Boron in Catalysis and Materials Chemistry

Boron in Catalysis and Materials Chemistry: A Themed Issue in Honor of Professor Todd B. Marder on the Occasion of His 65th Birthday

Editor

Ashok Kakkar

MDPI • Basel • Beijing • Wuhan • Barcelona • Belgrade • Manchester • Tokyo • Cluj • Tianjin

Editor
Ashok Kakkar
McGill University
Canada

Editorial Office
MDPI
St. Alban-Anlage 66
4052 Basel, Switzerland

This is a reprint of articles from the Special Issue published online in the open access journal *Molecules* (ISSN 1420-3049) (available at: https://www.mdpi.com/journal/molecules/special_issues/boron_marder).

For citation purposes, cite each article independently as indicated on the article page online and as indicated below:

LastName, A.A.; LastName, B.B.; LastName, C.C. Article Title. *Journal Name* **Year**, *Volume Number*, Page Range.

ISBN 978-3-0365-0812-2 (Hbk)
ISBN 978-3-0365-0813-9 (PDF)

© 2021 by the authors. Articles in this book are Open Access and distributed under the Creative Commons Attribution (CC BY) license, which allows users to download, copy and build upon published articles, as long as the author and publisher are properly credited, which ensures maximum dissemination and a wider impact of our publications.

The book as a whole is distributed by MDPI under the terms and conditions of the Creative Commons license CC BY-NC-ND.

Contents

About the Editor vii

Preface to "Boron in Catalysis and Materials Chemistry: A Themed Issue in Honor of Professor Todd B. Marder on the Occasion of His 65th Birthday" ix

Ashok Kakkar
Celebrating Todd Marder: 65th Birthday and His Contributions to Inorganic Chemistry
Reprinted from: *Molecules* **2021**, *26*, 776, doi:10.3390/molecules26040776 1

Ashanul Haque, Rayya A. Al-Balushi, Paul R. Raithby and Muhammad S. Khan
Recent Advances in π-Conjugated N^C-Chelate Organoboron Materials
Reprinted from: *Molecules* **2020**, *25*, 2645, doi:10.3390/molecules25112645 7

Oriol Salvadó and Elena Fernández
Tri(boryl)alkanes and Tri(boryl)alkenes: The Versatile Reagents
Reprinted from: *Molecules* **2020**, *25*, 1758, doi:10.3390/molecules25071758 29

Josep M. Oliva-Enrich, Ibon Alkorta and José Elguero
Hybrid Boron-Carbon Chemistry
Reprinted from: *Molecules* **2020**, *25*, 5026, doi:10.3390/molecules25215026 47

Ines Bennour, Francesc Teixidor, Zsolt Kelemen and Clara Viñas
m-Carborane as a Novel Core for Periphery-Decorated Macromolecules
Reprinted from: *Molecules* **2020**, *25*, 2814, doi:10.3390/molecules25122814 71

Martin Kellert, Imola Sárosi, Rajathees Rajaratnam, Eric Meggers, Peter Lönnecke and Evamarie Hey-Hawkins
Ruthenacarborane–Phenanthroline Derivatives as Potential Metallodrugs
Reprinted from: *Molecules* **2020**, *25*, 2322, doi:10.3390/molecules25102322 95

Hyunhee So, Min Sik Mun, Mingi Kim, Jea Ho Kim, Ji Hye Lee, Hyonseok Hwang, Duk Keun An and Kang Mun Lee
Deboronation-Induced Ratiometric Emission Variations of Terphenyl-Based *Closo-o*-Carboranyl Compounds: Applications to Fluoride-Sensing
Reprinted from: *Molecules* **2020**, *25*, 2413, doi:10.3390/molecules25102413 107

Li Ma, Xiaolin Zhang, Wenbo Ming, Shengxin Su, Xiaoyong Chang and Qing Ye
Reactions of Dihaloboranes with Electron-Rich 1,4-Bis(trimethylsilyl)-1,4-diaza-2,5-cyclohexadienes
Reprinted from: *Molecules* **2020**, *25*, 2875, doi:10.3390/molecules25122875 123

Maria Talavera, Silke Hinze, Thomas Braun, Reik Laubenstein and Roy Herrmann
A SF_5 Derivative of Triphenylphosphine as an Electron-Poor Ligand Precursor for Rh and Ir Complexes
Reprinted from: *Molecules* **2020**, *25*, 3977, doi:10.3390/molecules25173977 135

Arpita Saha, Hamdi Ben Halima, Abhishek Saini, Juan Gallardo-Gonzalez, Nadia Zine, Clara Viñas, Abdelhamid Elaissari, Abdelhamid Errachid and Francesc Teixidor
Magnetic Nanoparticles Fishing for Biomarkers in Artificial Saliva
Reprinted from: *Molecules* **2020**, *25*, 3968, doi:10.3390/molecules25173968 151

About the Editor

Ashok Kakkar is a Professor in the Department of Chemistry at McGill University in Montreal, Quebec, Canada. He obtained his PhD from the University of Waterloo (Canada, Professor Todd B. Marder), followed by post-doctoral studies at the University of Cambridge (UK, Professor the Lord Lewis) and Northwestern University (USA, Professor Tobin Marks). His research interests include developing methodologies for complex nanostructures with applications in a variety of areas including drug delivery and diagnostics.

Preface to "Boron in Catalysis and Materials Chemistry: A Themed Issue in Honor of Professor Todd B. Marder on the Occasion of His 65th Birthday"

Professor Todd B. Marder, a pioneer in boron chemistry, has had an illustrious career featuring many contributions to science in different fields, while mentoring students who are now academics and industrial scientists around the world. This special collection of articles in the themed issue "Boron in Catalysis and Materials Chemistry", on the occasion of his 65th birthday in November 2020, is a tribute to his accomplishments and a depiction of the diversity he has maintained and promoted in his science and his group members.

Ashok Kakkar
Editor

Editorial

Celebrating Todd Marder: 65th Birthday and His Contributions to Inorganic Chemistry

Ashok Kakkar

Department of Chemistry, McGill University, 801 Sherbrooke St. West, Montreal, QC H3A 0B8, Canada; ashok.kakkar@mcgill.ca

Academic Editor: Farid Chemat
Received: 28 January 2021; Accepted: 1 February 2021; Published: 3 February 2021

Professor Todd B. Marder grew up in Brooklyn, New York, and received his BSc from the Massachusetts Institute of Technology, Cambridge (USA, 1976), and his PhD from the University of California at Los Angeles (USA, 1981). This was followed by postdoctoral research at the University of Bristol (UK, 1981–1983). Working with the well-renowned inorganic chemists Professor Alan Davison, FRS (BSc), Professor Fred Hawthorne (PhD), and Professor Gordon Stone, CBE, FRS (postdoc) embedded the seeds of curiosity in inorganic chemistry in him, and it initiated his passion for organometallic and boron chemistry, homogeneous catalysis, dynamic NMR studies, and MO calculations. After two years as a visiting research scientist at DuPont Central Research, Wilmington, Delaware (USA), he began his independent academic career at the University of Waterloo (Canada) in 1985, where I first encountered this young, highly enthusiastic, gentleman inorganic chemist. I am proud to be the first PhD student to have graduated from his group.

Within a short span of about 8 years, he was promoted first to Associate Professor and then Full Professor in 1993. Diversity in chemistry was apparent in Todd's very first collection of young graduate students: indenyl-rhodium chemistry (Ashok Kakkar); Rh and Pt-acetylide chemistry: Pauline Chow, Davit Zargarian (Professor at Université de Montreal, Canada), M.J. Gerald Lesley (Professor at Southern Connecticut State University, USA), Dr. Graham Stringer and Dr. Ian R. Jobe (early postdocs from Oxford and the University of Western Ontario, respectively); metal boryl complexes and metal-catalyzed hydroboration chemistry: Steve Westcott (Professor at Mt. Allison University, Canada). As they say, there is always a woman behind every successful man, and Todd met his wife and a gem of a person, Anne, who was also teaching at the University of Waterloo. These two are a team that continue to support some of the best scientists in their care. In fact, their son, Ian, is an Assistant Professor in Criminology in the Department of Law at Maynooth University, Ireland, so academics run in the family.

With the excellence in chemistry that Todd demanded from himself and his crew, awards started to follow, the first being the Rutherford Memorial Medal for Chemistry (1995), given by the Royal Society of Canada to the leading chemist in Canada under 40 years of age.

After spending 12 years in Canada, Todd was lured by the British way of life, and decided to move from the colonies to the kingdom in 1997, when he accepted the Chair in Inorganic Chemistry at the University of Durham in England. He received the Royal Society of Chemistry Award in Main Group Element Chemistry in 2008 for his contributions to the chemistry of boron and its organometallic compounds and their applications, and a Japan Society for the Promotion of Science (JSPS) Invitation Fellowship (Japan), a Humboldt (Senior) Research Award (Germany), and the Royal Society (UK) Wolfson Research Merit Award, all in 2010. It was finally the lure of German beer, sausages and boron chemistry that brought him to the Institut für Anorganische Chemie, Julius-Maximilians-Universität Würzburg (Germany) in 2012, as Professor and Chair I of Inorganic Chemistry, as well as co-Head of the Institute for Sustainable Chemistry & Catalysis with Boron. Todd also won the Royal Society of Chemistry Organometallic Chemistry Award and was elected to the Bayerische Akademie der Wissenschaften (Bavarian Academy of Sciences) in 2015, Fellow of the American Association for the Advancement of Science (AAAS) in 2018, and Fellow of the European Academy of Sciences (EurASc) in 2019. In 2018, he was also awarded Docteur Honoris Causa, by the Université de Rennes 1, France.

Todd loves varied cultures, food, and his many friends scattered around the world and, prior to the current pandemic, would never miss an opportunity to travel both to contribute and to learn. This has led to several Visiting and similar Professorships in Europe (England, France), Asia (China, Hong Kong, India, Japan), and Australia (David Craig Visiting Professor at the Australian National University, 2014). He has served on the editorial board of several journals including Inorganic Chemistry, Organometallics, the Journal of Organometallic Chemistry, Applied Organometallic Chemistry, Canadian Journal of Chemistry, Crystal Engineering, and Chinese Journal of Chemistry.

Todd has given over 400 invited lectures worldwide, published over 385 papers with an h-index of 88 and around 25,000 citations (WoS). The co-workers he has trained occupy academic (over 35 of them) as well as industrial and other positions all over the world, and their success is his proudest achievement. Indeed, he refers to us as his children, and our academic offspring as his grandchildren, and has even had the pleasure of supervising a postdoc who was also his academic great granddaughter, Marie-Hélène Thibault (Professeure adjointe, Université de Moncton, Canada)!

Todd embraces diversity not only in his group members and numerous local and international collaborators, but also in his research interests [1–20]. For over 30 years, a significant part of his research has focused on metal-catalyzed borylation reactions, metal boryl complexes, mechanisms of borylation and other catalytic reactions [1–6], and the development of diboron(4) compounds and their chemistry [7–9]. One of the diboron(4) compounds he developed in collaboration with Prof. N. C. Norman (Bristol University), namely B_2neop_2, is now available commercially world-wide and is produced in multi-ton quantities. He also works in the fields of conjugated organometallic [10], organoboron [11–16] and organic materials [17–19] for linear and nonlinear optics, 2-photon excited fluorescent cell imaging [15], DNA/RNA/protein sensing [16], and on crystal engineering using perfluoroarene-arene interactions [20], among other areas, often combining several areas of his research in a single project.

Boron chemistry continues to offer a diverse platform for new directions in catalysts [21–23], as well as new cluster [24,25] and other materials [26–30] with varied applications, as indicated in these reviews, some by authors of papers in this issue. Professor Todd B. Marder has contributed significantly to expand the scope of boron chemistry, and it is an honor to celebrate his achievements, and mark his 65th birthday with a Special Issue of *Molecules*: "Boron in Catalysis and Materials Chemistry: A Themed Issue in Honor of Professor Todd B. Marder on the Occasion of His 65th Birthday". It includes excellent review articles on conjugated NˆC-organoboron materials, and tri(boryl)-alkanes and -alkenes as versatile reagents in organic synthesis [31,32], and research articles depicting diversity [33–39], as noted in the research interests of Professor Marder. We thank all the authors for joining us to congratulate Todd, and for making this issue a successful one through their excellent scientific contributions.

Funding: This research received no external funding.

Conflicts of Interest: The author declares no conflict of interest.

References

1. Tian, Y.M.; Guo, X.N.; Krummenacher, I.; Wu, Z.; Nitsch, J.; Braunschweig, H.; Radius, U.; Marder, T.B. Visible-Light-Induced Ni-Catalyzed Radical Borylation of Chloroarenes. *J. Am. Chem. Soc.* **2020**, *142*, 18231–18242. [CrossRef] [PubMed]
2. Liu, X.; Ming, W.; Friedrich, A.; Kerner, F.; Marder, T.B. Copper-Catalyzed Triboration of Terminal Alkynes Using B_2pin_2: Efficient Synthesis of 1,1,2-Triborylalkenes. *Angew. Chem. Int. Ed.* **2020**, *59*, 304–309. [CrossRef] [PubMed]
3. Tian, Y.-M.; Guo, X.-N.; Kuntze-Fechner, M.; Krummenacher, I.; Braunschweig, H.; Radius, U.; Steffen, A.; Marder, T.B. Selective Photocatalytic C-F Borylation of Polyfluoroarenes by Rh/Ni Dual Catalysis Providing Valuable Fluorinated Arylboronate Esters. *J. Am. Chem. Soc.* **2018**, *140*, 17612–17623. [CrossRef] [PubMed]
4. Budiman, Y.P.; Jayaraman, A.; Friedrich, A.; Kerner, F.; Radius, U.; Marder, T.B. Palladium-Catalyzed Homocoupling of Highly Fluorinated Arylboronates: Studies of the Influence of Strongly vs. Weakly Coordinating Solvents on the Reductive Elimination Process. *J. Am. Chem. Soc.* **2020**, *142*, 6036–6050. [CrossRef] [PubMed]
5. Bose, S.K.; Fucke, K.; Liu, L.; Steel, P.G.; Marder, T.B. Zinc-Catalyzed Borylation of Primary, Secondary and Tertiary Alkyl Halides with Alkoxy Diboron Reagents at Room Temperature. *Angew. Chem. Int. Ed.* **2014**, *53*, 1799–1803. [CrossRef] [PubMed]
6. Batsanov, A.S.; Cabeza, J.A.; Crestani, M.G.; Fructos, M.R.; García-Álvarez, P.; Gille, M.Z.Z.; Marder, T.B. Fully Borylated Methane and Ethane by Ruthenium Mediated Cleavage and Coupling of CO. *Angew. Chem. Int. Ed.* **2016**, *55*, 4707–4710. [CrossRef] [PubMed]
7. Eck, M.; Würtemberger-Pietsch, S.; Eichhorn, A.; Berthel, J.H.J.; Bertermann, R.; Paul, U.; Schneider, H.; Friedrich, A.; Kleeberg, C.; Radius, U.; et al. B–B Bond Activation and NHC Ring-expansion Reactions of Diboron(4) Compounds, and Accurate Molecular Structures of $B_2(NMe_2)_4$, B_2eg_2, B_2neop_2 and B_2pin_2. *Dalton Trans.* **2017**, *46*, 3661–3680. [CrossRef] [PubMed]
8. Dewhurst, R.D.; Neeve, E.C.; Braunschweig, H.; Marder, T.B. sp^2-sp^3 Diboranes: Astounding Structural Variability and Mild Sources of Nucleophilic Boron for Organic Synthesis. *Chem. Commun.* **2015**, *51*, 9594–9607. [CrossRef] [PubMed]
9. Neeve, E.C.; Geier, S.J.; Mkhalid, I.A.I.; Westcott, S.A.; Marder, T.B. Diboron(4) Compounds: From Structural Curiosity to Synthetic Workhorse. *Chem. Rev.* **2016**, *116*, 9091–9161. [CrossRef] [PubMed]
10. Sieck, C.; Tay, M.G.; Thibault, M.-H.; Edkins, R.M.; Costuas, K.; Halet, J.-F.; Batsanov, A.S.; Haehnel, M.; Edkins, K.; Lorbach, A.; et al. Reductive Coupling of Diynes at Rhodium Gives Fluorescent Rhodacyclopentadienes or Phosphorescent Rhodium 2,2′-Biphenyl Complexes. *Chem. Eur. J.* **2016**, *22*, 10523–10532. [CrossRef]
11. Ji, L.; Griesbeck, S.; Marder, T.B. Recent Developments in and Perspectives on Three-Coordinate Boron Materials: A Bright Future. *Chem. Sci.* **2017**, *8*, 846–863. [CrossRef] [PubMed]
12. He, J.; Rauch, F.; Finze, M.; Marder, T.B. (Hetero)Arene-Fused Boroles: A Broad Spectrum of Applications. *Chem. Sci.* **2021**, *12*, 128–147. [CrossRef]
13. Stennett, T.E.; Bissinger, P.; Griesbeck, S.; Ullrich, S.; Krummenacher, I.; Auth, M.; Sperlich, A.; Stolte, M.; Radacki, K.; Yao, C.-J.; et al. Near-Infrared Quadrupolar Chromophores Combining Three-Coordinate Boron-Based Superdonor and Superacceptor Units. *Angew. Chem. Int. Ed.* **2019**, *58*, 6449–6454. [CrossRef] [PubMed]
14. Wu, Z.; Nitsch, J.; Schuster, J.; Friedrich, A.; Edkins, K.; Loebnitz, M.; Dinkelbach, F.; Stepanenko, V.; Würthner, F.; Marian, C.M.; et al. Persistent Room Temperature Phosphorescence from Triarylboranes: A Combined Experimental and Theoretical Study. *Angew. Chem. Int. Ed.* **2020**, *59*, 17137–17144. [CrossRef] [PubMed]
15. Griesbeck, S.; Michail, E.; Wang, C.; Ogasawara, H.; Lorenzen, S.; Gerstner, L.; Zhang, T.; Nitsch, J.; Sato, Y.; Bertermann, R.; et al. Tuning the π-Bridge of Quadrupolar Triarylborane Chromophores for One- and Two-Photon Excited Fluorescence Imaging of Lysosomes in Live Cells. *Chem. Sci.* **2019**, *10*, 5405–5422. [CrossRef] [PubMed]

16. Ban, Ž.; Griesbeck, S.; Tomić, S.; Nitsch, J.; Marder, T.B.; Piantanida, I. A Quadrupolar bis-Triarylborane Chromophore as a Fluorimetric and Chirooptic Probe for Simultaneous and Selective Sensing of DNA, RNA and Proteins. *Chem. Eur. J.* **2020**, *26*, 2195–2203. [CrossRef] [PubMed]
17. Merz, J.; Fink, J.; Friedrich, A.; Krummenacher, I.; Al Mamari, H.H.; Lorenzen, S.; Hähnel, M.; Eichhorn, A.; Moos, M.; Holzapfel, M.; et al. Pyrene Molecular Orbital Shuffle—Controlling Excited State and Redox Properties by Changing the Nature of the Frontier Orbitals. *Chem. Eur. J.* **2017**, *23*, 13164–13180. [CrossRef]
18. Merz, J.; Steffen, A.; Nitsch, J.; Fink, J.; Schürger, C.B.; Friedrich, A.; Krummenacher, I.; Braunschweig, H.; Moos, M.; Mims, D.; et al. Synthesis, Photophysical and Electronic Properties of Tetra- Donor- or Acceptor-Substituted ortho-Perylenes Displaying Four Reversible Oxidations or Reductions. *Chem. Sci.* **2019**, *10*, 7516–7534. [CrossRef]
19. Ma, X.; Maier, J.; Wenzel, M.; Friedrich, A.; Steffen, A.; Marder, T.B.; Mitrić, R.; Brixner, T. Direct Observation of o-Benzyne Formation in Photochemical Hexadehydro-Diels–Alder (hν-HDDA) Reactions. *Chem. Sci.* **2020**, *11*, 9198–9208. [CrossRef]
20. Friedrich, A.; Collings, I.E.; Dziubek, K.F.; Fanetti, S.; Radacki, K.; Ruiz-Fuertes, J.; Pellicer-Porres, J.; Hanfland, M.; Sieh, D.; Bini, R.; et al. Pressure-Induced Polymerization of Polycyclic Arene-Perfluoroarene Cocrystals: Single Crystal X-ray Diffraction Studies, Reaction Kinetics, and Design of Columnar Hydrofluorocarbons. *J. Am. Chem. Soc.* **2020**, *142*, 18907–18923. [CrossRef]
21. Cuenca, A.B.; Shishido, R.; Ito, H.; Fernández, E. Transition-Metal-Free B–B and B–Interelement Reactions with Organic Molecules. *Chem. Soc. Rev.* **2017**, *46*, 415–430. [CrossRef] [PubMed]
22. Mkhalid, I.A.I.; Barnard, J.H.; Marder, T.B.; Murphy, J.M.; Hartwig, J.F. C–H Activation for the Construction of C–B Bonds. *Chem. Rev.* **2010**, *110*, 890–931. [CrossRef] [PubMed]
23. Wang, M.; Shi, Z. Methodologies and Strategies for Selective Borylation of C−Het and C−C Bonds. *Chem. Rev.* **2020**, *120*, 7348–7398. [CrossRef] [PubMed]
24. Quan, Y.; Xie, Z. Controlled Functionalization of o-Carborane via Transition Metal Catalyzed B–H Activation. *Chem. Soc. Rev.* **2019**, *48*, 3660–3673. [CrossRef] [PubMed]
25. Núñez, R.; Tarrés, M.; Ferrer-Ugalde, A.; Fabrizi de Biani, F.; Teixidor, F. Electrochemistry and Photoluminescence of Icosahedral Carboranes, Boranes, Metallacarboranes, and their Derivatives. *Chem. Rev.* **2016**, *116*, 14307–14378. [CrossRef] [PubMed]
26. Jäkle, F. Lewis Acidic Organoboron Polymers. *Coord. Chem. Rev.* **2006**, *250*, 1107–1121. [CrossRef]
27. Wakamiya, A.; Yamaguchi, S. Designs of Functional π-Electron Materials based on the Characteristic Features of Boron. *Bull. Chem. Soc. Jpn.* **2015**, *88*, 1357–1377. [CrossRef]
28. Wade, C.R.; Broomsgrove, A.E.; Aldridge, S.; Gabbaï, F.P. Fluoride Ion Complexation and Sensing Using Organoboron Compounds. *Chem. Rev.* **2010**, *110*, 3958–3984. [CrossRef]
29. Hudson, Z.M.; Wang, S. Metal-containing Triarylboron Compounds for Optoelectronic Applications. *Dalton Trans.* **2011**, *40*, 7805–7816. [CrossRef]
30. von Grotthuss, E.; John, A.; Kaese, T.; Wagner, M. Doping Polycyclic Aromatics with Boron for Superior Performance in Materials Science and Catalysis. *Asian J. Org. Chem.* **2018**, *7*, 37–53. [CrossRef]
31. Haque, A.; Al-Balushi, R.A.; Raithby, P.A.; Khan, M.S. Recent Advances in π-Conjugated N^C-chelate Organoboron Materials. *Molecules* **2020**, *25*, 2645. [CrossRef] [PubMed]
32. Salvadó, O.; Fernández, E. Tri(boryl)alkanes and Tri(boryl)alkenes: The Versatile Reagents. *Molecules* **2020**, *25*, 1758. [CrossRef] [PubMed]
33. Oliva-Enrich, J.M.; Alkorta, I.; Elguero, J. Hybrid Boron-Carbon Chemistry. *Molecules* **2020**, *25*, 5026. [CrossRef] [PubMed]
34. Talavera, M.; Hinze, S.; Braun, T.; Laubenstein, R.; Herrmann, R. A SF_5 Derivative of Triphenylphosphine as an Electron-poor Ligand Precursor for Rh and Ir Complexes. *Molecules* **2020**, *25*, 3977. [CrossRef]
35. Saha, A.; Halima, H.B.; Saini, A.; Gallardo-Gonzalez, J.; Zine, N.; Viñas, C.; Elaissari, A.; Errachid, A.; Teixidor, F. Magnetic Nanoparticles Fishing for Biomarkers in Artifical Saliva. *Molecules* **2020**, *25*, 3968. [CrossRef]
36. Ma, L.; Zhang, X.; Ming, W.; Su, S.; Chang, X.; Ye, Q. Reactions of Dihaloboranes with Electron-Rich 1,4-Bis(trimethylsilyl)-1,4-diaza-2,5-cyclohexadienes. *Molecules* **2020**, *25*, 2875. [CrossRef]

37. Bennour, I.; Teixidor, F.; Keleman, Z.; Viñas, C. *m*-Carborane as a Novel Core for Periphery-Decorated Macromolecules. *Molecules* **2020**, *25*, 2814. [CrossRef]
38. So, H.; Mun, M.S.; Kim, M.; Kim, J.H.; Lee, J.H.; Hwang, H.; An, D.K.; Lee, K.M. Deboronation-Induced Ratiometric Emission Variations of Terphenyl-Based Closo-o-Carboranyl Compounds: Applications to Fluoride Sensing. *Molecules* **2020**, *25*, 2413. [CrossRef]
39. Kellert, M.; Sárosi, I.; Rajaratnam, R.; Meggers, E.; Lönnecke, P.; Hey-Hawkins, E. Ruthenacarborane-Phenanthroline Derivatives as Potential Metallodrugs. *Molecules* **2020**, *25*, 2322. [CrossRef]

© 2021 by the author. Licensee MDPI, Basel, Switzerland. This article is an open access article distributed under the terms and conditions of the Creative Commons Attribution (CC BY) license (http://creativecommons.org/licenses/by/4.0/).

Review

Recent Advances in π-Conjugated N^C-Chelate Organoboron Materials

Ashanul Haque [1,*], Rayya A. Al-Balushi [2], Paul R. Raithby [3,*] and Muhammad S. Khan [4,*]

1. Department of Chemistry, College of Science, University of Hail, Ha'il 81451, Saudi Arabia
2. Department of Basic Sciences, College of Applied and Health Sciences, A'Sharqiyah University, P.O. Box 42, Ibra 400, Sultanate of Oman; rayya.albalushi@asu.edu.om
3. Department of Chemistry, University of Bath, Claverton Down BA2 7AY, UK
4. Department of Chemistry, Sultan Qaboos University, P.O. Box 36, Al-Khod 123, Sultanate of Oman
* Correspondence: a.haque@uoh.edu.sa (A.H.); p.r.raithby@bath.ac.uk (P.R.R.); msk@squ.edu.om (M.S.K.)

Academic Editor: Ashok Kakkar
Received: 17 May 2020; Accepted: 4 June 2020; Published: 6 June 2020

Abstract: Boron-containing π-conjugated materials are archetypical candidates for a variety of molecular scale applications. The incorporation of boron into the π-conjugated frameworks significantly modifies the nature of the parent π-conjugated systems. Several novel boron-bridged π-conjugated materials with intriguing structural, photo-physical and electrochemical properties have been reported over the last few years. In this paper, we review the properties and multi-dimensional applications of the boron-bridged fused-ring π-conjugated systems. We critically highlight the properties of π-conjugated N^C-chelate organoboron materials. This is followed by a discussion on the potential applications of the new materials in opto-electronics (O-E) and other areas. Finally, attempts will be made to predict the future direction/outlook for this class of materials.

Keywords: boron; π-conjugated materials; opto-electronics; tetracoordinated

1. Introduction

The continuing development in the area of π-conjugated materials during the last few decades is the result of the combined interdisciplinary research interests of chemists, physicists, and materials scientists [1]. This class of materials provides an excellent platform to merge the features of organics with inorganics to obtain hybrid materials [2,3]. π-Conjugated materials are of considerable interest to academic and industrial researchers compared to conventional inorganic semiconductor materials, which have involve costs and labor-intensive fabrication procedures. The synthetic versatility, solution processability, and ability to tune their photophysical properties render conjugated organic materials as attractive candidates for applications in various domains of materials science [4].

To realize a material with superior optical and electronic properties for real-life applications, one or more strategies, such as ring expansion, π-length extension, and metal or main-group element incorporation into the π-conjugated organic system are being adopted. Among these, the incorporation of a non-carbon dopant is an emerging strategy, as it induces a network of covalent and non-covalent interactions, and allows fine tuning of the frontier orbitals (FOs) and the intramolecular charge transfer (ICT) processes. In the last two decades, a plethora of molecular and polymeric systems incorporating heteroatoms (*viz.* N, Si, S, Se, B, and P) have been reported with intriguing properties and applications. In particular, research on π-conjugated organoboron materials has seen a tremendous upsurge due to high usability in several real-life applications. Both three and four coordinated B-based conjugated materials have been reported. In this review, we wish to present properties and applications of N^C-chelated tetra-coordinated B-based conjugated oligo- and polymeric materials.

2. Boron-Bridged π-Conjugated Materials: Properties and Features

Boron (B, atomic number = 5, electronic configuration = $1s^2\,2s^2\,2p^1$) is a group 13 element with three valence electrons and two vacant p-orbitals. Depending upon the chelating and auxiliary fragments present, a π-conjugated organic backbone can be attached to the B-center to realize trigonal planar and tetrahedral organoboron compounds. The chelation of organic and inorganic fragments induces rigidity into the framework, along with enhanced π-conjugation and improved photoluminescence (PL) properties [5,6]. Consequently, physio-chemical properties, stability, and applications of the resulting materials can be significantly controlled and tuned by variation of the coordination environments around the B-center.

The trigonal planar B-fragment is considered to be isoelectronic with the carbocation (sp^2 hybridization with one unoccupied p_z orbital); thus, it is formally electron deficient (Lewis acid). Therefore, when it is incorporated into a π-conjugated framework, there is an overlap between the p orbital of boron and π-orbitals of the organic segment, leading to significant delocalization and PL property modulation [7]. It is interesting that the vacant p-orbital can be both advantageous and detrimental. The benefit of this vacant orbital is that it can be exploited to tune PL properties through Lewis base coordination. On the other hand, low stability against moisture (or other nucleophiles) is also attributed to this vacant orbital. To overcome this challenge, one or more bulky substituents such as 2,4,6-trimethylphenyl (Mes), 2,4,6-tri-iso-propylphenyl (Tip), and 2,4,6-tris(trifluoromethyl)phenyl (FMes) are often employed. These substituents not only modulate electron density over the material but also sterically shield the boron atom and provide kinetic stability towards air and moisture. Overall, a trigonal planar boron system functions as a strong π-acceptor and a σ-donor [8]. Exploiting these features, a number of main/side chain, cyclic, and Lewis acid-base type B-functionalized materials are available with intriguing properties and applications.

As mentioned above, coordination of vacant p_z orbital of the B-atom paves a unique way to develop new materials with tunable properties. A range of neutral tetracoordinated organoboranes can be obtained by forming covalent bonds with mono-anionic chelate ligands. Tetracoordinated B-units can act as electron-donating substituents. A large number of chelating ligands based on C, N, and O donors have been synthesized and reported by different groups. Unlike three-coordinated compounds, boron atoms here play a more significant role in tetracoordinated boron compounds with the opportunity to fine tune the properties via modulation of electronic and steric effects. The introduction of a boron atom in tetracoordinated disposition within a conjugated framework enhances the planarity of the host (conjugated) structure, leading to improved charge transfer and other properties [9]. Besides, tetracoordinated organoboron structures possess high stability and rigid structure, leading to high fluorescence quantum yields. B ← N coordinate bond is labile compared to the B-F bond; this feature can be exploited to develop molecular switches with on-off PL properties [10]. In the literature, synthesis, characterization, and applications of numerous C^O- [11], N^O [12], N^N [13], N^C [14], C^C [15], and O^O [16] chelated B-containing π-conjugated materials have appeared. These chelating sites come from pyridine, thiazole, iso-quinoline, pyrimidine/quinoline, pyrazine, imidazole, and other cores.

In this review, we wish to present the features and properties of N^C-chelate tetracoordinated π-conjugated oligomeric and polymeric materials (Figure 1). In the last two decades, such organoboron-functionalized materials have been extensively studied by the research groups led by Wang, Pischel, and Yam, among others. They discovered several new features and properties associated with these materials. In this review we highlight the recent advances in the new area of N^C chelate compounds because these materials display promising PL features and a specific area has not been reviewed previously. Readers interested in related topics are referred to the reviews [17–23]. We have divided this review into five main headings. Following this introduction (Section 1), we have briefly compared the features of tricoordinated and tetracoordinated Boron-bridged π-conjugated materials in Section 2. In Section 3, we have discussed the properties of different N^C-chelate-based organoboron compounds. We have reviewed and compared properties of some classical compounds in conjunction with recent ones. Following this, we have outlined some applications of N^C-chelate π-conjugated oligomeric and polymeric organoboron materials in Section 4. In this section, applications such as O-E, imaging, sensing, and other areas have been covered. At the end, in Section 5, we discuss the opportunities and challenges existing in this area of research.

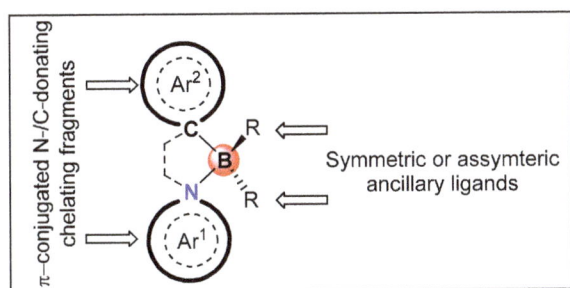

Figure 1. General structure of N^C-chelate organoboron materials discussed in this review. Ar^1 and Ar^2 represent aromatic N/C-donating fragments while R represents symmetric or asymmetric ancillary ligands.

3. π-Conjugated N^C-Chelate Organoboron Materials

Developed around a decade and a half ago, tetracoordinated N^C-chelate organoboron compounds are a unique class of photoresponsive materials. Figure 2 depicts a simple 2-phenylpyridine (ppy) based N^C donor ligand and the effect on the lowest unoccupied molecular orbital (LUMO) energy level upon borylation [24]. As is clear, borylation of a ppy ligand causes polarization of the LUMO. For instance, in non-borylated material, pyridine and phenyl rings contributed ~43% and ~55%, respectively to the LUMO level. Upon B-N coordination, the major contribution was from pyridine ring (~74%) indicating the polarization of the orbitals. This, in turn, led to modulated energy levels and stability (*viz.* destabilization of phenyl π* and the stabilization of pyridine π* orbital) of the compound. In addition, it has also been demonstrated that borylated unit possesses a significantly lower LUMO energy than their non-coordinated counterparts. The replacement of ancillary ligand such as bromine by an aromatic ring such as a thienyl or a phenyl ring has been found to increase the thermal and chemical stability of N^C-chelate organoboron materials [25]. The above-mentioned features motivated researchers to develop new materials with modified C and N-based fully color tunable materials [26,27]. In the subsections below, we discuss properties of different N^C-donor based organoboron compounds.

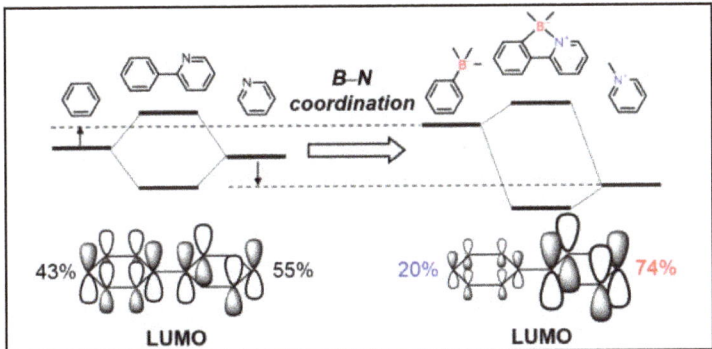

Figure 2. Schematic representation of molecular orbital polarization induced by B-N coordination. Reproduced with permission from reference [24].

3.1. 2-Arylpyridine-Derived N^C-Chelates

In 2008, Wang and coworkers discovered that the ppy-BMes$_2$ organoboron compound (**1**, Figure 3) undergoes intramolecular C-C/C-B bond rearrangement to produce reversible isomerized **1a** and an irreversible isomerized **1b** (Figure 3) products accompanied by a dramatic change in color [28]. Structure property relationship studies using symmetrical and unsymmetrical systems indicated that a bulky substituent (viz. mesityl) on boron fragment is required to initiate the photoisomerization process. In asymmetrical systems (**2**, Figure 3), borirane ring formation takes place regioselectively on the less bulky substituent attached to boron (**2a**; Figure 3). Further thermally assisted transformation produced 4bH-azaborepins (**2b**; Figure 3) [21]. It has been found that in a multi-borylated π-conjugates systems, isomerization takes place at one boron center only while others remain intact and assist this process via ICT [29]. Despite this fascinating process, the rearranged "dark" products significantly limit their usability for real-life applications [30].

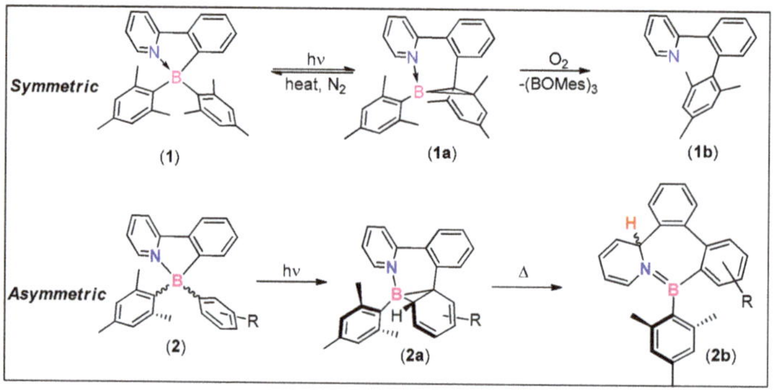

Figure 3. Photoisomerization and oxidation of symmetric and asymmetric organoboron compound ppy-BMes$_2$ (Mes = mesityl).

In order to control such process and to increase photochemical stability, several modifications (such as modulation of electronic factors and steric congestion) have been suggested [31,32]. It is shown that the functionalities attached to the chelating/auxiliary units significantly affect the isomerization process [33]. For instance, when a π-conjugated metal acetylide fragment was attached to the pyridine ring, a significant quenching of the photoisomerization (at boron unit) resulted, owing to a low-lying intra-ligand charge transfer (ILCT)/ metal to ligand charge transfer (MLCT) triplet state [34]. On the other hand, conjugation extension using an olefin was found to be an effective strategy as it inhibits photoisomerization (at boron unit) via dissipating energy through the alternative cis-trans isomerization pathway (**3** and **4**, Figure 4) [31]. This cis–trans isomerization of the olefinic bond has also been found to be modulated by metal chelation in such systems [35]. In contrast to this study, authors found that when a bithienyl unit is attached via an alkynyl ligand, it completely turned off photoisomerization arising from the N,C-chelate boryl core (**5** and **6**, Figure 4). They attributed this to relatively low-lying π → π* transition state of the bithienyl unit along with its effective competition with the CT transition of the boryl unit [36]. A similar observation was made by Yam et al. [37] Here, the authors merged photoactive diarylethene-functionalized N∧C chelated thienylpyridine with bis-alkynyl borane complexes.

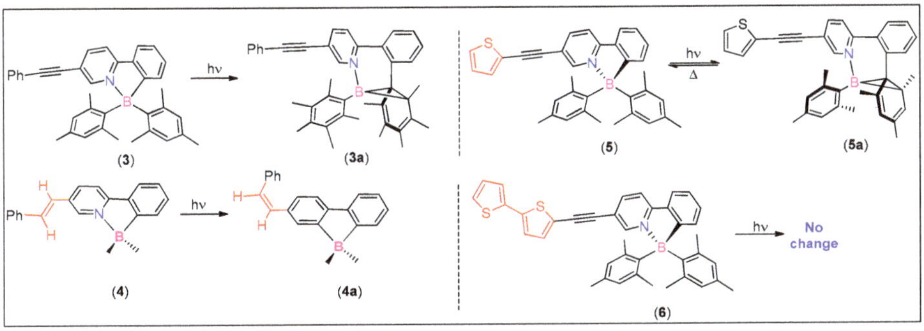

Figure 4. Photoisomerization processes in alkyne and olefin-containing organoboron compounds.

In order to examine the competitive photoisomerization processes, Wang and coworkers [38] prepared a series of cis and trans Pt(II) acetylides containing two photochromic units; i.e., dithienylethene (DTE) and B(ppy)Mes$_2$ (**7a–c**, Figure 5). Interestingly, in such systems, DTE showed preferential reversible photochromism over the boryl unit while the latter enhanced photoisomerization quantum efficiency of the DTE via antenna effect. The quantum efficiency (open → closed) of the system follows the order **7a** < **7b** < **7c**. In order to realize a swift reversible isomerization process, the presence of bulky groups and electron-donating groups has been found to be favorable while the opposite was found for the presence of electron withdrawing groups on the N^C-chelate [39]. Similarly, cyclometallation of the N^C ligands also quenched the photoisomerization process of the B(ppy)Mes$_2$ chromophore [40]. Several other studies have been carried out to obtain an in-depth knowledge of such chromophores and more can be found in references [38,41–46].

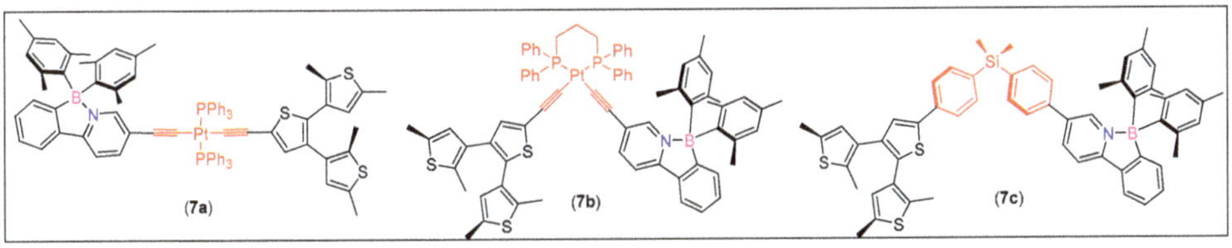

Figure 5. Examples of borylated materials in which two photochromic units are separated by cis, trans-Pt(II)-acetylide, and Si-containing spacers.

Through various structural modifications and analogues, it has been established that photoisomerization could also be extended to N^C-chelate in addition to the ppy systems [47]. For instance, in a recent work, Wang and coworkers [47] replaced pyridine by a pyrazole as the N-donor (**8**, Figure 6) and reported its two-stage photoreactivity. These new systems undergo photoisomerization to produce thermally stable aza borata bisnorcaradienes (**8a**), which upon prolonged irradiation led to isomerize products 14aH-diazabor-epins or BNN-benzotropilidene (**8b**). Although irreversible, formation of such interesting isomeric products is rare and opens up new vistas for researchers.

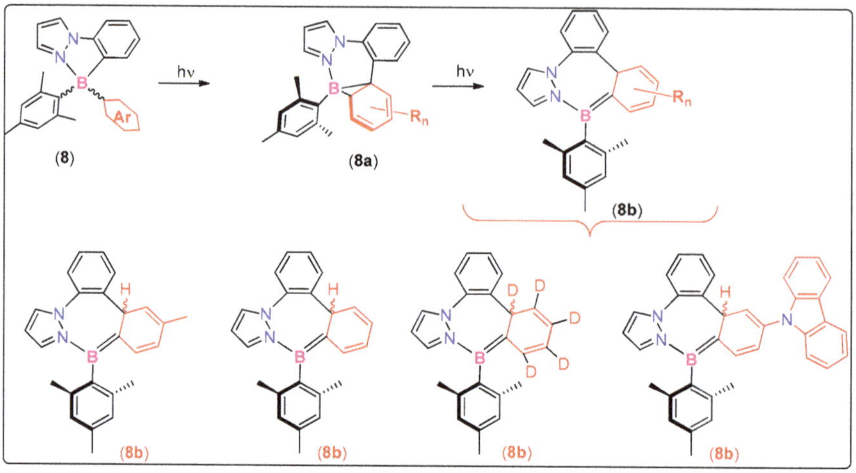

Figure 6. New generation organoboron compounds with rare two-stage photoreactivity.

3.2. 2-Arylthiazole and Aminobenzothiadiazole Derived N^C-Chelates

Research work on 2-arylthiazole as the N^C-chelates dates back to 2006 when Yagamuchi and coworkers reported the synthesis and hole mobility properties of mono, di, and oligomeric thienylthiazole-based tetracoordinated B-materials [48]. They found that the incorporation of boryl units into the thienylthiazole core promotes a planar configuration of the π-conjugated framework by intramolecular B-N coordination, which reduces the LUMO level and increases the electron-accepting ability (electron affinity) of the materials. Later they found that the electron affinity along with the thermal stability of the materials can be further improved by replacing thienyl by a C_6F_4 moiety and by coordination with $(C_6F_5)_2B$ [49]. For example, Figure 7 shows the effect of borylation on thienylthiazole-based N^C-chelates [50]. As can be seen, the LUMO level of the chelated system reduces upon borylation and further dips upon the modification of ancillary ligands. Similarly, Wang and co-workers reported a decrease in HOMO/LUMO energy level (by 0.6 eV) of a conjugated system upon B-N coordination [51].

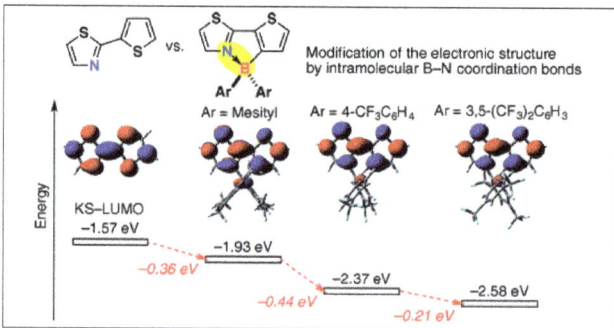

Figure 7. Effect of merging pyridine with phenyl followed by B-N coordination on the LUMO levels. Reproduced with permission from reference [50].

Benzothiadiazole (BTD) is a well-known strong electron-acceptor and is used to develop high-performance donor-acceptor (D-A) materials. BTD-containing materials often possess low band gaps, intense absorption, and emission that extends to the visible to near infrared (NIR) region, and thus are considered as unique materials for opto-electronics (O-E) and other applications [52].

Besides, it has also been reported that HOMO-LUMO energy levels of N^C-chelate organoboron compounds can be tuned via changing the exocyclic boron substituents [53]. Ingleson and coworkers reported simple routes to realize (N^C-chelate)B(aryl)$_2$ species (**9–15**, Figure 8) [54,55]. They found that the incorporation of borylated aryl amine donor units in compound **9** (Ar = C$_6$F$_5$) raised the HOMO level, thereby lowering the band gap (~1.5 eV). Despite the fact that the borylated D–A–D material possesses $\lambda_{max.}$ > 700 nm, low PL quantum yields was the main drawback to these systems.

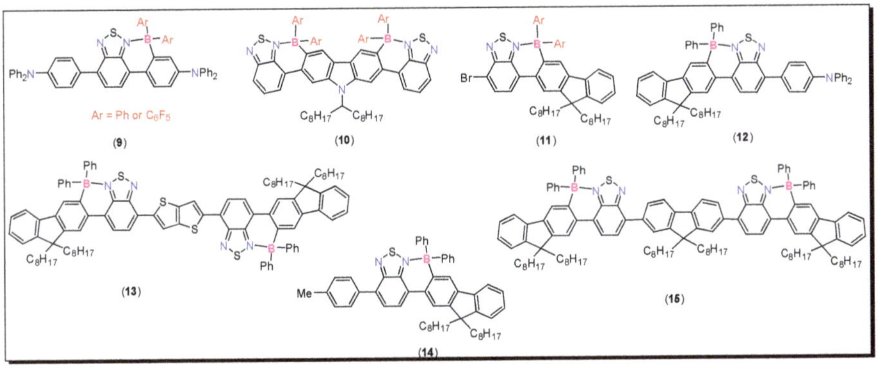

Figure 8. Some examples of BTD-containing N^C-chelate organoboron compounds.

An analysis of the molecular structure of such systems indicated the presence of strain arising from the steric bulkiness of the groups attached to the B-center and the BTD unit. Such a strain can lead to the formation of B-N bonds that can be broken reversibly. Based on this idea, Shimogawa et al. [56] reported a near infrared (NIR) emitting Mes$_2$B-substituted BTD material **16** (Figure 9a) having the ability to reversibly form intramolecular B-N coordination bonds. Such reversible bond breaking/formation has a direct effect on the electronic properties and color that are visible to the naked eye. For example, compound **16a**, depending upon its bond formation/cleavage exhibited interesting thermo, solvo, and mechanochromism features (Figure 9b–d).

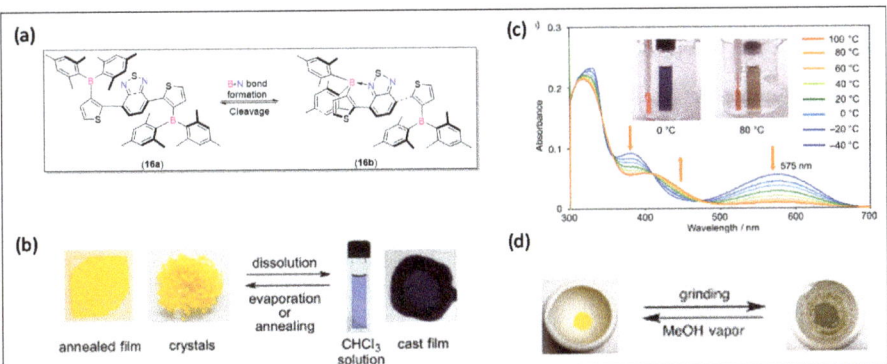

Figure 9. (a) Reversible formation of an intramolecular B-N coordination bond in **16a**. (b) Photographs of **16a** as an annealed film, yellow crystals, solution in CHCl$_3$, and cast film; (c) thermochromism of **16a** in toluene; and (d) yellow crystals of **16a** before and after grinding. Reproduced with permission from reference [56].

3.3. 2-Arylquinolines-Derived N^C-Chelates

Molecular fluorophores capable of absorbing and emitting light at different wavelengths are highly desirable for OE and bioimaging applications. In this quest, Shaikh et al. [57] reported molecular fluorophores based on 2-arylquinoline chelates (**17a–f**, Figure 10). The substituents present on the quinolines or on the boron center control the color of the emission. Because of their intense luminescence covering the whole visible region, the fluorophores were used for the imaging of breast cancer cells (MCF 7). Cell viability assays indicated the safe nature (at 1 µM) of organoboron compounds. Extensive bioimaging studies revealed its potential for targeted imaging. Bachollet et al. [58] carried out an extensive structure–optical property relationship study on bipyridine-based tetracoordinated B-materials (BOBIPYs). The new BOBIPYs showed promising properties, such as solid-state luminescence, blue to green emission, and high quantum yields. One of the derivatives based on compound **18** (Figure 10) has the potential to be used as an imaging probe. Other quinoline-based systems with aggregation-induced emission (AIE) properties have also been reported [59].

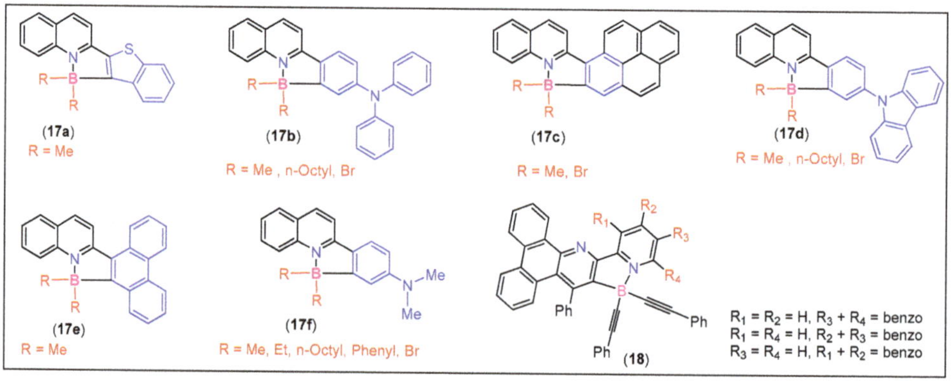

Figure 10. Some examples of 2-arylquinoline-derived N^C-chelates.

3.4. 1-Arylisoquinolines-Derived N^C-Chelates

N^C-chelate organoboron dyes, including those with arylisoquinoline ligands (**19–20**, Figure 11), display high stability in air-equilibrated media, and an excited state intramolecular charge-transfer (ICT) character leading to high Stokes shifts, solvatofluorochromic behavior, and high photochemical stability [60,61]. Pischel and coworkers [61] found that the π-extended system **20a–c** (Figure 11) exhibits unique two-photon absorption (TPA) properties in the NIR region. In such compounds, the presence of D/A moieties on the termini significantly affected the two-photon absorption cross-section and emission properties. For example, the compound having the donor-π-acceptor (D-π-A) configuration **20a** (X = NMe$_2$) showed TPA cross Section 59 GM (960 nm) and 95 GM (700 nm). On the other hand, compound **20b**, having the acceptor-π-acceptor (A-π-A) configuration showed TPA cross Section 14 GM (900 nm) and 223 GM (700 nm). The values higher than their non-extended counterpart **20d** (20 GM at 710 nm and 13 GM at 840 nm) clearly indicated the usefulness of the extension of the π-conjugation via the alkynyl core.

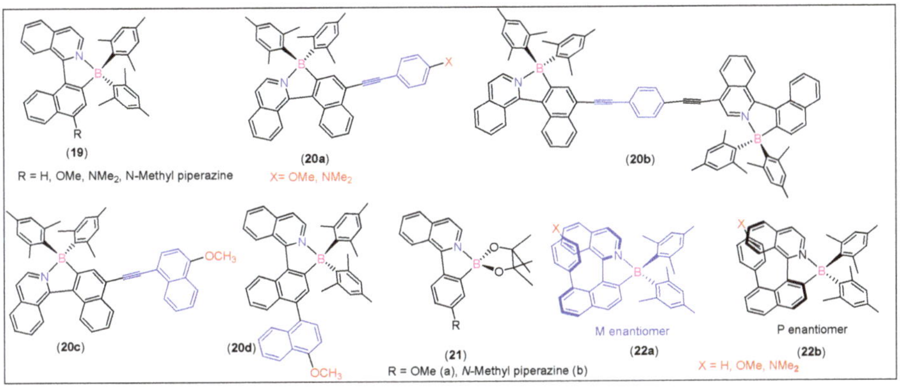

Figure 11. Some examples of 1-arylisoquinoline-derived N^C-chelates.

Similarly to **16a** (Figure 9), arylisoquinoline-based borylated compounds **21a and b** (Figure 11) exhibited a unique steric hindrance-dependent response against temperature and PL properties. These systems exhibited a different geometry in the ICT excited states and thus acted as molecular thermometers [62]. For instance, **21a** has a three-times higher emission quantum yield than the sterically hindered compound. Likewise, the sterically hindered compound showed no temperature effect. In contrast to this, the emission properties of compounds with helicene-type substituents on a borylated arylisoquinoline skeleton **22a and b** (Figure 11) were found to be controlled by the electron-donating substituents on the helicene-type cores [63].

4. Applications

Design and development of electronic devices based on conjugated organic materials have seen a huge upsurge in the last few decades. This is due to their synthetic flexibility, excellent film-forming properties, and tunable electronic properties which gained them popularity [64]. As highlighted in the sub-sections above, the PL properties, stability, and applications of B-containing conjugated polymers are greatly determined by the lability of the B-N coordination bond, N^C-donating, π-conjugated cores, and the ancillary ligands. Based on this, several tetradentate B-based small, large, and polymeric π-conjugated materials have been tested for a range of applications. Since we restricted ourselves to N^C based conjugated materials, we highlight some pertinent examples and applications of such materials.

4.1. O-E Applications

4.1.1. Organic Light Emitting Diodes (OLEDs)

For applications as OE components, and sensory or imaging probes, both oligomeric and polymeric systems have been tested. Among these, polymeric materials are favored over oligomeric systems due to manifold benefits. The intense emission extending to the Vis-NIR region, high quantum yield, and charge carrier mobility are some of the important features that make organoborons promising candidates for full-color tunable light emitters [5]. It has been shown that B-N coordination improves EL properties of the luminogens [65]. As discussed before, tetracoordinated B-materials act as electron-donating substituents, and in the presence of a suitable acceptor (such as BT), they create D-A systems with low lying LUMOs. Ingleson and co-workers found that BTD-based D-A materials **23a–c** (Figure 12) possess low LUMO energy levels, minimally changed HOMO energy levels, far red/NIR emission with solid state quantum yields of up to 34%, and good stability towards moisture [66]. OLEDs fabricated using these materials showed low turn-on and operation voltages. Among the reported compounds, OLED fabricated using **23a** showed the maximum external quantum efficiency

(EQE) value of 0.46% (Table 1) with the EL emission maxima (λ_{max}) at 678 nm. The EL value improved slightly (0.48%, λ_{max} = 679 nm) upon changing the device architecture (95:5 wt% PF8BT/**23a**).

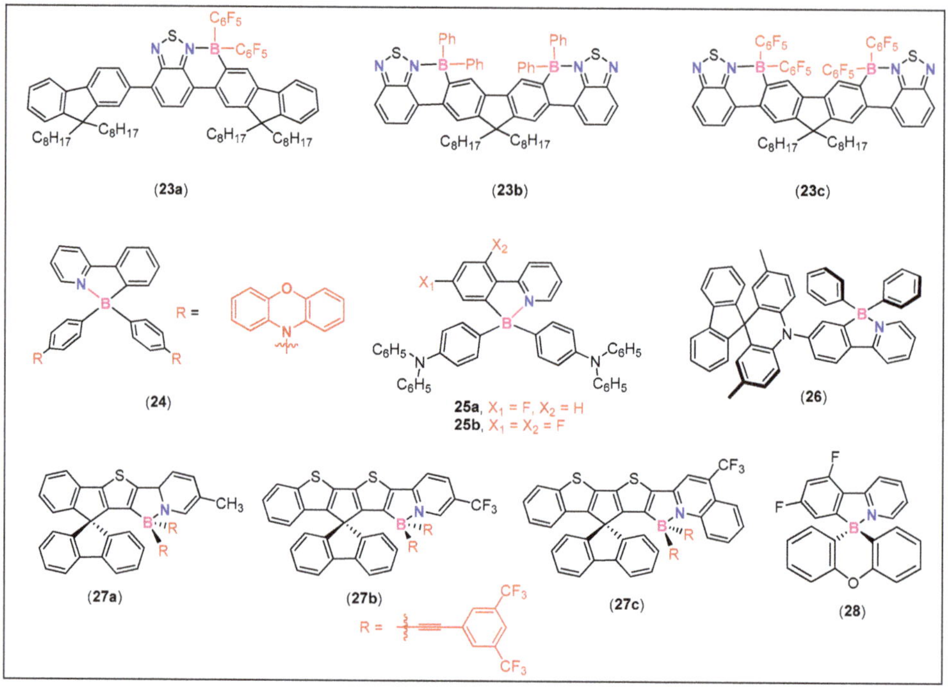

Figure 12. Examples of N^C-donating, π-conjugated cores as OLEDs materials.

Table 1. OLED performances of different π–conjugated N^C-chelate organoboron materials.

Comp. #	Device Architecture	EQE (%)	Current Efficiency (Cd/A)	Power Efficiency (lm/W)	Ref.
23a	ITO/Plexcore OC/PF8-TFB/PF8-BT/PF8-TFB/**23a**/Ba	0.46	-	-	[66]
23b	ITO/Plexcore OC/PF8-TFB/PF8-BT/PF8-TFB/**23b**/Ba	0.14	-	-	[66]
23c	ITO/Plexcore OC/PF8-TFB/PF8-BT/PF8-TFB/**23c**/Ba	0.13	-	-	[66]
25a	ITO/PEDOT:PSS/TAPC/mCP/mCPCN doped with **25a**)/3TPYMB/LiF/Al	20.2	63.9 [a]	66.9 [a]	[67]
25b	ITO/PEDOT:PSS/TAPC/mCP/mCPCN doped with **25b**)/3TPYMB/LiF/Al	26.6	88.2 [b]	81.5 [b]	[67]
26	ITO/HAT-CN/α-NPD/CCP/EML/PPF/TPBi/Liq/Al	22.7	56.4	44.3	[24]
27a	ITO/PEDOT:PSS/**27a**/MCP/3TPYMB/TmPyPB/LiF/Al	1.1	1.6	1.0	[68]
27b	ITO/PEDOT:PSS/**27b**/MCP/3TPYMB/TmPyPB/LiF/Al	1.3	4.8	3.0	[68]
27c	ITO/PEDOT:PSS/**27b**/MCP/3TPYMB/TmPyPB/LiF/Al	0.9	1.4	0.9	[68]

[a] = Peak value at 8 wt% concentration; [b] = Peak value at 25 wt% concentration; ITO = indium tin oxide; OC = organic conductive; PF8-TFB = poly[(9,9-dioctyl-fluorenyl-2,7-diyl)-co-(4,40-(N-(4-sec-butylphenyl)-diphenylamine)]; PF8-BT = poly[(9,9-di-n-octyl-fluorenyl-2,7-diyl)-alt-(benzo[2,1,3]-thiadiazol-4,8-diyl)]; PEDOT:PSS = poly (3,4-ethylenedioxythiophene): poly(styrenesulfonate); TAPC = di-[4-(N,N-ditolylamino) phenyl] cyclohexane; mCP = N,N-dicarbazolyl-3,5-benzene; mCPCN = 9-(3-(9H-carbazol-9-yl)phenyl)-9H-carbazole-3-carbonitrile; 3TPYMB = tris-[3-(3-pyridyl)mesityl]borane; LiF = lithium fluoride; HAT-CN = 1,4,5,8,9,11-hexaaza triphenylenehexacarbonitrile; α-NPD = 4,4′-bis-[N-(1-naphthyl)-N-phenylamino]-1,1′-biphenyl; CCP = 9-phenyl-3,9′-bicarbazole; EML = emission/emitting layer; PPF = 2,8-iso (diphenylphosphoryl) dibenzo[b,d]furan; TPBi = 1,3,5-tris(N-phenylbenzimidazol-2-yl)-benzene; Liq = 8-hydroxyquinoline lithium; MCP = N,N′-dicarbazolyl-3,5-benzene; TmPyPB = 1,3,5-tri[(3-pyridyl)-phen-3-yl]benzene.

Due to their well separated FO, high solid-state quantum efficiency, and small $\Delta E_S \to {}_T$, Wang and coworkers [69] fabricated OLEDs based on compound **24** (Figure 12). However, this material failed dramatically and degraded under operational condition, thereby displaying very poor performance. They suggested that such systems undergo photo-oxidation or exciton-driven transformations within OLEDs, and thus N,C-chelate organoboron compounds bearing an amino group are not suitable for use as emitters in OLEDs. However, another study proved this finding wrong. By replacing phenoxazine by a N,N-diphenylamino core (present over ancillary boron) and introducing one or more F-atom on the C-donating ppy yielded high quantum yield (56.4–100%) materials **25a–b** (Figure 12) with significant delayed fluorescence (Figure 13a,b) [67]. OLEDs that contain dopant **25a** (8 wt%) exhibited an EQE value 20.2%, while those fabricated using **25b** (25 wt%) showed nearly 27% EQE with high current (~64 and 88 Cd/A) and power efficiency (~67 and 82 lm/W) (Figure 13c,d and Table 1). These values (~27%) are considered as some of the best values for thermally activated delayed fluorescence (TADF) materials-based OLEDs. Fluorination not only improved the OLED performance, but also its stability.

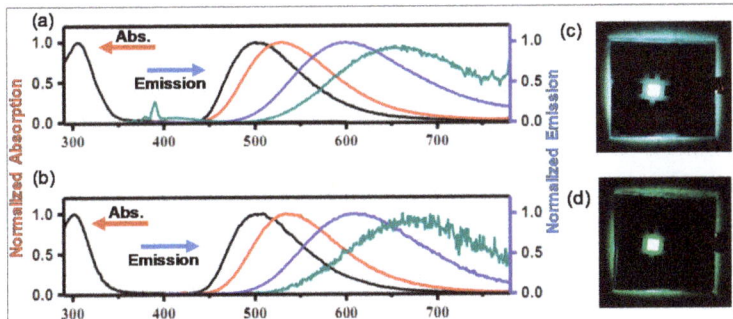

Figure 13. Absorption (arrow towards left) and emission (arrow towards right) spectra of (**a**) **25a**, and (**b**) **25b** in solid (black), and in toluene (red), DCM (blue), and ACN solution (green) at 298 K. Fabricated TADF OLED of (**c**) **25a** (8 wt%) and (**d**) **25b** (25 wt%) devices. Reproduced with permission from reference [67].

Bipolar molecules are considered as a unique class of materials for OE applications as they can transport both the electrons and holes. Despite the fact that bipolar molecules bearing D and A fragments are well known for their ability to enhance TADF, one main challenge in their use as host material is the compression of the band gap via ICT processes, leading to reduction in the device performance. Fortunately, in tetrahedral N^C coordinated materials, the B-atom acts as a node as it separates the FOs, thereby making TADF materials excellent candidates as colorful emissive materials in OLEDs. Matsio and Asuda [24] reported a twisted organoboron TADF molecule **26** (Figure 12), bearing ppy-BPh$_2$ as the acceptor and a spiro[2,7-dimethylacridan-9,9′-fluorene] (MFAc) as the D units. OLED fabricated using **26** showed EL an emission maxima (λ_{max}) at 494 nm (i.e., intense green color) leading to 22.7% EQE with high current (56.4 Cd/A) and power efficiency (44.3 lm/W, Table 1). On the other hand, devices doped with spiro compounds **27a–c** (Figure 12) exhibited tunable EL emission color [68]. Clearly, the presence of different N-donating substituents had a marked effect on the emission. For example, a device with dopant **27a** exhibits a blue emission (λ_{max} = 468 nm); **27b** showed a green emission (λ_{max} = 519 nm); and **27c** showed red EL (λ_{max} = 605 nm, Table 1). Recently, Adachi and coworkers found that the OLEDs fabricated using dopant **28** (Figure 12) exhibit a very high efficiency (up to 10.5%), and surpasses the theoretical limit (5%) for conventional fluorescence emitters [70].

4.1.2. Organic Field-Effect Transistors (OFETs)

Organic field-effect transistors (OFETs) have great potential to be developed as next-generation flexible displays, and e-skins. However, compared to p-type semiconductors, a limited number of investigations has been carried out on the n-type materials. Based on the idea that borylation significantly rigidifies the backbone and increases the electron affinity of the resulting materials, new generation OFETs have been developed. For example, Wang and coworkers [71] designed A-π-A type small organic frameworks incorporating indandione and 1,1-dicyanomethylene-3-indandione as acceptors bridged by thienylthiazyl-boron-core (**29a–c**, Figure 14). They found that OFETs fabricated using **29a** did not work, which was attributed to the steric hindrance and small size π-core. On the other hand, **29b** and **29c** based semiconducting layers exhibit high performance. In fact, **29c** showed the highest electron mobility value (~1.4×10^{-2} cm^2 V^{-1} s^{-1}) reported for small organoboron molecules. They attributed this high performance to increased interchromophoric interactions imparted by the acceptor units with an extended π-surface. Additionally, devices based on **29c** showed much more defined structures and high crystallinity. Besides, control experiments (using C-analogue) indicated that the borylation is an effective way to realize n-type materials [72].

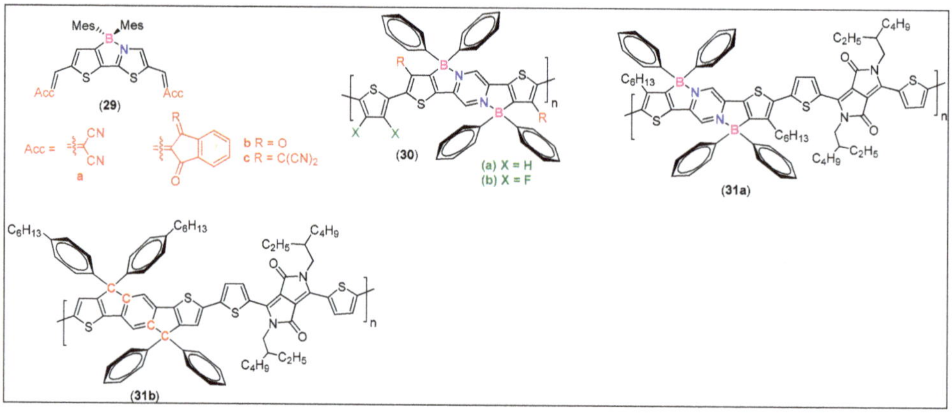

Figure 14. Examples of N^C-donating, π-conjugated cores as OFETs materials.

Similar results (structure dependent electron/hole mobilities) were found for semiconductors such as **30** (Figure 14) [73]. It was noted that the transport characteristics (n-type) of fluorinated systems (**30b**) were much better than non-fluorinated ones (**30a**). OFETs studies (top gate/bottom contact TGBC configuration) indicated ambipolar transport characteristics of these materials. Because of the coplanar backbone conformation of the polymers, both **30a** and **30b** show impressive hole and electron mobilities (10^{-3}–10^{-2} cm^2 V^{-1} s^{-1}), with **30b** being more effective (~2×) for electron mobility than **30a**. On the other hand, OFET tests indicate a well-defined p-type characteristic for both **31a–b** (Figure 14). The hole mobilities of **31b** and **31a** tested by OFETs are 0.059 and 0.035 cm^2 V^{-1} s^{-1}, respectively [74].

4.1.3. Photovoltaics

Oligo and polymeric π-conjugated N^C-chelate organoboron materials have also been tested as component (donor, acceptor, or dye) in bulk heterojunction (BHJs) and dye sensitized solar cells (DSSCs). As discussed before, science is advancing towards the development of the next-generation organic electronic devices, also called plastic electronics. In the design and development of light to electricity converting devices (solar cells), all-polymer solar cells (all-PSCs) are in great demand. To realize all-PSCs, development of new acceptor materials is one of the most challenging tasks [75]. This is why compared to donor materials, fewer research initiatives have been carried out in the area of acceptor materials.

Fortunately, N^C-chelate organoboron materials fulfill the requirement of promising acceptor materials due to their unique structures and excellent optoelectronic properties [76]. Several tetracoordinate organoboron materials are available with low band gaps and LUMO levels comparable to $PC_{61}BM$'s LUMO level [72]. While developing materials with high efficiency, Wang and workers [77] found that the electron mobility of a borylated polymer can be significantly modulated by reducing the steric hindrance effect. Using this knowledge, they developed acceptors **32a–c** (Figure 15) exhibiting power conversion efficiency (PCE) up to 4.95% (Table 2). Similarly, halogenation is an effective strategy to regulate the performance of PV materials. To understand the effect of halogenation on OE properties, Huang and co-workers [76] reported **33** (Figure 15) with different halogens installed on the donor fragment. This arrangement (BDT as acceptor and halogenated CN borane as acceptor) led to weakened ICT transitions. Depending upon the presence and the type of halogen, downshifted energy levels and varying efficiency were observed. For example, polymer **33** ($X = Cl$, $\eta = 4.10\%$, Table 2) showed higher efficiency than **33** ($X = F$, $\eta = 3.65\%$, Table 2), which in turn were better than non-halogenated **33** ($X = H$, $\eta = 1.54\%$, Table 2). A further significant improvement in the efficiency was noted when benzodithiophene fragment from **33** was replaced by a thiophene or 3,4-difluorothiophene **34** (Figure 15) [73]. Good coplanarity, narrow bandgap, downshifted energy levels, and extended absorption profiles (350–800 nm) were some of the features found with **34**. A BHJ device fabricated using acceptor **34** ($X = F$) exhibited a very high PCE (8.42%, Table 2) attributed to the favorable film morphology, balanced charge carriers, and collection among others.

Table 2. PV (all-PSCs and DSSCs) performances of selected π-conjugated N^C-chelate organoboron materials.

Comp. #	Device Architecture	V_{oc} (V)	J_{sc} (mA/cm^{-2})	FF (%)	PCE [a] (%)	Ref.
32a	ITO/PEDOT:PSS/**PTB7-Th:32a**/Ca/Al	0.92	11.37	48	4.95	[77]
	ITO/PEDOT:PSS/**PTB7:32a**/Ca/Al	0.93	9.05	45	3.71	
32c	ITO/PEDOT:PSS/**PTB7-Th:32c**/Ca/Al	1.08	0.51	22	0.10	
	ITO/PEDOT:PSS/**PTB7:32c**/Ca/Al	1.00	2.48	30	0.63	
33	ITO/PEDOT:PSS/**PBDB-T:33**/PDINO/Al	0.97 (X = H)	4.15	38.12	1.54	[76]
		0.95 (X = F)	8.74	43.66	3.65	
		0.95 (X = Cl)	9.19	46.55	4.10	
34	ITO/PEDOT:PSS/**PBDB-T:34**/PNDIT-F3N/Al	0.92 (X = H)	8.01	48.7	3.79	[73]
		0.92 (X = F)	13.01	69.8	8.42	
36	ITO/PEDOT:PSS/**36:PC$_{71}$BM**/LiF/Al	0.82	9.89	46.1	3.62	[78]
37	37/TiO$_2$	0.51–0.73	10.3–14.2	54–72	3.9–6.1	[50]
38	38TBP/TiO$_2$ or 38/TiO$_2$ or 38+DCA-TBP/TiO$_2$ or 38+DCA/TiO$_2$	0.44–0.68	7.8–19.8	49–68	3.2–6.1	[79]

[a] = average PCE value. PDINO = 2,9-Bis[3-(dimethyloxido amino)propyl]anthra[2,1,9-def:6,5,10-d'e'f']diisoquinoline-1,3,8,10(2H,9H)-tetrone; PNDIT-F3N = poly[[2,7-bis(2-ethylhexyl)-1,2,3,6,7,8-hexahydro-1,3,6,8-tetraoxobenzo[lmn][3,8]phenanthroline-4,9-diyl]-2,5-thiophenediyl[9,9-bis[3-(dimethylamino)propyl]-9H-fluorene-2,7-diyl]-2,5-thiophenediyl]; PTB7-Th = poly[4,8-bis(5-(2-ethylhexyl)thiophen-2-yl)benzo[1,2-b:4,5-b']-dithiophene-co-3-fluorothieno[3,4-b]thiophene-2-carboxylate]; PTB7 = poly({4,8-bis[(2-ethylhexyl)oxy]-benzo[1,2-b:4,5-b']dithiophene-2,6-diyl}{3-fluoro-2-[(2-ethylhexyl)carbonyl] thieno[3,4-b]thiophenediyl}); PBDB-T= poly[(2,6-(4,8-bis(5-(2-ethylhexyl)thiophen-2-yl)-benzo[1,2-b:4,5-b']dithiophene))-alt-(5,5-(1',3'-di-2-thienyl-5',7'-bis(2-ethylhexyl)benzo[1',2'-c:4',5'-c']dithiophene-4,8-dione)]; PC$_{71}$BM = [6,6]-phenyl-C$_{71}$-butyric acid methyl ester; TiO$_2$ = titanium(IV) oxide or titanium dioxide; DCA = deoxycholic acid; TBP = 4-tert-butylpyridine.

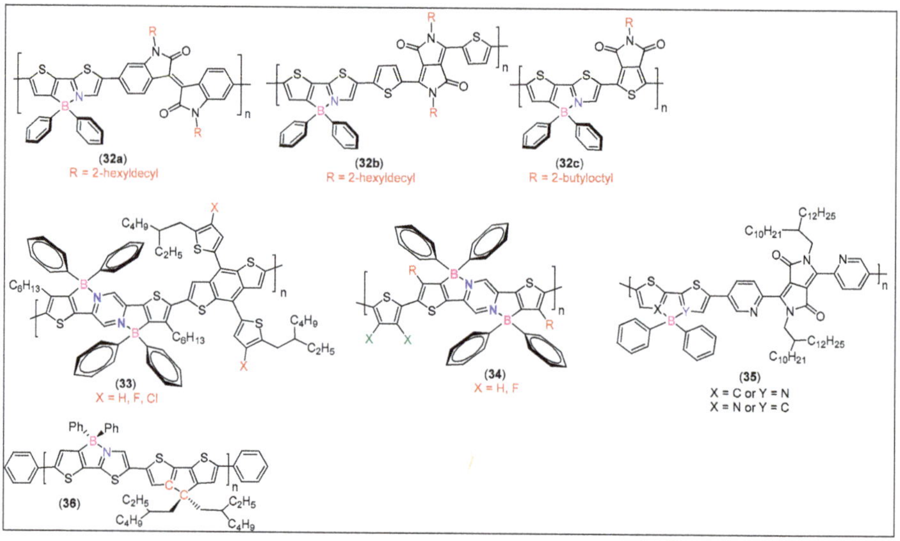

Figure 15. Examples of N^C-donating, π-conjugated cores as BHJ active layers.

To get in-depth knowledge about different side chains on phase separation morphology and all-PSC device performance, Wang and co-workers [80,81] assessed the performance of regio-random amorphous polymer with large steric hindrance (**35**, Figure 15). The reported acceptor has been found to possess energy level matching that of several donor materials. leading to high performance devices (V_{oc} = 0.89–1.98 V, J_{sc} = 10.18–14.24 mA/cm^2, FF = 3.78–6.55%, and PCE = 3.8–6.6%). Among six donor polymers studied, the active layer containing J91:**35** blend showed the most optimal phase separation morphology and the best performance (Figure 16). On the other hand, J51:**35** blend exhibits sub-optimal active layer morphology and poor PV performance (Figure 16). These results indicate that the aggregation tendency in solution of polymer donor is the dominant factor in the phase separation of semi-crystalline polymer donor/amorphous polymer acceptor blend in all-PSCs.

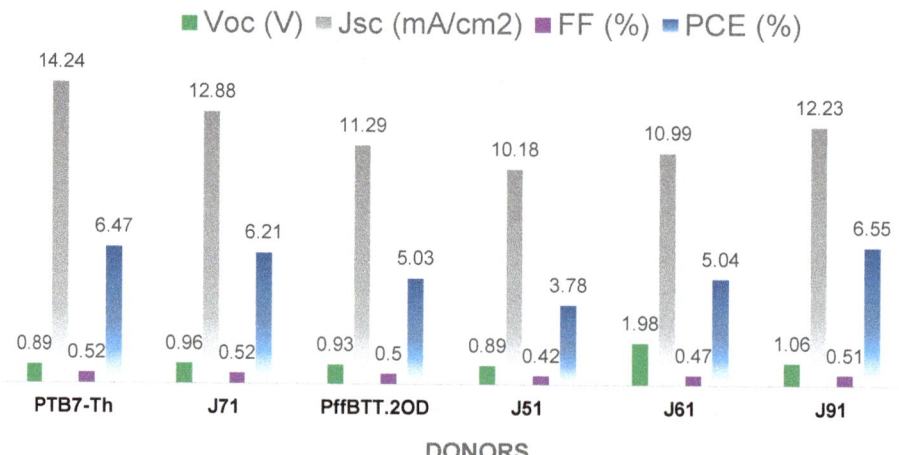

Figure 16. All-PSC device performance of compound **35** with different donors [80,81].

To be used as donor materials, a common strategy is to prepare D-A molecular architecture, which offers tunable LUMO/HOMO energy levels as a low FO level is helpful for harvesting sunlight. Since B-N based materials offer such opportunity, Wang et al. [78] compared the efficiency of polymers **36** (Figure 15). Clearly, the replacement of C–C unit by a B ← N unit significantly altered the frontier energy levels and PV performance. They found that control polymer (all carbon analogue) exhibited higher FOs energy levels, which is detrimental for use in BHJ as donor materials. On the other hand, borylation led to reduction in FOs energy levels, delocalization of FMO throughout the structure, and enhancement in the device performance (PCE up to 3.62%, Table 2).

Despite the fact that the conversion efficiency of DSSCs has reached >10%, further improvement is necessary. This is especially because of two reasons: one is practical applications on a commercial scale and the second is to develop metal free dyes. In the quest for metal free dyes, Shimogawa and coworkers [50] prepared a series of D-π-A dyes (**37**, Figure 17) as DSSC sensitizers, which contain triphenylamine as electron donors (D), bithiophene as π-spacer, boryl-substituted thienylthiazoles as electron acceptors (A), and carboxylic acid derivatives as anchor groups. DSSCs based on **37** exhibited good PCE values (5.1–6.1%). Based on these results, they replaced the bithiophene unit by diketopyrrolopyrrole to develop D-π-A dyes **38** (Figure 17). DSSCs containing these dyes exhibited high J$_{SC}$ values (≤19.8 mA cm^{-2}) as the absorption of **38** extends into the NIR region. [79]

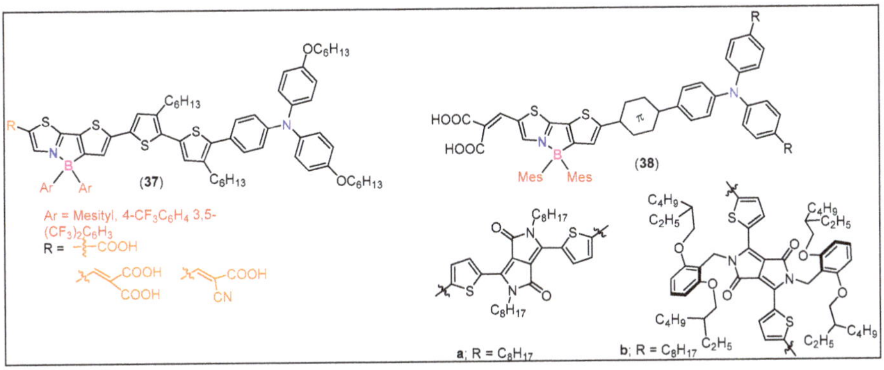

Figure 17. Examples of N^C-donating, π-conjugated cores as DSSC dyes.

4.2. Sensing

Due to the highly labile intramolecular N → B-Lewis pair formation dependent PL properties, N^C complexes have also been utilized for the detection of ions in solution and gaseous analytes. In a recent work, Liu et al. [82,83] found that, in the presence of light, red colored BN-functionalized anthracene **39** (Figure 18) rapidly reacts with oxygen to form colorless endoperoxides. Generally, for this type of reaction on typical diarylanthracenes, a photosensitizer is typically required to promote singlet O$_2$ generation. However, **39** did not require any catalyst and exhibited self-sensitizing properties. This high, catalyst-free O$_2$ reactivity was attributed to strong absorption of visible light, the small singlet-triplet gaps, and the release of steric strain upon peroxide formation. In another report, Schraff et al. [84] showed that in triazole-appended boranes, N-B-coordinate bonds are weak in solution and exist in close (coordinated)-open (non-coordinated) forms. The strength of such bonds (close-open equilibria) can be significantly modulated by changing the substituents present on the triazole ring. The same group showed that this process can be used to monitor on/off emission features upon the addition of different anions [85]. Yan et al. [10] exploited the labile nature of the B ← N coordinate bond and developed a fluorescent probe **40** (Figure 18), which selectively detected fluoride anions (F$^-$) in organic solvent. The authors proposed that in the presence of F$^-$ anions a competitive reaction (labile B ← N *vs* stable B-F bond) takes place leading to the formation of an open form with quenched emission.

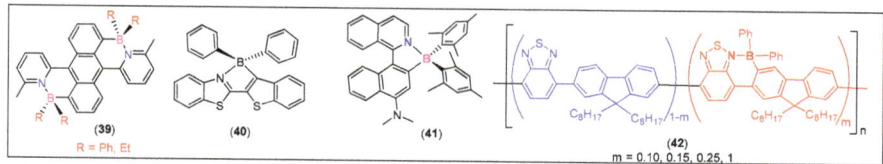

Figure 18. Examples of N^C-donating, π-conjugated cores for sensing and bioimaging.

4.3. Bioimaging

Bioimaging using a fluorescent chemical probe is an emerging area of research. Several organic, inorganic, and organometallic probes have been reported for the imaging of cellular organelles and biological processes [86–88]. Due to their high demand in non-invasive early disease detection and progression, a number of borylated molecular fluorophores have been reported. For example, due to their characteristic features (high chemical and photostability, absorption/emission spanning visible/NIR/FIR regions, biocompatibility, etc.), arylquinolie-based molecular fluorophores emerged as potential candidates [57]. Pischel and coworkers [89] utilized an aryl isoquinoline containing organo-boron N^C chelate dye **41** (Figure 18) for the imaging of the N13 mouse microglial cell line. Moderate absorption co-efficient, large Stokes shift, polarity sensitive fluorescence, and possibility of bioconjugation make them promising imaging probes. Moreover, significant TP absorption cross-sections (up to 61 GM) allow the use of excitation wavelengths in the NIR region (>800 nm). Borylated poly(9,9-dioctylfluorene-alt-benzothiadiazole), **42** (m = 1, Figure 18), is a deep red/near-IR absorbing and highly emissive polymer tested for both OLED and bioimaging applications [90]. Conjugated polymer nanoparticles prepared using pegylated poly(lactic-co-glycolic acid) encapsulating **42** (m = 1) were found to be bright, photostable, low toxicity bioimaging agents for in vivo optical imaging. This includes, but is not limited to, high signal to background ratios, quantum yield (2.3%), low photobleaching (less than 10%), low toxicity, and preferential accumulation in the liver.

4.4. Others

In addition to the above-mentioned applications, other applications have been reported too. This includes the activation of small molecules [91,92], switchable chiral anions [93], NLO materials [94,95], among others. In a recent example, Wang et al. [96] reported blue fluorescent polymers exhibiting thermally reversible photochromism. These photochromic polymers, based on photochromic boron chromophores, showed monomer-ratio-dependent photoisomerization quantum efficiencies and boron dependent fluorescence quenching efficiency. The developed material can be used as switchable/erasable ink and are promising candidate for optical device applications.

5. Opportunities and Challenges

We highlight the features, properties, and applications of N^C-chelate tetracoordinate π-conjugated borylated materials. As discussed, different types of organic backbone can be utilized as N and C donors in combination with various ancillary ligands to prepare small, medium, and large π-conjugated materials. A thorough review of the literature indicated that a minor change in structure significantly modulates properties; therefore, several new synthetic pathways have been reported. The labile B-N bond formation offers coordination-induced stability to the structure and luminescence enhancement. Complexes such as those based on the phenylpyridine core show interesting reversible "bright" and "dark" photochromic states, while those on pyrazole display rare isomerism (two-stage photoreactivity). Such features can be exploited to develop smart optical devices. Due to their excellent charge transport abilities, especially hole transfer, they are being extensively used for the development of acceptor materials in all-PSCs. Compared to imide and cyano-based polymers, organoboron materials exhibit low performance, which is mainly attributed to the lack of well-designed studies [76]. It is envisioned that the selection of suitable co-units and halogenation at ideal positions are important to further improve the performance [76], which should be considered in future studies.

Some researchers also assessed boron containing complexes as donor materials for BHJs; more research is needed to improve the device performance. Similarly, by modifying π-skeletons, intrinsic D/A fragments, a range of dyes could also be achieved. For the biological applications, it is essential to carry out in vivo studies to explore the pros and cons of the materials in the human body. Besides, to achieve targetability, probes bearing a targeting unit (such as peptides) should also be prepared and assessed.

6. Conclusions

This article provides an overview of π-conjugated N^C-chelate organoboron materials. We delineated the important properties, features, and applications of different classes of such materials. We discussed here the effects of different structural components on the properties of the materials. The examples selected in this article clearly demonstrate that tetracoordinate N^C-chelate organoboron materials are potential candidates for the development of next generation materials. We have also suggested some future directions which might prove helpful for the design and development of new boron-based materials for modern applications.

Funding: This research received no external funding. The APC was funded by MDPI.

Acknowledgments: A.H. acknowledges the Department of Chemistry, University of Hail, Kingdom of Saudi Arabia for providing infrastructure support during the preparation of this manuscript. R.A. acknowledges The Research Council (TRC), Oman (project number BFP/RGP/EI/18/076) and A'Sharqiyah University, Oman (ASU-FSFR/CAS/FSHN-01/2019). MSK acknowledges Sultan Qaboos University, Oman (EG/SQU-BP/SCI/CHEM/19/01) and The Research Council (TRC), Oman (BFP/RGP/EI/20/010) for support.

Conflicts of Interest: The authors declare no conflict of interest.

References

1. Tour, J.M. Molecular electronics. Synthesis and testing of components. *Acc. Chem. Res.* **2000**, *33*, 791–804. [CrossRef] [PubMed]
2. Cheng, Y.J.; Yang, S.H.; Hsu, C.S. Synthesis of Conjugated Polymers for Organic Solar Cell Applications. *Chem. Rev.* **2009**, *109*, 5868–5923. [CrossRef] [PubMed]
3. Duan, C.; Zhang, K.; Zhong, C.; Huang, F.; Cao, Y. Recent advances in water/alcohol-soluble π-conjugated materials: New materials and growing applications in solar cells. *Chem. Soc. Rev.* **2013**, *42*, 9071–9104. [CrossRef] [PubMed]
4. Haque, A.; Al-Balushi, R.A.; Al-Busaidi, I.J.; Khan, M.S.; Raithby, P.R. Rise of Conjugated Poly-ynes and Poly(Metalla-ynes): From Design Through Synthesis to Structure-Property Relationships and Applications. *Chem. Rev.* **2018**, *118*, 8474–8597. [CrossRef] [PubMed]
5. Li, D.; Zhang, H.; Wang, Y. Four-coordinate organoboron compounds for organic light-emitting diodes (OLEDs). *Chem. Soc. Rev.* **2013**, *42*, 8416–8433. [CrossRef] [PubMed]
6. Pang, S.; Mas-Montoya, M.; Xiao, M.; Duan, C.; Wang, Z.; Liu, X.; Janssen, R.A.J.; Yu, G.; Huang, F.; Cao, Y. Adjusting Aggregation Modes and Photophysical and Photovoltaic Properties of Diketopyrrolopyrrole-Based Small Molecules by Introducing B ← N Bonds. *Chem. Eur. J.* **2019**, *25*, 564–572. [CrossRef]
7. Hertz, V.M.; Bolte, M.; Lerner, H.W.; Wagner, M. Boron-Containing Polycyclic Aromatic Hydrocarbons: Facile Synthesis of Stable, Redox-Active Luminophores. *Angew. Chem. Int. Ed. Engl.* **2015**, *54*, 8800–8804. [CrossRef]
8. Glogowski, M.; Williams, J. Boron Photochemistry: XV. Determination of the Hammett Substituent Constant for the *p*-Dimesitylboryl Group. *J. Organomet. Chem.* **1981**, *218*, 137–146. [CrossRef]
9. Santos, F.M.; Rosa, J.N.; Candeias, N.R.; Carvalho, C.P.; Matos, A.I.; Ventura, A.E.; Florindo, H.F.; Silva, L.C.; Pischel, U.; Gois, P.M. A Three-Component Assembly Promoted by Boronic Acids Delivers a Modular Fluorophore Platform (BASHY Dyes). *Chem. Eur. J.* **2016**, *22*, 1631–1637. [CrossRef]

10. Yan, N.; Wang, F.; Wei, J.; Song, J.; Yan, L.; Luo, J.; Fang, Z.; Wang, Z.; Zhang, W.; He, G. Highly emissive B←N unit containing four-coordinate C, N-Chelated organoboron compound for the detection of fluoride ions. *Dyes Pigm.* **2019**, *166*, 410–415. [CrossRef]
11. Shi, Y.G.; Wang, J.W.; Li, H.; Hu, G.F.; Li, X.; Mellerup, S.K.; Wang, N.; Peng, T.; Wang, S. A simple multi-responsive system based on aldehyde functionalized amino-boranes. *Chem. Sci.* **2018**, *9*, 1902–1911. [CrossRef] [PubMed]
12. Qi, Y.; Xu, W.; Ding, N.; Chang, X.; Shang, C.; Peng, H.; Liu, T.; Fang, Y. A film-based fluorescent device for vapor phase detection of acetone and related peroxide explosives. *Mater. Chem. Front.* **2019**, *3*, 1218–1224. [CrossRef]
13. Li, X.; Tang, P.; Yu, T.; Su, W.; Li, Y.; Wang, Y.; Zhao, Y.; Zhang, H. Two N, N-chelated difluoroboron complexes containing phenanthroimidazole moiety: Synthesis and luminescence properties. *Dyes Pigm.* **2019**, *163*, 9–16. [CrossRef]
14. Dhanunjayarao, K.; Sa, S.; Aradhyula, B.P.R.; Venkatasubbaiah, K. Synthesis of phenanthroimidazole-based four coordinate organoboron compounds. *Tetrahedron* **2018**, *74*, 5819–5825. [CrossRef]
15. Nagura, K.; Saito, S.; Fröhlich, R.; Glorius, F.; Yamaguchi, S. N-Heterocyclic Carbene Boranes as Electron-Donating and Electron-Accepting Components of π-Conjugated Systems. *Angew. Chem. Int. Ed. Engl.* **2012**, *51*, 7762–7766. [CrossRef]
16. Oda, S.; Shimizu, T.; Katayama, T.; Yoshikawa, H.; Hatakeyama, T. Tetracoordinate Boron-Fused Double [5] Helicenes as Cathode Active Materials for Lithium Batteries. *Org. Lett.* **2019**, *21*, 1770–1773. [CrossRef]
17. Cinar, M.E.; Ozturk, T. Thienothiophenes, dithienothiophenes, and thienoacenes: Syntheses, oligomers, polymers, and properties. *Chem. Rev.* **2015**, *115*, 3036–3140. [CrossRef]
18. Wakamiya, A.; Yamaguchi, S. Designs of functional π-electron materials based on the characteristic features of boron. *Bull. Chem. Soc. Jpn.* **2015**, *88*, 1357–1377. [CrossRef]
19. Mellerup, S.K.; Wang, S. Boron-doped molecules for optoelectronics. *Trends Chem.* **2019**, *1*, 77–89. [CrossRef]
20. Rao, Y.-L.; Amarne, H.; Wang, S. Photochromic four-coordinate N, C-chelate boron compounds. *Coord. Chem. Rev.* **2012**, *256*, 759–770. [CrossRef]
21. Mellerup, S.K.; Wang, S. Isomerization and rearrangement of boriranes: From chemical rarities to functional materials. *Sci. China Mat.* **2018**, *61*, 1249–1256. [CrossRef]
22. Huang, Z.; Wang, S.; Dewhurst, R.D.; Ignat'ev, N.V.; Finze, M.; Braunschweig, H. Boron: Its role in energy related research and applications. *Angew. Chem. Int. Ed. Engl.* **2020**, *59*, 2–19.
23. Entwistle, C.D.; Marder, T.B. Boron chemistry lights the way: Optical properties of molecular and polymeric systems. *Angew. Chem. Int. Ed. Engl.* **2002**, *41*, 2927–2931. [CrossRef]
24. Matsuo, K.; Yasuda, T. Enhancing thermally activated delayed fluorescence characteristics by intramolecular B–N coordination in a phenylpyridine-containing donor-acceptor π-system. *Chem. Commun.* **2017**, *53*, 8723–8726. [CrossRef] [PubMed]
25. Li, Y.; Pang, B.; Meng, H.; Xiang, Y.; Li, Y.; Huang, J. Synthesis of aromatic substituted B ← N embedded units with good stability and strong electron-affinity. *Tetrahedron Lett.* **2019**, *60*, 151286. [CrossRef]
26. Patil, N.T.; Shaikh, A.C. N, C-Chelate Four-Coordinate Organoborons with Full Colourtunability. U.S. Patent 10,301,330, 28 May 2019.
27. Wang, S.; Amarne, H.Y.; Rao, Y. Boron Compounds and Uses Thereof. U.S. Patent 8,697,872 B2, 15 April 2014.
28. Rao, Y.L.; Amarne, H.; Zhao, S.B.; McCormick, T.M.; Martic, S.; Sun, Y.; Wang, R.Y.; Wang, S. Reversible intramolecular C-C bond formation/breaking and color switching mediated by a N,C-chelate in (2-ph-py)BMes2 and (5-BMes2-2-ph-py)BMes2. *J. Am. Chem. Soc.* **2008**, *130*, 12898–12900. [CrossRef]
29. Baik, C.; Murphy, S.K.; Wang, S. Switching of a single boryl center in pi-conjugated photochromic polyboryl compounds and its impact on fluorescence quenching. *Angew. Chem. Int. Ed. Engl.* **2010**, *49*, 8224–8227. [CrossRef]
30. Rao, Y.L.; Amarne, H.; Chen, L.D.; Brown, M.L.; Mosey, N.J.; Wang, S. Photo- and thermal-induced multistructural transformation of 2-phenylazolyl chelate boron compounds. *J. Am. Chem. Soc.* **2013**, *135*, 3407–3410. [CrossRef]
31. Baik, C.; Hudson, Z.M.; Amarne, H.; Wang, S. Enhancing the photochemical stability of N,C-chelate boryl compounds: C-C bond formation versus C=C bond cis,trans-isomerization. *J. Am. Chem. Soc.* **2009**, *131*, 14549–14559. [CrossRef]

32. Novoseltseva, P.; Li, H.; Wang, X.; Sauriol, F.; Wang, S. Structural dynamics and stereoselectivity of chiral benzylidene-amine N, C-chelate borane photo-thermal isomerization. *Chem. Eur. J.* **2020**, *26*, 2276–2284. [CrossRef]
33. Zeng, C.; Yuan, K.; Wang, N.; Peng, T.; Wu, G.; Wang, S. The opposite and amplifying effect of B←N coordination on photophysical properties of regioisomers with an unsymmetrical backbone. *Chem. Sci.* **2019**, *10*, 1724–1734. [CrossRef] [PubMed]
34. Wang, N.; Ko, S.B.; Lu, J.S.; Chen, L.D.; Wang, S. Tuning the photoisomerization of a N^C-chelate organoboron compound with a metal-acetylide unit. *Chem. Eur. J.* **2013**, *19*, 5314–5323. [CrossRef] [PubMed]
35. Baik, C.; Wang, S. Inhibiting olefin cis, trans-photoisomerization and enhancing electron-accepting ability of a diboryl compound by metal chelation. *Chem. Commun.* **2011**, *47*, 9432–9434. [CrossRef] [PubMed]
36. Rao, Y.L.; Amarne, H.; Lu, J.S.; Wang, S. Impact of a dithienyl unit on photostability of N,C-chelating boron compounds. *Dalton Trans.* **2013**, *42*, 638–644. [CrossRef]
37. Wong, H.-L.; Wong, W.-T.; Yam, V.W.-W. Photochromic thienylpyridine–bis (alkynyl) borane complexes: Toward readily tunable fluorescence dyes and photoswitchable materials. *Org. Lett.* **2012**, *14*, 1862–1865. [CrossRef]
38. Li, X.; Shi, Y.; Wang, N.; Peng, T.; Wang, S. Photoisomerization of Pt(II) Complexes Containing Two Different Photochromic Chromophores: Boron Chromophore versus Dithienylethene Chromophore. *Chem. Eur. J.* **2019**, *25*, 5757–5767. [CrossRef]
39. Amarne, H.; Baik, C.; Murphy, S.K.; Wang, S. Steric and electronic influence on photochromic switching of N,C-chelate four-coordinate organoboron compounds. *Chem. Eur. J.* **2010**, *16*, 4750–4761. [CrossRef]
40. Rao, Y.-L.; Wang, S. Impact of cyclometalation and π-conjugation on photoisomerization of an N, C-chelate organoboron compound. *Organometallics* **2011**, *30*, 4453–4458. [CrossRef]
41. Mellerup, S.K.; Yousefalizadeh, G.; Wang, S.; Stamplecoskie, K.G. Experimental Evidence for a Triplet Biradical Excited-State Mechanism in the Photoreactivity of N,C-Chelate Organoboron Compounds. *J. Phys. Chem. A* **2018**, *122*, 9267–9274. [CrossRef]
42. Hudson, Z.M.; Ko, S.B.; Yamaguchi, S.; Wang, S. Modulating the photoisomerization of N,C-chelate organoboranes with triplet acceptors. *Org. Lett.* **2012**, *14*, 5610–5613. [CrossRef]
43. Wang, N.; Wang, J.; Zhao, D.; Mellerup, S.K.; Peng, T.; Wang, H.; Wang, S. Lanthanide Complexes with Photochromic Organoboron Ligand: Synthesis and Luminescence Study. *Inorg. Chem.* **2018**, *57*, 10040–10049. [CrossRef] [PubMed]
44. Yusuf, M.; Liu, K.; Guo, F.; Lalancette, R.A.; Jäkle, F. Luminescent organoboron ladder compounds via directed electrophilic aromatic C–H borylation. *Dalton Trans.* **2016**, *45*, 4580–4587. [CrossRef] [PubMed]
45. Li, F.-P.; Zhu, H.-Y.; Li, Q.-S.; Li, Z.-S. Theoretical study on the regioselective photoisomerization of asymmetric N, C-chelate organoboron compounds. *Phys. Chem. Chem. Phys.* **2019**, *21*, 8376–8383. [CrossRef] [PubMed]
46. Li, Q.-S.; Zhu, H.-Y. Insights into the Photo-induced Isomerization Mechanisms of a N, C-Chelate Organoboron Compound: A Theoretical Study. *ChemPhysChem* **2020**, *21*, 510–517.
47. Li, C.; Mellerup, S.K.; Wang, X.; Wang, S. Accessing Two-Stage Regioselective Photoisomerization in Unsymmetrical N, C-Chelate Organoboron Compounds: Reactivity of B (ppz)(Mes) Ar. *Organometallics* **2018**, *37*, 3360–3367. [CrossRef]
48. Wakamiya, A.; Taniguchi, T.; Yamaguchi, S. Intramolecular B–N Coordination as a Scaffold for Electron-Transporting Materials: Synthesis and Properties of Boryl-Substituted Thienylthiazoles. *Angew. Chem. Int. Ed. Engl.* **2006**, *45*, 3170–3173. [CrossRef] [PubMed]
49. Job, A.; Wakamiya, A.; Kehr, G.; Erker, G.; Yamaguchi, S. Electronic tuning of thiazolyl-capped pi-conjugated compounds via a coordination/cyclization protocol with B(C_6F_5)$_3$. *Org. Lett.* **2010**, *12*, 5470–5473. [CrossRef]
50. Shimogawa, H.; Endo, M.; Taniguchi, T.; Nakaike, Y.; Kawaraya, M.; Segawa, H.; Murata, Y.; Wakamiya, A. D–π–A dyes with an intramolecular B-N coordination bond as a key scaffold for electronic structural tuning and their application in dye-sensitized solar cells. *Bull. Chem. Soc. Jpn.* **2017**, *90*, 441–450. [CrossRef]
51. Dou, C.; Ding, Z.; Zhang, Z.; Xie, Z.; Liu, J.; Wang, L. Developing Conjugated Polymers with High Electron Affinity by Replacing a C—C Unit with a B← N Unit. *Angew. Chem. Int. Ed. Engl.* **2015**, *54*, 3648–3652. [CrossRef]
52. Khan, M.S.; Al-Suti, M.K.; Maharaja, J.; Haque, A.; Al-Balushi, R.; Raithby, P.R. Conjugated Poly-ynes and Poly (metalla-ynes) Incorporating Thiophene-based Spacers for Solar Cell (SC) Applications. *J. Organomet. Chem.* **2015**, *812*, 13–33. [CrossRef]

53. Crossley, D.L.; Goh, R.; Cid, J.; Vitorica-Yrezabal, I.; Turner, M.L.; Ingleson, M.J. Borylated arylamine–benzothiadiazole donor–acceptor materials as low-LUMO, low-band-gap chromophores. *Organometallics* **2017**, *36*, 2597–2604. [CrossRef]
54. Crossley, D.L.; Cid, J.; Curless, L.D.; Turner, M.L.; Ingleson, M.J. Facile Arylation of Four-Coordinate Boron Halides by Borenium Cation Mediated Boro-desilylation and-destannulation. *Organometallics* **2015**, *34*, 5767–5774. [CrossRef] [PubMed]
55. Crossley, D.L.; Vitorica-Yrezabal, I.; Humphries, M.J.; Turner, M.L.; Ingleson, M.J. Highly Emissive Far Red/Near-IR Fluorophores Based on Borylated Fluorene–Benzothiadiazole Donor–Acceptor Materials. *Chem. Eur. J.* **2016**, *22*, 12439–12448. [CrossRef] [PubMed]
56. Shimogawa, H.; Yoshikawa, O.; Aramaki, Y.; Murata, M.; Wakamiya, A.; Murata, Y. 4,7-Bis[3-(dimesitylboryl)thien-2-yl]benzothiadiazole: Solvato-, Thermo-, and Mechanochromism Based on the Reversible Formation of an Intramolecular B-N Bond. *Chemistry* **2017**, *23*, 3784–3791. [CrossRef] [PubMed]
57. Shaikh, A.C.; Ranade, D.S.; Thorat, S.; Maity, A.; Kulkarni, P.P.; Gonnade, R.G.; Munshi, P.; Patil, N.T. Highly emissive organic solids with remarkably broad color tunability based on N,C-chelate, four-coordinate organoborons. *Chem. Commun. (Camb.)* **2015**, *51*, 16115–16118. [CrossRef] [PubMed]
58. Bachollet, S.P.; Volz, D.; Fiser, B.; Munch, S.; Ronicke, F.; Carrillo, J.; Adams, H.; Schepers, U.; Gomez-Bengoa, E.; Brase, S.; et al. A Modular Class of Fluorescent Difluoroboranes: Synthesis, Structure, Optical Properties, Theoretical Calculations and Applications for Biological Imaging. *Chem. Eur. J.* **2016**, *22*, 12430–12438. [CrossRef]
59. Zhao, Z.; Chang, Z.; He, B.; Chen, B.; Deng, C.; Lu, P.; Qiu, H.; Tang, B.Z. Aggregation-Induced Emission and Efficient Solid-State Fluorescence from Tetraphenylethene-Based N, C-Chelate Four-Coordinate Organoborons. *Chem. Eur. J.* **2013**, *19*, 11512–11517. [CrossRef]
60. Boscá Mayans, F.; Cuquerella Alabort, M.C.; Fernandes Pais, V.C.; Ros Lao, A.; Pischel, U. Excited-State Pathways of Four-Coordinate N, C-Chelate Organoboron Dyes. *ChemPhotoChem* **2018**, *2*, 34–41. [CrossRef]
61. Domínguez, Z.; Pais, V.N.F.; Collado, D.; Vázquez-Domínguez, P.; Albendín, F.N.; Pérez-Inestrosa, E.; Ros, A.; Pischel, U. π-Extended Four-Coordinate Organoboron N, C-Chelates as Two-Photon Absorbing Chromophores. *J. Org. Chem.* **2019**, *84*, 13384–13393. [CrossRef]
62. Pais, V.F.; Lassaletta, J.M.; Fernandez, R.; El-Sheshtawy, H.S.; Ros, A.; Pischel, U. Organic fluorescent thermometers based on borylated arylisoquinoline dyes. *Chemistry* **2014**, *20*, 7638–7645. [CrossRef]
63. Domínguez, Z.; López-Rodríguez, R.; Álvarez, E.; Abbate, S.; Longhi, G.; Pischel, U.; Ros, A. Azabora [5] helicene Charge-Transfer Dyes Show Efficient and Spectrally Variable Circularly Polarized Luminescence. *Chem. Eur. J.* **2018**, *24*, 12660–12668. [CrossRef] [PubMed]
64. Yang, J.; Zhao, Z.; Wang, S.; Guo, Y.; Liu, Y. Insight into high-performance conjugated polymers for organic field-effect transistors. *Chem* **2018**, *4*, 2748–2785. [CrossRef]
65. He, B.; Chang, Z.; Jiang, Y.; Chen, B.; Lu, P.; Kwok, H.S.; Qin, A.; Zhao, Z.; Qiu, H. Impacts of intramolecular B–N coordination on photoluminescence, electronic structure and electroluminescence of tetraphenylethene-based luminogens. *Dyes Pigm.* **2014**, *101*, 247–253. [CrossRef]
66. Crossley, D.; Cade, I.; Clark, E.R.; Escande, A.; Humphries, M.; King, S.; Vitorica-Yrezabal, I.; Ingleson, M.; Turner, M. Enhancing electron affinity and tuning band gap in donor–acceptor organic semiconductors by benzothiadiazole directed C-H borylation. *Chem. Sci.* **2015**, *6*, 5144–5151. [CrossRef]
67. Shiu, Y.-J.; Chen, Y.-T.; Lee, W.-K.; Wu, C.-C.; Lin, T.-C.; Liu, S.-H.; Chou, P.-T.; Lu, C.-W.; Cheng, I.-C.; Lien, Y.-J. Efficient thermally activated delayed fluorescence of functional phenylpyridinato boron complexes and high performance organic light-emitting diodes. *J. Mat. Chem. C* **2017**, *5*, 1452–1462. [CrossRef]
68. Wong, B.Y.; Wong, H.L.; Wong, Y.C.; Chan, M.Y.; Yam, V.W. Air-Stable Spirofluorene-Containing Ladder-Type Bis(alkynyl)borane Compounds with Readily Tunable Full Color Emission Properties. *Chem. Eur. J.* **2016**, *22*, 15095–15106. [CrossRef]
69. Mellerup, S.K.; Yuan, K.; Nguyen, C.; Lu, Z.H.; Wang, S. Donor-Appended N,C-Chelate Organoboron Compounds: Influence of Donor Strength on Photochromic Behaviour. *Chem. Eur. J.* **2016**, *22*, 12464–12472. [CrossRef]
70. Mamada, M.; Tian, G.; Nakanotani, H.; Su, J.; Adachi, C. The Importance of Excited-State Energy Alignment for Efficient Exciplex Systems Based on a Study of Phenylpyridinato Boron Derivatives. *Angew. Chem. Int. Ed. Engl.* **2018**, *57*, 12380–12384. [CrossRef]

71. Zhao, R.; Min, Y.; Dou, C.; Lin, B.; Ma, W.; Liu, J.; Wang, L. A Conjugated Polymer Containing a B ← N Unit for Unipolar n-Type Organic Field-Effect Transistors. *ACS Appl. Polym. Mater.* **2020**, *2*, 19–25. [CrossRef]
72. Hecht, R.; Kade, J.; Schmidt, D.; Nowak-Krol, A. n-Channel Organic Semiconductors Derived from Air-Stable Four-Coordinate Boron Complexes of Substituted Thienylthiazoles. *Chem. Eur. J.* **2017**, *23*, 11620–11628. [CrossRef]
73. Li, Y.; Meng, H.; Liu, T.; Xiao, Y.; Tang, Z.; Pang, B.; Li, Y.; Xiang, Y.; Zhang, G.; Lu, X.; et al. 8.78% Efficient All-Polymer Solar Cells Enabled by Polymer Acceptors Based on a B ← N Embedded Electron-Deficient Unit. *Adv. Mater.* **2019**, *31*, e1904585. [CrossRef] [PubMed]
74. Pang, B.; Tang, Z.; Li, Y.; Meng, H.; Xiang, Y.; Li, Y.; Huang, J. Synthesis of Conjugated Polymers Containing B ← N Bonds with Strong Electron Affinity and Extended Absorption. *Polymers* **2019**, *11*, 1630. [CrossRef] [PubMed]
75. Zhao, C.; Wang, J.; Jiao, J.; Huang, L.; Tang, J. Recent advances of polymer acceptors for high-performance organic solar cells. *J. Mat. Chem. C* **2020**, *8*, 28–43. [CrossRef]
76. Meng, H.; Li, Y.; Pang, B.; Li, Y.; Xiang, Y.; Guo, L.; Li, X.; Zhan, C.; Huang, J. Effects of Halogenation in B←N Embedded Polymer Acceptors on Performance of All-polymer Solar Cells. *ACS Appl. Mater. Inter.* **2020**, *12*, 2733–2742. [CrossRef] [PubMed]
77. Zhao, R.; Dou, C.; Xie, Z.; Liu, J.; Wang, L. Polymer Acceptor Based on B ← N Units with Enhanced Electron Mobility for Efficient All-Polymer Solar Cells. *Angew. Chem. Int. Ed. Engl.* **2016**, *55*, 5313–5317. [CrossRef] [PubMed]
78. Zhang, Z.; Ding, Z.; Dou, C.; Liu, J.; Wang, L. Development of a donor polymer using a B ← N unit for suitable LUMO/HOMO energy levels and improved photovoltaic performance. *Poly. Chem.* **2015**, *6*, 8029–8035. [CrossRef]
79. Shimogawa, H.; Endo, M.; Nakaike, Y.; Murata, Y.; Wakamiya, A. D-π-A Dyes with Diketopyrrolopyrrole and Boryl-substituted Thienylthiazole Units for Dye-sensitized Solar Cells with High J SC Values. *Chem. Lett.* **2017**, *46*, 715–718. [CrossRef]
80. Zhang, L.; Ding, Z.; Zhao, R.; Jirui, F.; Ma, W.; Liu, J.; Wang, L. Effect of polymer donor aggregation on the active layer morphology of amorphous polymer acceptor-based all-polymer solar cells. *J. Mat. Chem. C* **2020**, *8*, 5613–5619. [CrossRef]
81. Zhao, R.; Lin, B.; Feng, J.; Dou, C.; Ding, Z.; Ma, W.; Liu, J.; Wang, L. Amorphous Polymer Acceptor Containing B← N Units Matches Various Polymer Donors for All-Polymer Solar Cells. *Macromolecules* **2019**, *52*, 7081–7088. [CrossRef]
82. Liu, K.; Lalancette, R.A.; Jäkle, F. B–N Lewis pair functionalization of anthracene: Structural dynamics, optoelectronic properties, and O_2 sensitization. *J. Am. Chem. Soc.* **2017**, *139*, 18170–18173. [CrossRef]
83. Liu, K.; Lalancette, R.A.; Jäkle, F. Tuning the structure and electronic properties of B-N fused dipyridylanthracene and implications on the self-sensitized reactivity with singlet oxygen. *J. Am. Chem. Soc.* **2019**, *141*, 7453–7462. [CrossRef] [PubMed]
84. Schraff, S.; Sun, Y.; Pammer, F. Tuning of electronic properties via labile N → B-coordination in conjugated organoboranes. *J. Mat. Chem. C* **2017**, *5*, 1730–1741. [CrossRef]
85. Koch, R.; Sun, Y.; Orthaber, A.; Pierik, A.J.; Pammer, F.D. Turn-on Fluorescence Sensors Based on Dynamic Intramolecular N → B-Coordination. *Org. Chem. Front.* **2020**. [CrossRef]
86. Haque, A.; Faizi, M.S.H.; Rather, J.A.; Khan, M.S. Next generation NIR fluorophores for tumor imaging and fluorescence-guided surgery: A review. *Bioorg. Med. Chem.* **2017**, *25*, 2017–2034. [CrossRef]
87. Haque, A.; Al-Balushi, R.A.; Khan, M.S. σ-Acetylide Complexes for Biomedical Applications: Features, Challenges and Future directions. *J. Organomet. Chem.* **2019**, *897*, 95–106. [CrossRef]
88. DeRosa, C.A.; Fraser, C.L. *Tetracoordinate Boron Materials for Biological Imaging*; John Wiley & Sons Ltd.: Hoboken, NJ, USA, 2017.
89. Fernandes Pais, V.C.; Alcaide, M.M.; Pischel, U. Strongly Emissive and Photostable Four-Coordinate Organoboron N, C Chelates and Their Use in Fluorescence Microscopy. *Chem. Eur. J.* **2015**, *21*, 15369–15376. [CrossRef]
90. Neumann, P.R.; Crossley, D.L.; Turner, M.; Ingleson, M.; Green, M.; Rao, J.; Dailey, L.A. In Vivo Optical Performance of a New Class of Near-Infrared-Emitting Conjugated Polymers: Borylated PF8-BT. *ACS Appl. Mater. Inter.* **2019**, *11*, 46525–46535. [CrossRef]

91. Chen, J.; Lalancette, R.A.; Jäkle, F. Stereoselective ortho borylation of pyridylferrocenes. *Organometallics* **2013**, *32*, 5843–5851. [CrossRef]
92. Chen, J.; Lalancette, R.A.; Jakle, F. Chiral organoborane Lewis pairs derived from pyridylferrocene. *Chem. Eur. J.* **2014**, *20*, 9120–9129. [CrossRef]
93. Chen, J.; Murillo Parra, D.A.; Lalancette, R.A.; Jäkle, F. Redox-Switchable Chiral Anions and Cations Based on Heteroatom-Fused Biferrocenes. *Organometallics* **2015**, *34*, 4323–4330. [CrossRef]
94. Jiménez, C.C.; Enríquez-Cabrera, A.; González-Antonio, O.; Ordóñez-Hernández, J.; Lacroix, P.G.; Labra-Vázquez, P.; Farfán, N.; Santillan, R. State of the Art of Boron and Tin Complexes in Second-and Third-Order Nonlinear Optics. *Inorganics* **2018**, *6*, 131. [CrossRef]
95. Mukundam, V.; Sa, S.; Kumari, A.; Das, R.; Venkatasubbaiah, K. B–N coordinated triaryl pyrazole: Effect of dimerization, and optical and NLO properties. *J. Mat. Chem. C* **2019**, *7*, 12725–12737. [CrossRef]
96. Wang, J.; Jin, B.; Wang, N.; Peng, T.; Li, X.; Luo, Y.; Wang, S. Organoboron-based photochromic copolymers for erasable writing and patterning. *Macromolecules* **2017**, *50*, 4629–4638. [CrossRef]

© 2020 by the authors. Licensee MDPI, Basel, Switzerland. This article is an open access article distributed under the terms and conditions of the Creative Commons Attribution (CC BY) license (http://creativecommons.org/licenses/by/4.0/).

Review

Tri(boryl)alkanes and Tri(boryl)alkenes: The Versatile Reagents

Oriol Salvadó and Elena Fernández *

Department Química Física i Inorgànica, University Rovira i Virgili, 43007 Tarragona, Spain; oriol.salvado@urv.cat
* Correspondence: mariaelena.fernandez@urv.cat

Received: 26 March 2020; Accepted: 8 April 2020; Published: 10 April 2020

Abstract: The interest of organoboron chemistry in organic synthesis is growing, together with the development of new and versatile polyborated reagents. Here, the preparation of 1,1,1-tri(boryl)alkanes, 1,2,3-tri(boryl)alkanes, 1,1,2-tri(boryl)alkanes, as well as 1,1,2-tri(boryl)alkenes as suitable and accessible polyborated systems is demonstrated as being easily applied in the construction of new carbon-carbon and carbon-heteroatom bonds. Synthetic procedures and limitations have been collected to demonstrate the powerful strategies to construct selective molecules, taking advantages of the easy transformation of carbon-boron bond in multiple functionalities, under the total control of chemo- and stereoselectivity.

Keywords: 1,1,1-tri(boryl)alkanes; 1,2,3-tri(boryl)alkanes; 1,1,2-tri(boryl)alkanes; 1,1,2-tri(boryl)alkenes; synthetic approaches; synthetic applications

1. Introduction

The possibility of preparing a compound that contains three C-B bonds within its formula enhances the power towards polyfunctionalization strategies in a sequential manner. The fact that the three boryl units can be bonded to the same C or different carbons of the same single compound opens a large number of possibilities to construct a new functional product, especially from the perspective that the presence of more than one boryl group contributes to the stabilization of the intermediate carbanions formed, facilitating the transformation of the C-B bond, even under room temperature conditions. Both tri(boryl)alkanes and tri(boryl)alkenes preserve the stereoselectivity along the transformations and the current approaches to functionalize chemoselectively one boryl unit versus the other two, which nowadays open an unlimited opportunity to stimulate the creativity of synthetic chemists towards accessible protocols for useful applications.

2. 1,1,1-Tri(boryl)alkanes

2.1. Synthetic Approaches

Triborylmethane compounds have originally been prepared by Matteson and co-workers, developing a reliable synthetic procedure by mixing dimethoxyboron chloride, Cl-B(OMe)$_2$, with CHCl$_3$ and Li (Scheme 1) [1,2]. By using the appropriate chlorinated reagents, related triboryl compounds can be easily prepared, giving access to a new generation of organopolyborated species.

Scheme 1. Synthesis of H(B(OMe)$_2$)$_3$.

The lability of dimethoxyboryl moieties contributes to several difficulties in isolating the corresponding organotriboronates, and consequently, cyclic boronic esters have been prepared, as they are generally more stable towards hydrolysis. The transesterification of tris(dimethoxyboryl)methane with ethylenglycol or pinacol provides a direct access to the cyclic esters ethylenglycolboryl and pinacolboryl [3], with the expected degree of stability (Scheme 2) [4].

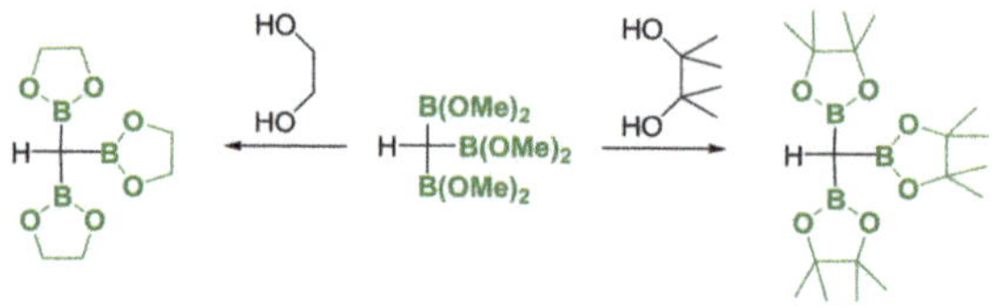

Scheme 2. Synthesis of H(Bpin)$_3$ from H(B(OMe)$_2$)$_3$.

In particular, the synthesis of tris(pinacolboryl)methane has been recently redesigned by Suginome and co-workers, via the α,α-diborylation of pyrazolylaniline-modified methylboronic acid with B$_2$pin$_2$ in the presence of [Ir(μ-OMe)(COD)]$_2$ as catalyst (Scheme 3) [5].

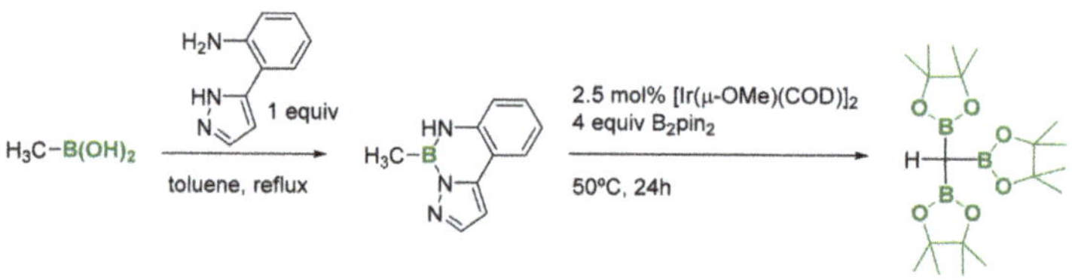

Scheme 3. Synthesis of H(Bpin)$_3$ from Me-B(OH)$_2$.

Triborylalkane derivatives have been prepared via the triple borylation of terminal C(sp^3)–H bonds with transition-metal complexes. Sato, Mita and co-workers determined the first catalytic C(sp^3)–H triborylation at a single carbon with the assistance of a nitrogen directing group [6]. Model substrate 2-ethylpyridine forms a five-membered metallacycle intermediate, via C(sp^3)–H activation with Ir from [Ir(μ-Cl)(COD)]$_2$ catalyst, that subsequently reacts with 4 equiv of B$_2$pin$_2$ towards site-selective C(sp^3)–H borylation leading to the desired triborylated product (Scheme 4a). Triborylation of terminal C(sp^3)–H bonds in toluene was also observed by Chirik and co-workers by using 50 mol% of α-diimine cobalt bis(carboxylate) precatalysts and 4 equiv of B$_2$pin$_2$ (Scheme 4b) [7]. Similarly, α-diimine nickel bis(carboxylate) complex enables the selective preparation of benzyltriboronates via the triborylation of benzylic C–H bonds (Scheme 4c) [8].

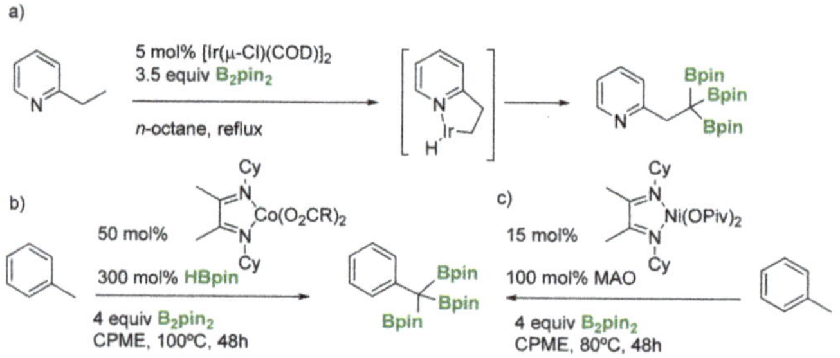

Scheme 4. Transition metal catalytized C(sp^3)–H triborylation at a single carbon.

The sequential dehydrogenative borylation/hydroboration of terminal alkynes with Cu complexes contributes to the expansion of the triborylalkane derivatives synthesis. Marder and co-workers have developed a one-step synthesis via the dehydrogenative borylation of alkynes followed by the double hydroboration of 1-alkynylboronate intermediate with HBpin (HBpin = pinacolborane) catalyzed by Cu(OAc)$_2$ (Scheme 5a) [9]. Interestingly, other copper sources, such as CuCl$_2$, CuCl, and Cu(OTf)$_2$, resulted in being inactive in this transformation. The base KF was essential in a stoichiometric manner, presumably by promoting the Cu-H formation in an early step of the catalytic cycle (Scheme 5b). This strategy complements a pioneer attempt by Marder and co-workers to perform the Rh-catalyzed reaction of ethynylarenes with HBcat (HBcat = catecholborane), to obtain the corresponding tris(catecholboryl)alkane, via vinylboronate intermediate [10]. Catechol-substituted monoborylacetylene could be regioselectively hydroborted via the transition-metal free syn-addition of catecholborane favoring the formation of the 1,1-diborylalkene, that can be further hydroborated selectively with HBCl$_2$ to provide the corresponding 1,1,1-triborylethane. The subsequent substitution of Cl atoms by cathecol leads to the desired tris(catecholboryl)ethane [11]. Further treatment of tris(catecholboryl)ethane with ^{t}BuLi results in the formation of 1,1,1-tris[di(tert-butyl)boryl]ethane (Scheme 6).

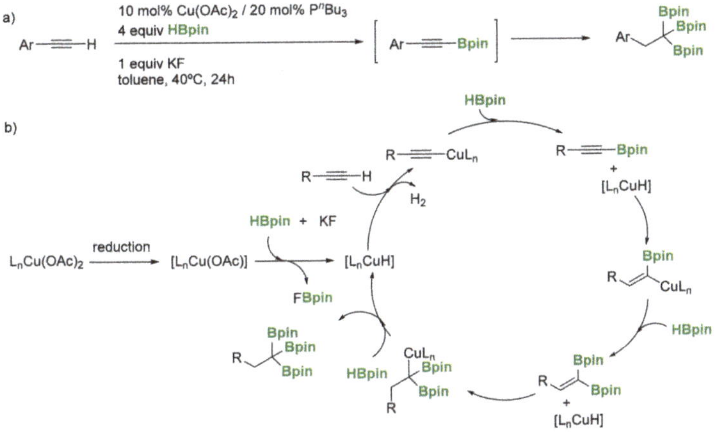

Scheme 5. (a) sequential dehydrogenative borylation/hydroboration of terminal alkynes; (b) mechanistic proposal.

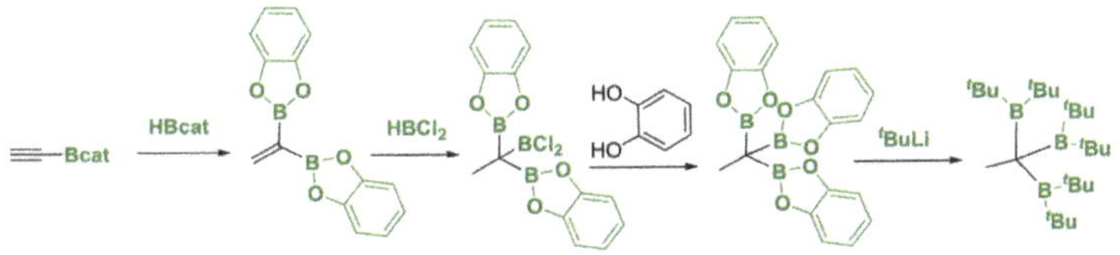

Scheme 6. Synthesis of 1,1,1-tris(catecholboryl)alkane and 1,1,1-tris[di(tert-butyl)boryl]ethane.

Alternatively, Chirik and co-workers have demonstrated that a cyclohexylsubstituted pyridine(diimine) cobalt methyl complex allows the dehydrogenative 1,1-diboration of terminal alkynes with B$_2$pin$_2$ at room temperature. This 1,1-diborylalkene intermediate can accomplish a consecutive cobalt-catalyzed hydroboration with HBpin to generate the desired of 1,1,1-triborylalkane (Scheme 7) [12]. The methodology is described as a one pot-sequential 1,1-diboration of terminal alkynes with B$_2$pin$_2$, followed by hydroboration with HBpin, although two different types of cobalt complexes are used for each step.

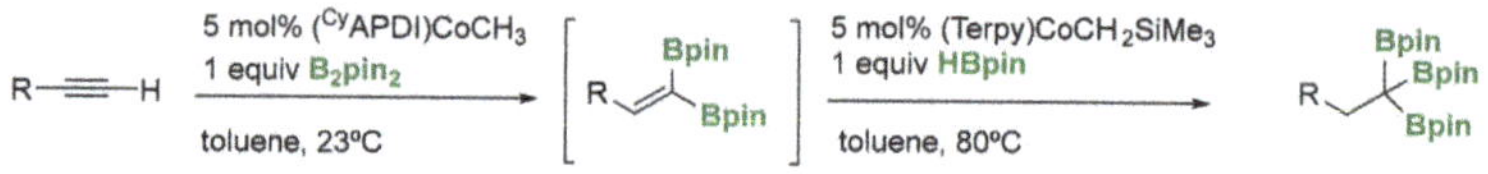

Scheme 7. Cobalt-catalyzed dehydrogenative 1,1-diboration of terminal alkynes.

Huang and co-workers have developed a Co(I)-catalyzed double dehydrogenative borylation of vinylarenes with B$_2$pin$_2$ to generate a 1,1-diborylalkene intermediate, which undergoes hydroboration with pinacolborane to give 1,1,1-tris(boronates) in high yield (Scheme 8a) [13]. The dehydrogenative borylation of alkenes leading to vinylic boron compounds was first reported in 1992 [14], and it is understood as being the result of a catalytic hydroboration of an alkene with a monohydrideborane, followed by the elimination of hydrogen. However, Huang and co-workers have postulated a mechanism where the dehydrogenative borylation occurs by the insertion of vinylarenes into the Co–B bond to generate a β-boryl-substituted Co(I) species, which undergoes β-hydride elimination to give monoborylalkenes and the Cu-H intermediate that regenerates the Co-B catalytic species via σ-bond methatesis with B$_2$pin$_2$, maintaining the same type of reactivity along the catalytic cycle (Scheme 8b). As a limitation, this protocol cannot be applied for 1,2- and 1,1-disubstituted vinylarenes as well as vinylalkanes, since 1,1,1-tris(boronates) were not formed.

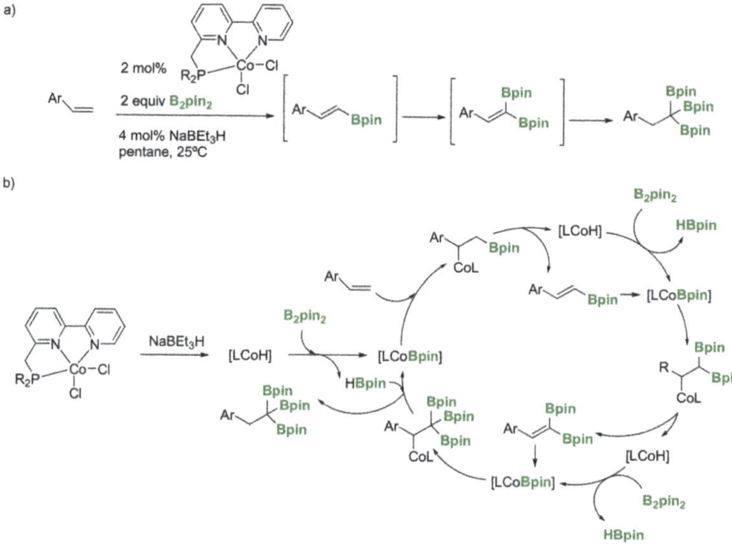

Scheme 8. (a) Co(I)-catalyzed double dehydrogenative borylation of vinylarenes; (b) mechanistic proposal.

2.2. Synthetic Applications

The reactivity of 1,1,1-tris(pinacolboryl)alkanes has been launched by alkoxide-assisted deborylative pathways, via α-bis(boryl) carbanion formation, that can be eventually trapped by different electrophiles [4]. The deborylative alkylation of 1,1,1-tris(boronate) products has been applied to the synthesis of internal geminal bis(boronates), by reaction with alkyl bromides in the presence of NaOMe (1.5 equiv), at 25 °C (Scheme 9) [13]. Interestingly, 1,1,1-tris(boronates) seems to be more reactive for deborylative alkylations than the analogue 1,1-bis(boronates), as the reactions with the latter require the use of an excess of NaO^{t}Bu (3 equiv) [15]. The reaction is compatible with diverse functionalities in the electrophiles, including cyanide, propylene epoxide, ester, halides and amines.

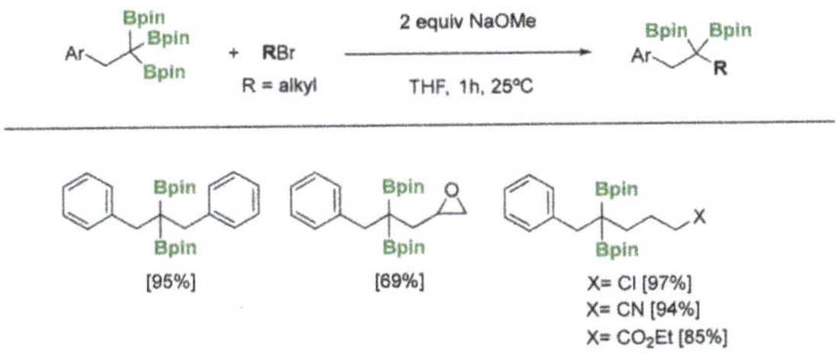

Scheme 9. Reactivity of 1,1,1-tris(pinacolboryl)alkanes with RBr.

Double deborylative alkylation delivers tertiary boronic esters, as well as carbocyclic derivatives. Using 1,n-dihalides as electrophiles and NaO^{t}Bu (4 equiv), the base-assisted deborylation acts twice on the triborylated reagent, to facilitate the formation of α-vinylboronates (when n = 0) and cyclic organoboronates (when n = 1–5) at room temperature within 6h (Scheme 10) [9]. Following the same synthetic strategy, different alkyl groups can be introduced in a stepwise manner through sequential deborylative alkylations, providing access to tertiary boronic esters, which, after in situ oxidation with H$_2$O$_2$/NaOH, furnished tertiary alcohols containing three different alkyl groups in moderate yield (Scheme 11) [9].

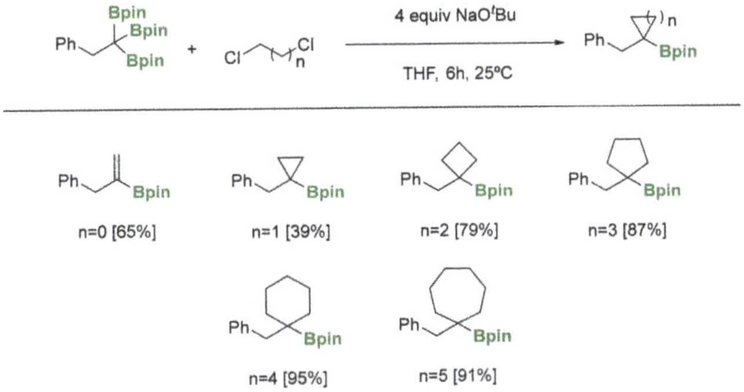

Scheme 10. Double deborylative alkylation of 1,1,1-tris(pinacolboryl)alkanes.

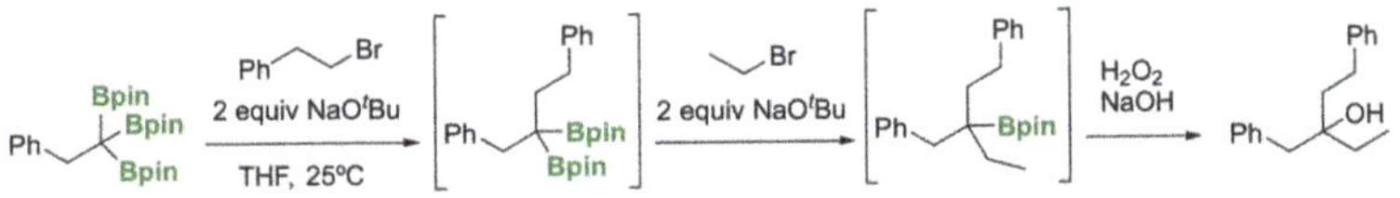

Scheme 11. Sequential deborylative alkylations, followed by in situ oxidation with H$_2$O$_2$/NaOH, towards tertiary alcohols.

The deborylative conjugate addition of 1,1,1-tris(pinacolboryl)alkanes has been pursued through a NaOEt (1 equiv) assisted reaction to generate benzylic α-diboryl anions that can act as soft nucleophiles to be added to (E)-methylcrotonate, at room temperature, followed by aqueous work up (Scheme 12) [8]. Products formed from the deborylative conjugate addition to phenyl-substituted (E)-methyl cinnamate, as well as α,β,γ,δ-unsaturated ethyl sorbate, suffered from a competitive 1,2-addition to the more electron deficient enoates (Scheme 12) [8]. The reaction can be diastereoselective for α-subtituted substrates, such as methyl (E)-2-methylbut-2-enoate, in favor of the anti diastereomer. The stereocontrol at the α-position of the ester group has been suggested to occur during the protonation when the reaction is quenched (Scheme 13a) [8]. Remarkably, when the deborylative conjugate addition of the benzylic α-diboryl anion to (E)-methylcrotonate was quenched with iodomethane, the resulting product was isolated in >20:1 as the syn diastereomer (Scheme 13b) [8], highlighting the exclusive ability of boronate groups to act as masking groups for soft nucleophilic anions and also as elements for substrate-controlled diastereoselectivity.

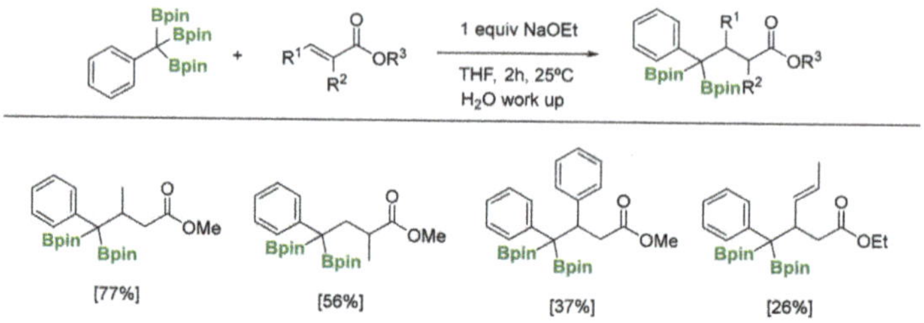

Scheme 12. Deborylative conjugate addition of 1,1,1-tris(pinacolboryl)alkanes.

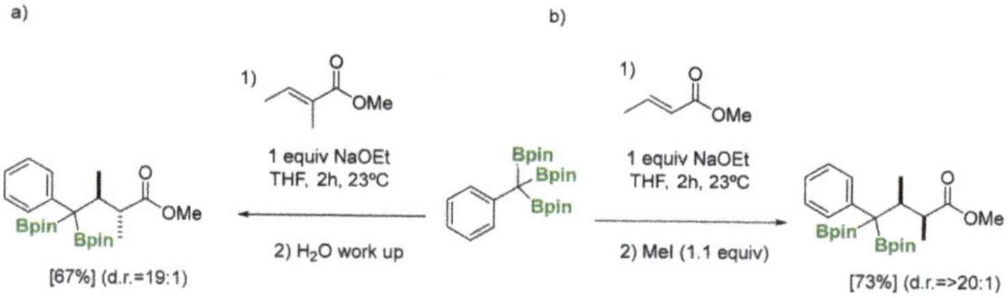

Scheme 13. (a) diastereoselective deborylative conjugate addition of 1,1,1-tris(pinacolboryl)alkanes; (b) diastereoselective sequential deborylative conjugate addition / electrophilic methylation.

The carboxylation of 1,1,1-tris(pinacolboryl)alkanes, using CsF as a mild activator for deborylation [16], has been explored under 1 atm of CO_2, providing the corresponding carboxylated products in moderate yields, after esterification with RI (Scheme 14) [6]. In addition, the sequence based on the trioxidation of 1,1,1-tris(pinacolboryl)alkanes using H_2O_2 in the presence of an acidic additive such as TsOH·H_2O prevents protodeborylation pathways, and n-C_6H_{13}-I allows the isolation of hexyl 2-(4-methylpyridin-2-yl)acetate in moderate yield. Both methods' results are useful, since carboxylic acids with complementary chain lengths can be prepared by selecting the appropriate reagents to interact with the tris(boryl)alkane.

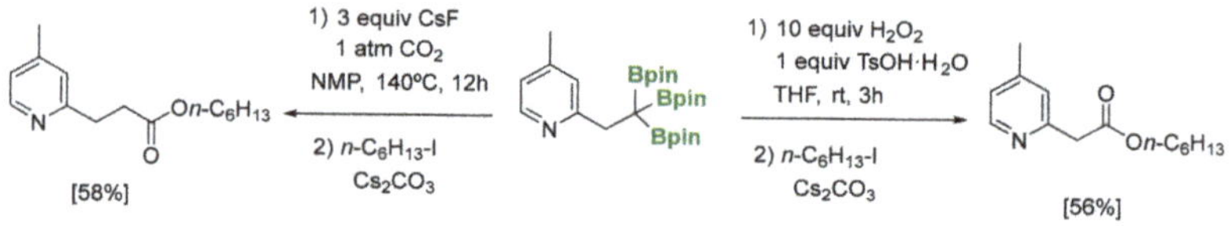

Scheme 14. Carboxylation of 1,1,1-tris(pinacolboryl)alkanes.

Other interesting reactivity in C-C bond formation from tris(boryl)alkane involves deborylative Boron–Wittig olefination, via the B-O elimination pathway. Originally, the proof of concept was established by the addition of methyllithium to tris(ethylenedioxyboryl)methane, to generate in situ the corresponding bis(ethylenedioxyboryl)methide salt, that interacts with aldehydes or ketones to give the corresponding alkene monoboronic esters (Scheme 15) [17,18].

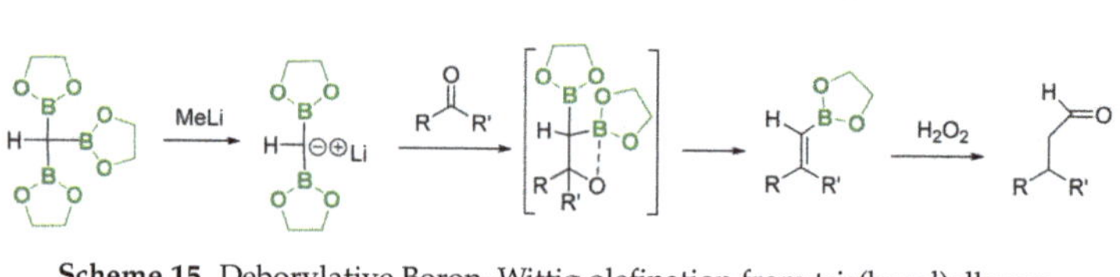

Scheme 15. Deborylative Boron–Wittig olefination from tris(boryl)alkanes.

The analogue 1,1,1-tris(boryl)methide salt has exclusively been prepared via base-assisted deborylation of the corresponding tetra(boryl)methane (C(BO$_2$C$_2$Me$_4$)$_4$) [19], promoting the condensation with aldehydes and ketones to obtain 1,1-gem-diborylalkenes (Scheme 16) [20].

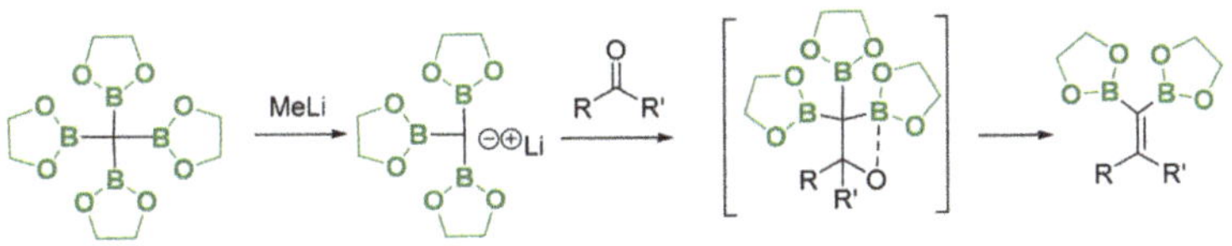

Scheme 16. Deborylation of tetra(boryl)methane and condensation with aldehydes and ketones.

The reactivity developed from 1,1,1-tris(boryl)alkanes offers a significant number of applications, since their activation in the presence of bases generates the corresponding carbanions, which are strongly stabilized by delocalization, with vacant p-orbitals of the remaining two boron atoms. The possibility to perform further functionalization of the remaining C-B bonds eventually leads to new strategic synthetic protocols, emphasizing the straightforward methodology, as well as the smooth reaction conditions.

3. 1,2,3-Tri(boryl)alkanes

3.1. Synthetic Approaches

The introduction of three vicinal boryl moieties in a one-pot protocol represents a challenging reaction. The triboration of alkynes towards 1,2,3-triborated compounds has been accomplished by Yoshida and co-workers [21] via the Cu–PCy$_3$ catalysed borylcupration of alkyl-substituted propargyl ethers (Scheme 17a). This approach demonstrates the easy access to multifunctinalized organoboron compounds that contain two alkenyl boryl and one allylboryl moiety. Mechanistically, it has been suggested that alkyl-substituted propargyl ethers insert regioselectively into Cu-B bonds generating the alkenyl copper where Cu is located next to OMe group. Subsequently, σ-bond methatesis with B$_2$pin$_2$ facilitates the formation of π-allyl species that after a second σ-bond methatesis with the diboron reagent provides the triborated product (Scheme 17b). However, phenyl-substituted propargyl ethers react towards the diborated product, presumably as a consequence of the regioselective alkenyl copper species formation, with Cu next to Ph group, and eventual σ-bond metathesis with B$_2$pin$_2$ (Scheme 17b).

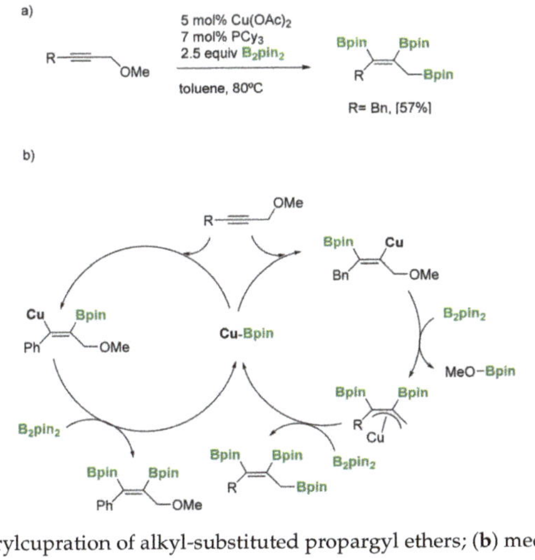

Scheme 17. (a) Borylcupration of alkyl-substituted propargyl ethers; (b) mechanistic proposal.

Alternatively, Ma and co-workers [22] developed a highly selective 1,2,3-triboration of propargylic carbonates towards (E)-propen-1,2,3-triboronates, by the catalytic systems Pd(PPh$_3$)$_4$ and CuCl as co-catalyst (Scheme 18a). Mechanistically, two catalytic cycles have been suggested. The first one is based on the Pd(0) catalyzed S$_N$2'-type oxidative addition of propargylic carbonates to give 1,2-allenyl palladium intermediate I1 (Scheme 18b), with the concomitant release of CO$_2$. In the following step, I1 undergoes transmetallation with Cu-Bpin species, previously formed from Cu-OMe and B$_2$pin$_2$, in order to generate the intermediate I2. Subsequent reductive elimination allows the formation of 1,2-allenyl boronate P1, that releases from the catalytic cycle 1. Interestingly, it has been suggested that P1 might insert in a Pd-B bond in catalytic cycle 2, producing η^1-allylic intermediate I3, which undergoes reductive elimination to afford the final triborated product with exclusive (E)-stereoselectivity (Scheme 18b).

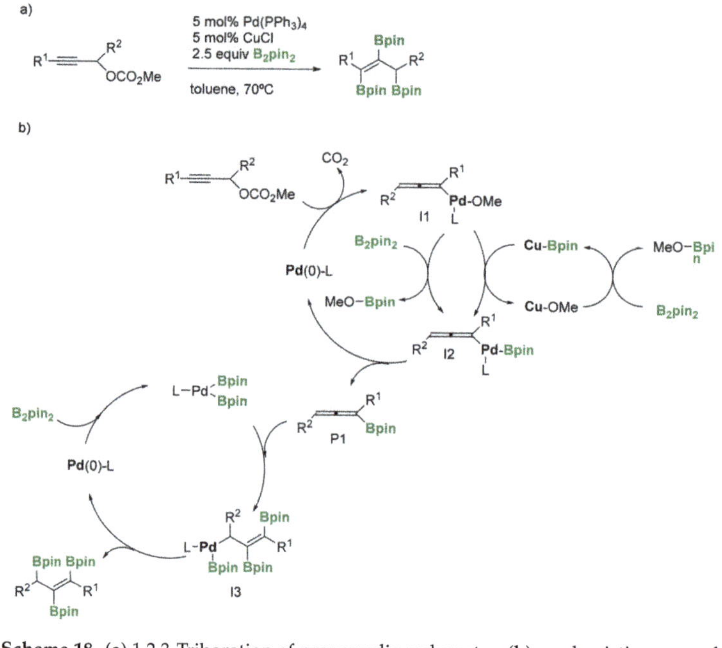

Scheme 18. (a) 1,2,3-Triboration of propargylic carbonates; (b) mechanistic proposal.

All the previous efforts devoted to introducing three vicinal C–B bonds into a one-pot sequential reaction required transition metal complexes to activate the borane reagent and facilitate the addition to the unsaturated substrates. However, Fernández, Szabó and co-workers [23] were first to observe the possibility of conducting the 1,2,3-triboration reaction in a transition-metal free context, using allylic alcohols as substrates and conducting a one-pot allylic borylation/diboration sequence (Scheme 19). The reaction takes place in the presence of catalytic amount of base Cs$_2$CO$_3$, that contributes to the generation of MeO$^-$ from MeOH. The methoxy group activates the B$_2$pin$_2$, through intermediate [Hbase][MeOB$_2$pin$_2$], and promotes the nucleophilic attack of the Bpin moiety to the terminal position of the tertiary allylic alcohol (Scheme 20).

The initial mechanism of this reaction can be described as an S$_N$2'-type process. In the absence of MeOH or base, the reaction does not take place. After the transition-metal free allylic borylation, the allylic boronate product can interact with the [Hbase][MeOB$_2$pin$_2$] adduct and complete the diboration reaction [24], to generate the 1,2,3-triborated product (Scheme 20). This unprecedented tandem allylic borylation/diboration reaction in the absence of transition-metal catalysts had not been observed before and served as the basis of a new methodology towards polyborated compounds.

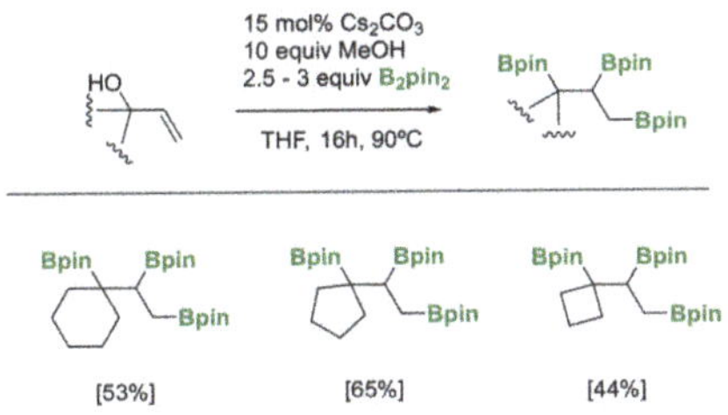

Scheme 19. Transition metal-free one-pot allylic borylation/diboration sequence.

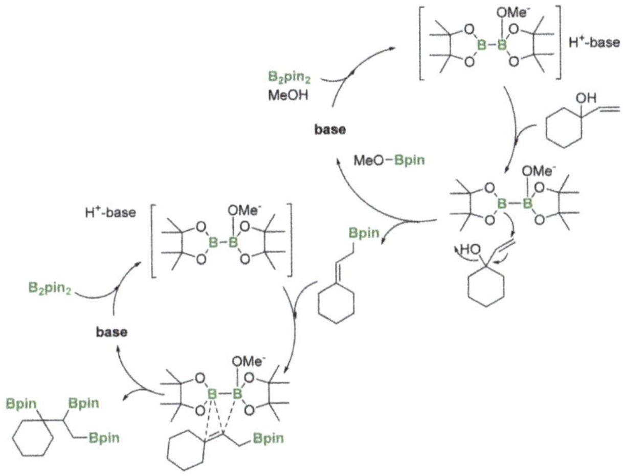

Scheme 20. Mechanistic proposal of transition metal-free 1,2,3-triboration of allylic alcohols.

Fernández and co-workers also explored the viability of performing a direct 1,2,3-triboration reaction in a transition-metal free scenario via the 1,4-hydroboration of 1,3-dienes [25], followed by the in situ diboration reaction of the resultant allylic boronate (Scheme 21) [26]. This work represents an unprecedented tandem 1,4-hydroboration/diboration reaction, that is generalized for terminal and internal 1,3-dienes.

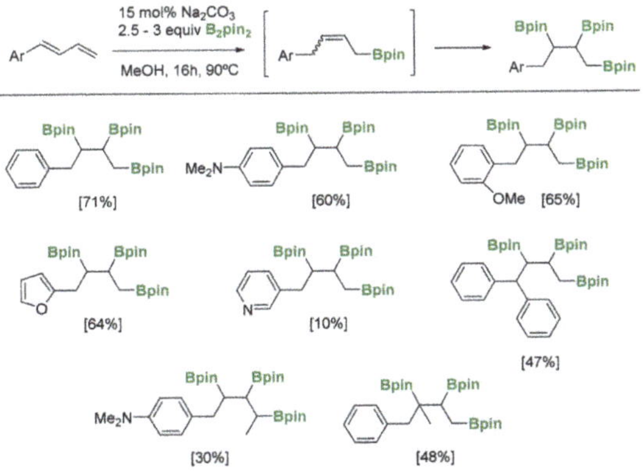

Scheme 21. Transition metal-free 1,2,3-triboration of conjugated dienes.

The nucleophilic attack of the Bpin moiety from [Hbase][MeOB$_2$pin$_2$] takes place to the less hindered position of the conjugated dienes throughout a S$_N$2'-borylative process. The resulting allylic boronate product conducts a transition-metal free diboration reaction of the internal C=C bond, giving access to the 1,2,3-triborated product (Scheme 22) [26].

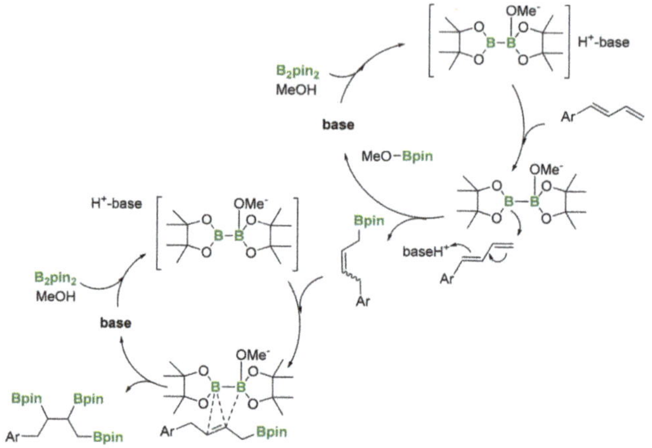

Scheme 22. Mechanistic proposal of transition metal-free 1,2,3-triboration of conjugated dienes.

3.2. Synthetic Applications

The polyfunctionalization of 1,2,3-triborated products can be accomplished throughout the cross-coupling reaction. Interestingly, in the presence of the 1 equiv of aryliodide and Pd complex, it was possible to exclusively functionalize the internal C-B bond of the triborated product (Scheme 23) [26]. The reaction was general for para- and ortho-substituted aryl systems, as well as for furyl groups. The in situ oxidation of the resulting 1,3-diborated products, in the presence of H_2O_2/NaOH, allowed access to 1,3-butanediol 2,3-diaryl systems in a straightforward manner (Scheme 23) [26].

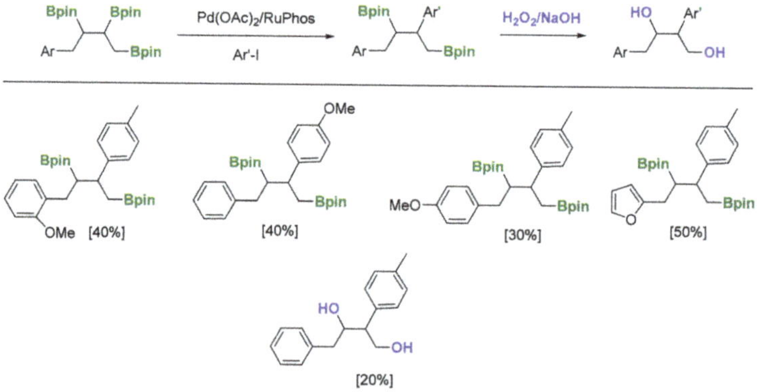

Scheme 23. Selective polyfunctionalization of 1,2,3-triborated products.

Aggarwal and co-workers have reported an alternative functionalization of 1,2,3-triborated compounds following a photoredoxcatalyzed mono-deboronation. Initially the reaction proceeds through the generation of primary β-boryl radicals that undergo a rapid 1,2-boron shift to form thermodynamically favored secondary followed by a subsequent 1,2-boron shift towards tertiary radicals (Scheme 24). This protocol allows for the selective transformation of the more hindered boronic ester from the 1,2,3-triborated substrate [23], conducting its radical addition to tert-butyl-acrylate [27]. The product is obtained with high selectivity, highlighting the thermodynamic control that favors the most stabilized radical center as the site of the reaction.

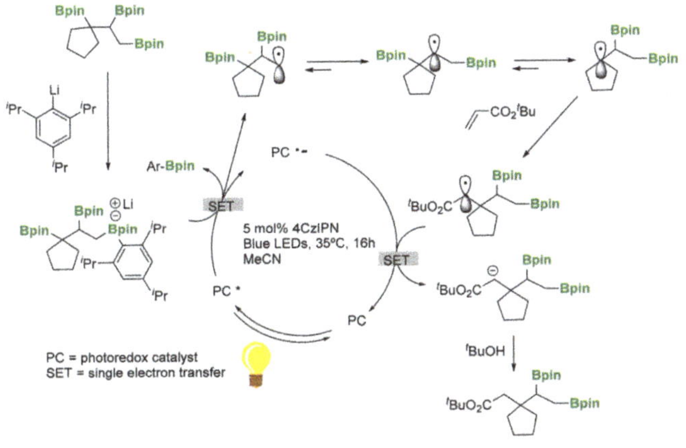

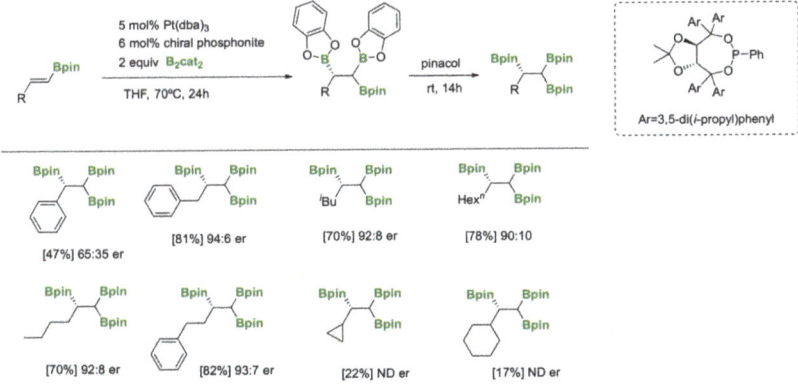

Scheme 24. Photoredoxcatalyzed mono-deboronation of 1,2,3-triborated products.

4. 1,1,2-Tri(boryl)alkanes

4.1. Synthetic Approaches

The preparation of versatile 1,1,2-tri(boryl)alkanes has been conducted via the Pt-catalyzed diboration of vinylboronates, (Scheme 25) [28]. Morken and co-workers employed chiral phosphonite to induce enantioselectivity along the internal C-B bond formation in the presence of Pt(dba)$_3$ and bis(catecholato)diboron. Further treatment of the triborated product with pinacol, allowed the purification of the desired stable products with moderate to high enantioselectivity. The reaction has been considered of general application, although limitations are related to α-substituted substrates that suffered from lower reactivity and selectivity [28].

Scheme 25. Synthesis of 1,1,2-tri(boryl)alkanes via bis(catecholato)diboron.

Another approach, launched by Song and co-workers [29], involves transition-metal free base-controlled 1,1,2-tris(boryl)alkane synthesis via diboration of alkenylboronates with B$_2$pin$_2$ (Scheme 26a). The reaction conditions for the synthesis of 1,1,2-tris(boryl)alkane were strictly optimized, since the corresponding 1,2-di(boryl)alkanes can also be formed as byproducts, depending on the base used. K$_2$CO$_3$ (1.5 equivalents) resulted in the optimized one, although 4.5 equivalents of B$_2$pin$_2$ were required. The substrate scope is broad and several functionalized terminal alkynes, such as propargyl amides and propargyl sulfonamides, are compatible. From a mechanistic point of view, a nucleophilic attack from the adduct [Hbase][MeOB$_2$pin$_2$], (generated in situ from B$_2$pin$_2$/base/MeOH) to the alkenylboronate substrate (Scheme 26b) has been suggested, as it was described on the transition-metal free diboration of alkenes developed by Fernández and co-workers in 2011 [24].

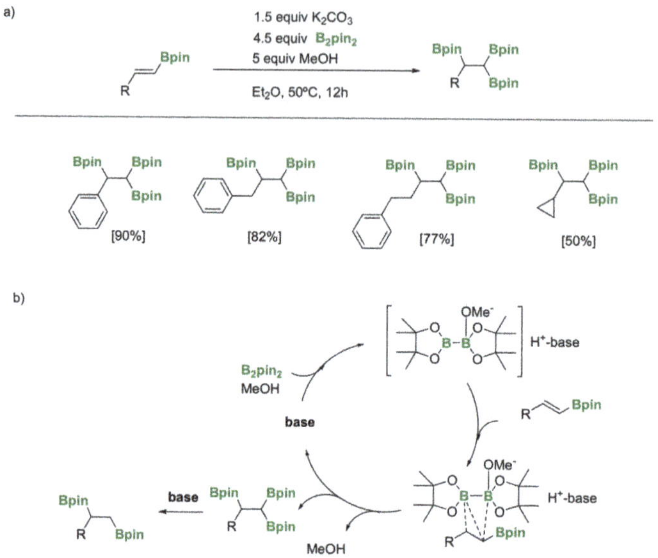

Scheme 26. (a) transition-metal free 1,1,2-tris(boryl)alkane synthesis; (b) mechanistic proposal.

4.2. Synthetic Applications

The versatility of 1,1,2-tri(boryl)alkanes is based on the construction of new C-C and C-heteroatom bonds, since the three boronate moieties can be distinguished in a stepwise manner. Song and co-workers proved the selective deprotonation on the geminal diboryl moiety, when NaO^{t}Bu is used as base (5 equiv) producing the 1,2-bis(boronate) product (Scheme 27) [29]. However, in the presence of Cs$_2$CO$_3$ (4 equiv) the 1,1,2-tri(boryl)alkane is converted into the linear alkyl monoboronate product (Scheme 27) [29].

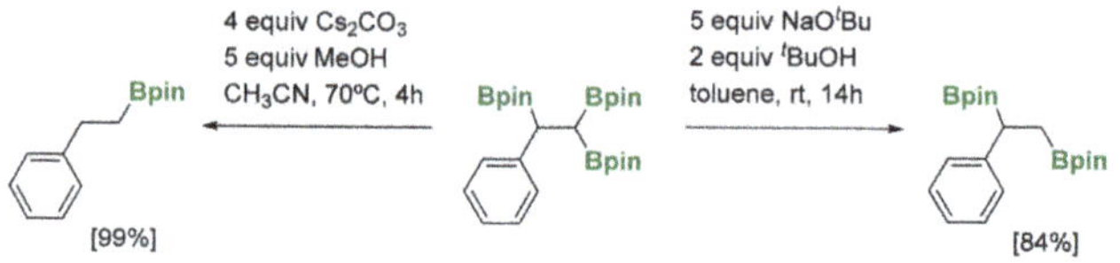

Scheme 27. Selective deprotonation on the geminal diboryl moiety in 1,1,2-tri(boryl)alkanes.

A simple transformation from 1,1,2-tri(boryl)alkane into mixed Bpin and B-MIDA triboronate compounds (MIDA=N-(methyliminodiacetic acid) has been developed by Sharma and co-workers [30]. However, the experiment showed that the treatment of 1,1,2-tri(pinacolboryl)alkane with MIDA provided a mixture (1:1 ratio by NMR) of the two mono-B(MIDA)-diBpin species, possessing the B(MIDA) group vicinal to Bpin or geminal to the Bpin moiety (Scheme 28). Unfortunately, the mixture resulted in being inseparable, using silica gel chromatography.

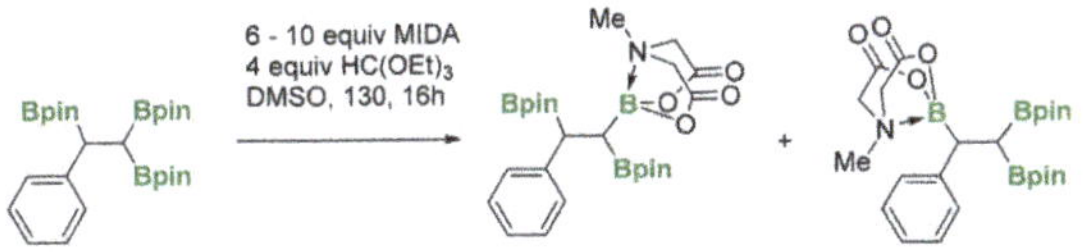

Scheme 28. Transformation from 1,1,2-tri(boryl)alkane into mixed Bpin and B-MIDA triboronate compounds.

Deborylative alkylation reactions with chiral 1,1,2-tris-(boronates) were performed in the group of Morken [28], and they found that five equivalents of NaO^{t}Bu in toluene were efficient to form the new C-C bond from one of the geminal boryl moieties. The syn diastereoisomer was the predominant product, obtained as 1,2-diol upon oxidative workup. The reaction was applied for primary and seconary electrophiles (Scheme 29). The deborylative alkylation with alkenyl electrophiles represents a suitable method for the synthesis of internal vicinal bis(boronates), in the presence of terminal alkenes. When the substrate contains alkyl halide groups, the vinyl boronate diboration enables the construction of 1,1,2-tri(boryl)alkanes that may undergo intramolecular alkylation to form cyclic 1,2-bis(boronates). This method allows the construction of five, six and seven membered rings, which, after in situ oxidation, provides anti-1,2-diol (Scheme 29).

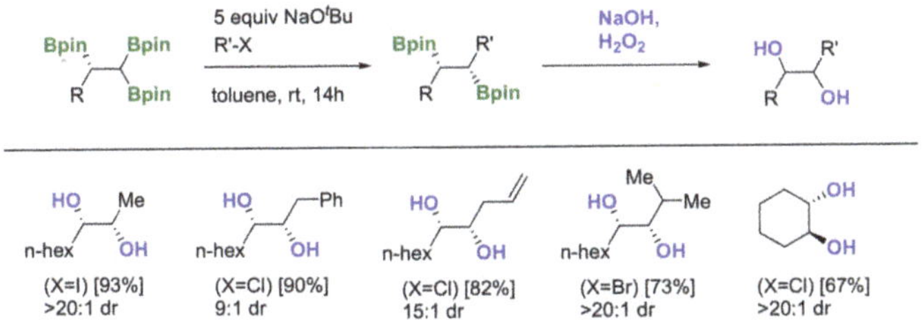

Scheme 29. Deborylative alkylation with chiral 1,1,2-tris-(boronates).

5. 1,1,2-Tri(boryl)alkenes

5.1. Synthetic Approaches

The preparation of polyborated alkenes has originally been accomplished by Marder and co-workers, via the tandem desilylative borylation of bis(trimethylsilyl)acetylene and the subsequent diboration with B$_2$pin$_2$ in the presence of [Pt(PPh$_3$)$_2$(η^2-C$_2$H$_4$)] as catalyst (Scheme 30a) [31]. Similarly, Srebnik and co-workers established the convenience of using [Pt(PPh$_3$)$_4$] as catalyst to diborate 1-alkynylboronates to easily obtain the expected 1,1,2-triborylalkenes (Scheme 30b) [32]. More extensively, Nishihara and co-workers launched a more general platinum catalyzed diboration of phenylethynyl MIDA boronate with B$_2$pin$_2$ that proceeds to yield 1,1,2-triboryl-2-phenylethene, with two different classes of the boron functionalities (Scheme 30c) [33]. This protocol uses the MIDA boronic group in the substrate, which is considered a masked group [34], towards strategic further selective C-B functionalization.

Moving from diboration to the hyroboration of alkynylboronates, Ozerov and co-workers discovered that pinacolborane (HBpin) can be efficiently added to terminal alkynes in the presence of iridium SiNC pincer complex, to efficiently prepare alkynylboronates. The following pathway involved the dehydrogenative diboration with HBpin and the same iridum complex under an atmosphere of CO (Scheme 31) [35]. The formation of 1,1,2-tri(boryl)alkanes through this protocol can be extended to a large number of substrates. However, the authors noted that the mechanism remains unclear, discarding the intermediacy of a hydroborated product or diboration with B$_2$pin$_2$.

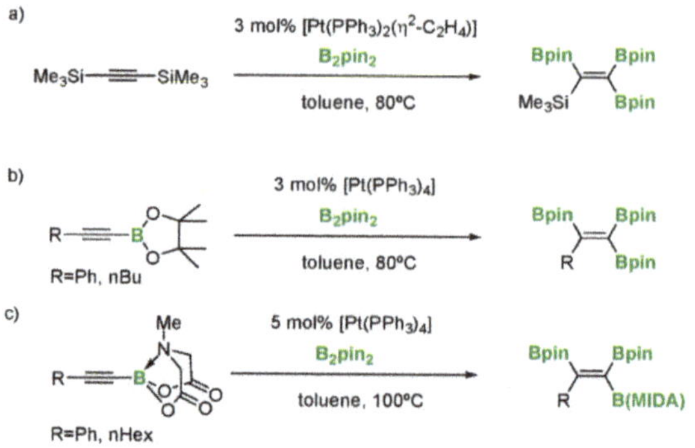

Scheme 30. (a) Tandem desilylative borylation of bis(trimethylsilyl)acetylene and the subsequent Pt-catalyzed diboration; (b) Pt-catalyzed diboration of 1-alkynylboronates; (c) platinum-catalyzed diboration of phenylethynyl MIDA boronate.

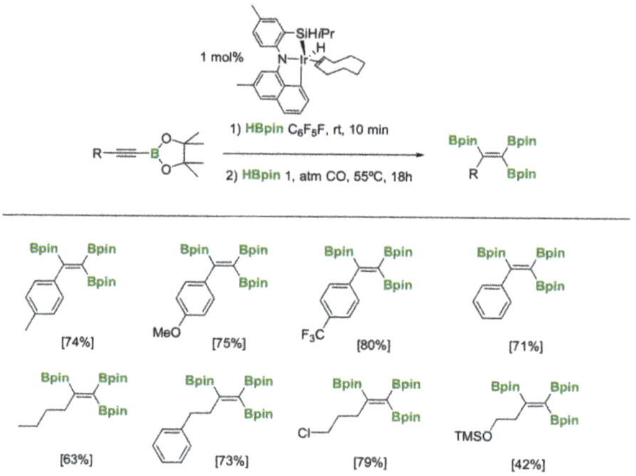

Scheme 31. Dehydrogenative diboration with HBpin of alkynylboronates.

Marder and co-workers have developed a copper catalyzed triboration protocol for the synthesis of 1,1,2-triborylalkenes, using Cu(OAc)$_2$ and P^{n}Bu$_3$ as suitable ligands, and with B$_2$pin$_2$ as a boron source (Scheme 32) [36]. The use of stoichiometric amounts of acrylonitrile as additive seems to be required as hydrogen (B-H) acceptor, in order to avoid by-products reactions. The reaction features mild reaction conditions, broad substrate scope and good functional group tolerance.

The mechanism has been suggested on the basis of a coordination of terminal alkynes to [LnCuOAc], which is formed from Cu(OAc)$_2$, the ligand and a reductive pathway (Scheme 33). Cu(OTf)$_2$ or CuCl$_2$ resulted inactive catalytic systems. Alternatively, the addition of KOAc to CuCl$_2$ and CuCl reactivated the system, indicating that the ⁻OAc plays an important role in the reaction. The reactivity of [LnCuOAc] with the terminal alkyne provides the intermediate I1, which interacts with B$_2$pin$_2$ through σ-bond methatesis to afford the alkynylboronate and the Cu-Bpin species I2. Consequently, insertion of the alkynylboronate into Cu-Bpin generates the alkenylcopper I3 that eventually reacts with B$_2$pin$_2$ through σ-bond methatesis, to generate the desired 1,1,2-tri(boryl)alkene (Scheme 33). Acrylonirile is added, because its hydroboration is faster than that of alkynes, avoiding the alkyne hydroboration side reaction [36].

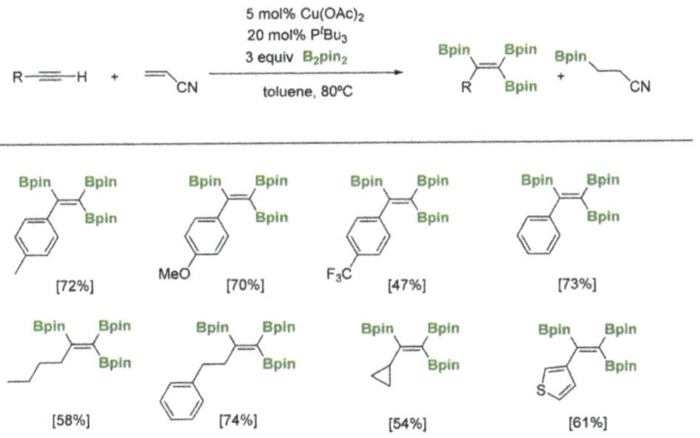

Scheme 32. Copper catalyzed triboration of alkynes.

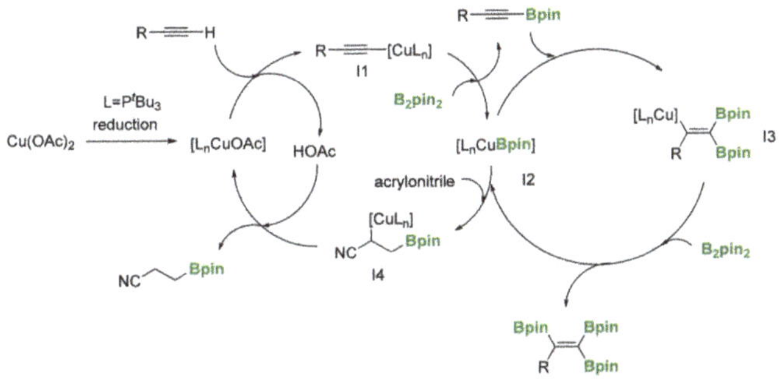

Scheme 33. Mechanistic proposal of copper catalyzed triboration of alkynes.

5.2. Synthetic Applications

The transformation of 1,1,2-tri(boryl)alkenes can be conducted by palladium-catalyzed arylation reaction with aryl iodides. Pd(dba)2 was found to be the best catalyst to give the arylated product and the new C-C is selectively formed in trans to the B(MIDA) moiety (Scheme 34). HP^{t}Bu$_3$BF$_4$ was selected as the best ligand to modify the palladium complex in the presence of three equivalents of K$_3$PO$_4$ as the base [33]. The limitation of this protocol concerns the aryl halide used, since aryl iodide results in being very efficient for the cross-coupling reaction, and the aryl bromide and aryl chloride show a low or null reactivity (Scheme 34). Additionally, other electrophiles such as ethyl iodide or (E)-octenyl iodide were subjected to the Suzuki–Miyaura coupling reactions, but no desired products were formed. Despite the fact that some limitations were observed, this method represents the first example of obtaining gem-diborylated olefins with two different boryl groups via a chemoselective arylation. The chemoselectivity can be explained simply by a steric effect, because even the addition of two equivalents of aryl iodides did not provide the diarylated products [33].

The versatility of 1,1,2-tri(pinacolboryl)alkenes was also demonstrated, due to complementarily stereoselective Suzuki-Miyaura coupling reactions with aryl halides catalyzed by Pd(PPh$_3$)$_4$ in the presence of three equivalents of K$_3$PO$_4$ as the base (Scheme 35) [35]. The (E)-stereochemistry of the diborated product becomes one of the few examples described in the literature, and becomes a potential synthetic strategy to obtain trans-diaryldiborylalkenes in a simple way.

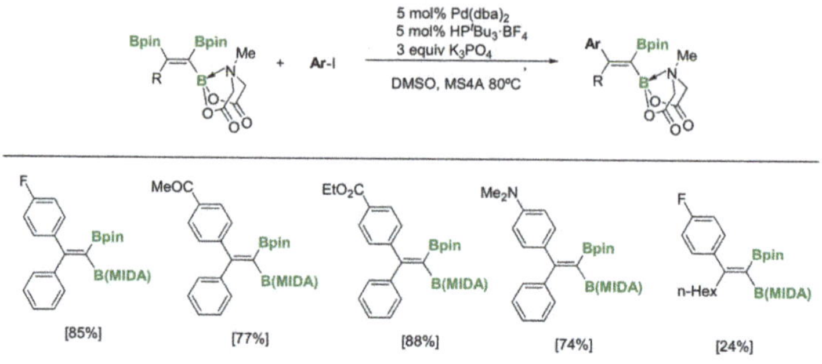

Scheme 34. Selective palladium catalyzed arylation of 1,1,2-tri(boryl)alkenes.

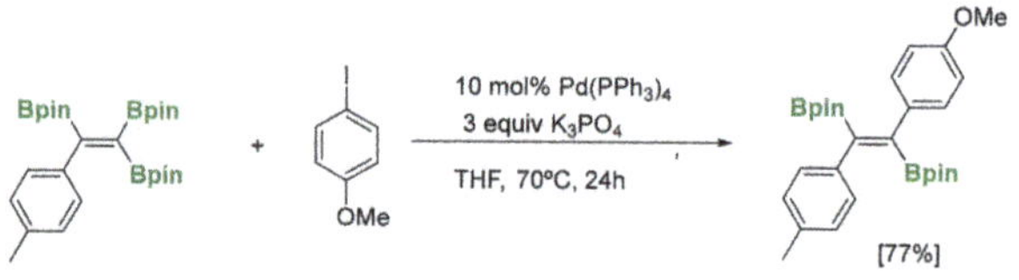

Scheme 35. Selective palladium catalyzed arylation of 1,1,2-tri(boryl)alkenes.

The utility of the 1,1,2-tri(pinacolboryl)alkenes has also explored different ways to be functionalized through difluorination (with Selectfluor) to afford gem-difluoroborylakenes (Scheme 36) [36]. Similarly, dibromination (with NBS) and dichlorination (with NCS) of 1,1,2-tri(pinacolboryl)alkenes facilitated the formation of gem-dihaloborylakenes (Scheme 36). The use of a lower amount of NBS or NCS, as well as lower reaction times, generated trans-stereoselectively towards the monohalo-diborylated alkene (Scheme 36).

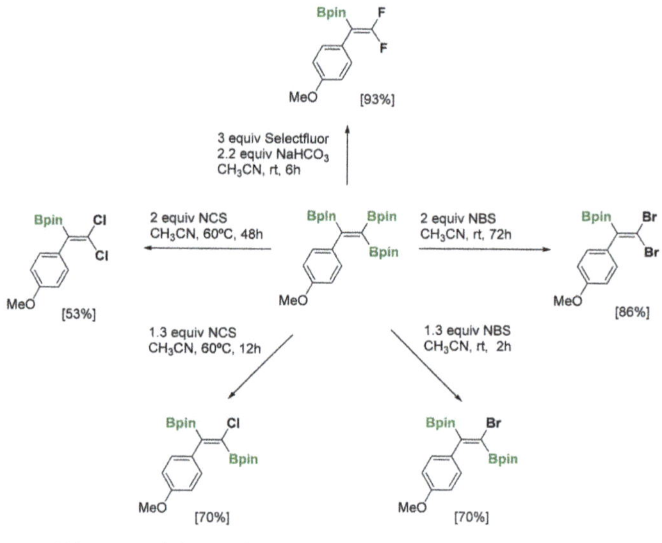

Scheme 36. Selective functionalization of 1,1,2-tri(boryl)alkenes.

6. Conclusions

The preparation of 1,1,1-tri(boryl)alkanes, 1,2,3-tri(boryl)alkanes, 1,1,2-tri(boryl)alkanes, as well as 1,1,2-tri(boryl)alkenes, appears to be a useful tool to synthesize organic compounds through selective

C-B functionalization. The assistance of vicinal or geminal boryl units to the selective functionalization of $C(sp^3)$-B or $C(sp^2$-B) fragments becomes crucial. In part, this advantage is due to the enhanced α-borylcarbanions stability as intermediates in synthesis. Within the last decade, intensive efforts are devoted to prepare tri(boryl)alkanes and tri(boryl)alkenes, and significant applications have illustrated the powerful tool to construct molecules in a selective way. We can only expect that their interest will keep growing, attending to their real utility.

Funding: This research was supported by MINECO through project CTQ2016-80328-P. O. Salvadó thanks Martí Franques grant.

Conflicts of Interest: The authors declare no conflict of interest.

References

1. Castle, R.B.; Matteson, D.S. A Methanetetraboronic Ester. *J. Am. Chem. Soc.* **1968**, *90*, 2194. [CrossRef]
2. Castle, R.B.; Matteson, D.S. Methanetetraboronic and methanetriboronic esters. *J. Organomet. Chem.* **1969**, *20*, 19–28. [CrossRef]
3. Matteson, D.S.; Thomas, J.R. C-alkylation of methanetetraboronic and methanetriboronic esters. *J. Organomet. Chem.* **1970**, *24*, 263–271. [CrossRef]
4. Matteson, D.S. Methanetetraboronic and methanetriboronic esters as synthetic intermediates. *Synthesis* **1975**, *1975*, 147–158. [CrossRef]
5. Yamamoto, Y.; Ishibashi, A.; Suginome, M. Boryl-Directed, Ir-Catalyzed $C(sp^3)$–H Borylation of Alkylboronic Acids Leading to Site-Selective Synthesis of Polyborylalkanes. *Org. Lett.* **2019**, *21*, 6235–6240. [CrossRef] [PubMed]
6. Mita, T.; Ikeda, Y.; Michigami, K.; Sato, Y. Iridium-catalyzed triple $C(sp^3)$–H borylations: Construction of triborylated sp^3-carbon centers. *Chem. Commun.* **2013**, *49*, 5601–5603. [CrossRef] [PubMed]
7. Palmer, W.N.; Obligacion, J.V.; Pappas, I.; Chirik, P.J. Cobalt-Catalyzed Benzylic Borylation: Enabling Polyborylation and Functionalization of Remote, Unactivated $C(sp^3)$–H Bonds. *J. Am. Chem. Soc.* **2016**, *138*, 766–769. [CrossRef]
8. Palmer, W.N.; Zarate, C.; Chirik, P.J. Benzyltriboronates: Building Blocks for Diastereoselective Carbon–Carbon Bond Formation. *J. Am. Chem. Soc.* **2017**, *139*, 2589–2592. [CrossRef]
9. Liu, X.; Ming, W.; Zhang, Y.; Friedrich, A.; Marder, T.B. Copper-Catalyzed Triboration: Straightforward, Atom-Economical Synthesis of 1,1,1-Triborylalkanes from Terminal Alkynes and HBpin. *Angew. Chem. Int. Ed.* **2019**, *58*, 18923–18927. [CrossRef]
10. Baker, R.T.; Nguyen, P.; Marder, T.B.; Westcott, S.A. Transition Metal Catalyzed Diboration of Vinylarenes. *Angew. Chem. Int. Ed. Engl.* **1995**, *34*, 1336–1338. [CrossRef]
11. Gu, Y.; Pritzkow, H.; Siebert, W. Synthesis and Reactivity of Monoborylacetylene Derivatives. *Eur. J. Inorg. Chem.* **2001**, *200*, 373–379. [CrossRef]
12. Krautwald, S.; Bezdek, M.J.; Chirik, P.J. *Cobalt-Catalyzed 1,1-Diboration of Terminal Alkynes: Scope, Mechanism, and Synthetic Applications*. *J. Am. Chem. Soc.* **2017**, *139*, 3868–3875. [PubMed]
13. Zhang, L.; Huang, Z. Synthesis of 1,1,1-Tris(boronates) from Vinylarenes by Co-Catalyzed Dehydrogenative Borylations–Hydroboration. *J. Am. Chem. Soc.* **2015**, *137*, 15600–15603. [CrossRef] [PubMed]
14. Brown, J.M.; Lloyd-Jones, G.C. Vinylborane Formation in Rhodium-catalysed Hydroborations; Ligand-free Homogeneous Catalysis. *J. Chem. Soc. Chem. Commun.* **1992**, 710–712. [CrossRef]
15. Hong, K.; Liu, X.; Morken, J.P. Simple Access to Elusive α-Boryl Carbanions and Their Alkylation: An Umpolung Construction for Organic Synthesis. *J. Am. Chem. Soc.* **2014**, *136*, 10581–10584. [CrossRef] [PubMed]
16. Nave, S.; Sonawane, R.P.; Elford, T.G.; Aggarwal, V.K. Protodeboronation of Tertiary Boronic Esters: Asymmetric Synthesis of Tertiary Alkyl Stereogenic Centers. *J. Am. Chem. Soc.* **2010**, *132*, 17096–17098. [CrossRef]
17. Matteson, D.S.; Moody, R.J.; Jesthi, P.K. Reaction of aldehydes and ketones with a boron-substituted carbanion, bis(ethylenedioxyboryl)methide. Simple aldehyde homologation. *J. Am. Chem. Soc.* **1975**, *97*, 5608–5609. [CrossRef]

18. Matteson, D.S.; Moody, R.J. Homologation of carbonyl compounds to aldehydes with lithium bis(ethylenedioxyboryl)methide. *J. Org. Chem.* **1980**, *45*, 1091–1095. [CrossRef]
19. Matteson, D.S.; Davies, R.A.; Hagelee, L.A. A bromomethanetriboronic ester. *J. Organomet. Chem.* **1974**, *69*, 45–52. [CrossRef]
20. Matteson, D.S.; Tripathy, P.B. Alkene-1,1-diboronic esters from the condensation of triborylmethide anions with ketones or aldehydes. *J. Organomet. Chem.* **1974**, *69*, 53–62.
21. Yoshida, H.; Kawashima, S.; Takemoto, Y.; Okada, K.; Ohshita, J.; Takaki, K. Copper-Catalyzed Borylation Reactions of Alkynes and Arynes. *Angew. Chem. Int. Ed.* **2012**, *51*, 253–258. [CrossRef] [PubMed]
22. Yang, Z.; Cao, T.; Han, Y.; Lin, W.; Liu, Q.; Tang, Y.; Zhai, Y.; Jia, M.; Zhang, W.; Zhu, T.; et al. Regio- and (E)-Stereoselective Triborylation of propargylic Carbonates. *Chin. J. Chem.* **2017**, *35*, 1251–1255. [CrossRef]
23. Miralles, N.; Alam, R.; Szabó, K.; Fernández, E. Transition-Metal-Free Borylation of Allylic and Propargylic Alcohols. *Angew. Chem. Int. Ed.* **2016**, *55*, 4303–4307. [CrossRef] [PubMed]
24. Bonet, A.; Pubill-Ulldemolions, C.; Bo, C.; Gulyás, H.; Fernández, E. Transition-metal-free diboration reaction by activation of diboron compounds with simple Lewis bases. *Angew. Chem. Int. Ed.* **2011**, *50*, 7158. [CrossRef]
25. Maza, R.J.; Davenport, E.; Miralles, N.; Carbó, J.; Fernández, E. Transition-Metal-Free Allylic Borylation of 1,3-Dienes. *Org. Lett.* **2019**, *21*, 2251–2254. [CrossRef]
26. Davenport, E.; Fernández, E. Transition-metal-free synthesis of vicinal triborated compounds and selective functionalisation of the internal C-B bond. *Chem. Commun.* **2018**, *54*, 10104–10107.
27. Kaiser, D.; Noble, A.; Fasano, V.; Aggarwal, V.K. 1,2-Boron Shifts of β-Boryl Radicals Generated from Bis-boronic Esters Using Photoredox Catalysis. *J. Am. Chem. Soc.* **2019**, *141*, 14104–14109. [CrossRef]
28. Coombs, J.R.; Zhang, L.; Morken, J.P. Enantiomerically Enriched Tris(boronates): Readily Accessible Conjunctive Reagents for Asymmetric Synthesis. *J. Am. Chem. Soc.* **2014**, *136*, 16140–16143. [CrossRef]
29. Gao, G.; Yan, J.; Yang, K.; Chen, F.; Song, Q. Base-controlled highly selective synthesis of alkyl 1,2-bis(boronates) or 1,1,2-tris(boronates) from terminal alkynes. *Green Chem.* **2017**, *19*, 3997–4001. [CrossRef]
30. Lin, S.; Wang, L.; Aminoleslami, N.; Lao, Y.; Yagel, C.; Sharma, A. A modular and concise approach to MIDA acylboronates via chemoselective oxidation of unsymmetrical geminal diborylalkane: Unlocking access to a novel class of acylborons. *Chem. Sci.* **2019**, *10*, 4684–4687. [CrossRef]
31. Lesley, G.; Nguyen, P.; Taylor, N.J.; Marder, T.B.; Scott, A.J.; Clegg, W.; Norman, N.C. Synthesis and Characterization of Platinum(II)-Bis(boryl) Catalyst Precursors for Diboration of Alkynes and Diynes: Molecular Structures of cis-[(PPh$_3$)$_2$Pt(B-4-Butcat)$_2$], cis-[(PPh$_3$)$_2$Pt(Bcat)$_2$], cis-[(dppe)Pt(Bcat)$_2$], cis-[(dppb)Pt(Bcat)$_2$], (E)-(4-MeOC$_6$H$_4$)C(Bcat)CH(Bcat), (Z)-(C$_6$H$_5$)C(Bcat)C(C$_6$H$_5$)(Bcat), and (Z,Z)-(4-MeOC$_6$H$_4$)C(Bcat)C(Bcat)C(4-MeOC$_6$H$_4$)(Bcat) (cat = 1,2-O$_2$C$_6$H$_4$; dppe) Ph$_2$PCH$_2$CH$_2$PPh$_2$; dppb) Ph$_2$P(CH$_2$)$_4$PPh$_2$. *Organometallics* **1996**, *15*, 5137–5154.
32. Ali, H.A.; Quntar, A.E.A.A.; Goldberg, I.; Srebnik, M. Platinum(0)-Catalyzed Diboration of Alkynylboronates and Alkynylphosphonates with Bis(pinacolato)diborane(4): Molecular Structures of [((Me$_4$C$_2$O$_2$)B)(C$_6$H$_5$)C=C(P(O)(OC$_2$H$_5$)$_2$)(B(O$_2$C$_2$Me$_4$))] and [((Me$_4$C$_2$O$_2$)B)(C$_4$H$_9$)C=C(B(O$_2$C$_2$Me$_4$))$_2$]. *Organometallics* **2002**, *21*, 4533–4539. [CrossRef]
33. Hyodo, K.; Suetsugu, M.; Nishihara, Y. Diborylation of Alkynyl MIDA Boronates and Sequential Chemoselective Suzuki–Miyaura Couplings: A Formal Carboborylation of Alkynes. *Org. Lett.* **2014**, *16*, 440–443. [CrossRef] [PubMed]
34. Mancilla, T.; Contreras, R.; Wrackmeyer, B.J. New bicyclic organylboronicesters derived from iminodiacetic acids. *J. Organomet. Chem.* **1986**, *307*, 1–6. [CrossRef]
35. Lee, C.-I.; Shih, E.-C.; Zhou, J.; Reibenspies, J.H.; Ozerov, O.V. Synthesis of Triborylalkenes from Terminal Alkynes by Iridium-Catalyzed Tandem C-H Borylation and Diboration. *Angew. Chem. Int. Ed.* **2015**, *54*, 14003–14007. [CrossRef]
36. Liu, X.; Ming, W.; Friedrich, A.; Kerner, F.; Marder, T.B. Copper-Catalyzed Triboration of Terminal Alkynes using B$_2$pin$_2$: Efficient synthesis of 1,1,2-triborylakenes. *Angew. Chem. Int. Ed.* **2020**, *59*, 304–309. [CrossRef]

© 2020 by the authors. Licensee MDPI, Basel, Switzerland. This article is an open access article distributed under the terms and conditions of the Creative Commons Attribution (CC BY) license (http://creativecommons.org/licenses/by/4.0/).

Article

Hybrid Boron-Carbon Chemistry

Josep M. Oliva-Enrich [1,*], Ibon Alkorta [2] and José Elguero [2]

1. Instituto de Química-Física "Rocasolano" (CSIC), Serrano 119, E-28006 Madrid, Spain
2. Instituto de Química Médica (CSIC), Juan de la Cierva, 3, E-28006 Madrid, Spain; ibon@iqm.csic.es (I.A.); iqmbe17@iqm.csic.es (J.E.)
* Correspondence: j.m.oliva@iqfr.csic.es; Tel.: +34-91-561-9400

Academic Editor: Ashok Kakkar
Received: 18 September 2020; Accepted: 26 October 2020; Published: 29 October 2020

Abstract: The recently proved one-to-one structural equivalence between a conjugated hydrocarbon C_nH_m and the corresponding borane B_nH_{m+n} is applied here to hybrid systems, where each C=C double bond in the hydrocarbon is consecutively substituted by planar $B(H_2)B$ moieties from diborane(6). Quantum chemical computations with the B3LYP/*cc*-pVTZ method show that the structural equivalences are maintained along the substitutions, even for non-planar systems. We use as benchmark aromatic and antiaromatic (poly)cyclic conjugated hydrocarbons: cyclobutadiene, benzene, cyclooctatetraene, pentalene, benzocyclobutadiene, naphthalene and azulene. The transformation of these conjugated hydrocarbons to the corresponding boranes is analyzed from the viewpoint of geometry and electronic structure.

Keywords: boron; conjugated hydrocarbon; isoelectronic molecule; electronic structure; quantum chemistry; singlet-triplet gap

1. Introduction

Planar conjugated hydrocarbons played a key role in the early days of quantum mechanics, when computers were not available, and physical models were needed in order to understand the electronic structure of molecules [1], the attractive nature of the chemical bond [2], the nature of ground and excited states in atoms and molecules [3–5], and mechanistic studies of chemical reactions [6].

On the other hand, a relevant quantity in chemistry is the energy gap between the lowest-lying singlet and triplet electronic states in a molecule, directly related to the reactivity of the system, and useful, e.g., for the design of photochemical molecular devices [7]. For instance, methylene CH_2 is a very reactive species with a triplet ground state and a singlet-triplet experimental energy gap of 38 kJ·mol^{-1} [8]. On the other hand, water H_2O has a singlet ground state with an experimental singlet-triplet energy gap of 675 kJ·mol^{-1} [9]. Hence, the larger the singlet-triplet energy gap the more stable a molecule.

The chemistry of boron compounds has evolved along the second half of the XXth century and beginning of the XXIst century in a generalised dual fashion since the original works of Stock [10] on B_nH_m boranes syntheses in the beginning of the XXth century: (i) organoboron [11] and metal-boron [12] chemistry and (ii) the chemistry of 3D polyhedral (metalla)heteroborane structures [13–17]. Given the huge and magnificent efforts done towards the synthesis of new boron derivatives in the last decades, a complete literature citation here is impractical.

The recent experimental isolation of borophane layers [18], a 2D (BH)$_1$ system structurally equivalent to graphene and the characterisation of chemical structures where one C=C double bond is substituted by one B(H$_2$)B moiety [19–23] calls for the possibility of creating a new field of research within boron chemistry, namely, the synthesis of finite planar neutral borane molecules. We have recently proved that there is a one-to-one structural correspondence between any planar conjugated hydrocarbon C_nH_m and the planar borane $B_nH_{(m+n)}$, indeed with the same number of electrons and n more hydrogen atoms in the latter [24]. This transformation can be carried out by substituting all C=C double bonds by a perpendicular planar $B(H_2)B$ moiety, which is the central rhombus of diborane(6). Up to this date we have not found any exception so far to this transformation and is extended here to non-planar systems.

The problems that we would like to tackle in this work, dedicated to Professor Todd B. Marder on the occasion of his 65th birthday, are related to the stability of hybrid boron-carbon isoelectronic chemical structures built by consecutive substitution of C=C by $B(H_2)B$ moieties in (poly)cyclic conjugated hydrocarbons. Particularly, we have chosen examples with $4n$ π electrons: cyclobutadiene, cyclooctatetraene, pentalene and benzocyclobutadiene, and examples with $(4n+2)$ π electrons: benzene, naphthalene and azulene. The question we would like to answer here is: Given a (poly)cyclic conjugated hydrocarbon C_nH_m with an even number of carbon atoms or π electrons, how similar or different are the hybrid systems $C_{(n-2k)}B_{(2k)}H_{(m+2k)}$, with $k = \{0, 1, 2, \ldots, n/2\}$, to the original hydrocarbon from the structural and electronic structure point of view? In Scheme 1 we gather the conjugated hydrocarbons considered in this work.

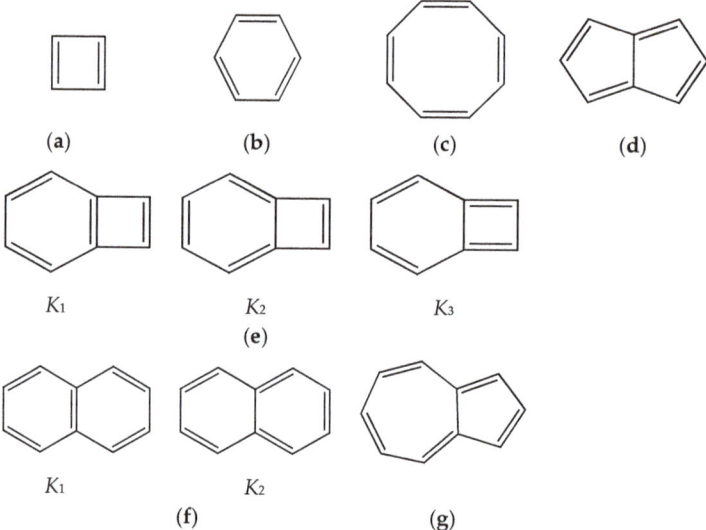

Scheme 1. Conjugated hydrocarbons included in this work. Kekulé structures K_i [25] are also included for those structures leading to different hybrid isomers on {C=C ↔ B(H2)B} substitutions. (**a**) cyclobutadiene, (**b**) benzene, (**c**) cyclooctatetraene, (**d**) pentalene, (**e**) benzocyclobutadiene with Kekulé structures K_1, K_2 and K_3, (**f**) naphthalene with Kekulé structures K_1 and K_2, and (**g**) azulene. All systems are planar, except for cyclooctatetraene (**c**) with a D_{2d} symmetry energy minimum.

2. Results

Tables 1–5 gather the formula, structure, electronic energy, vertical singlet-triplet energy gaps and point-group symmetry (PGS) for all the systems included in this work, derived from a conjugated hydrocarbon C_nH_m with an even number of carbon atoms or π electrons and the structures derived from consecutive k { C=C ↔ $B(H_2)B$ } substitutions, with $1 \leq k \leq n/2$. The triplet energy is computed with optimized geometry of the singlet ground state due to considerable geometrical changes in triplet optimizations, and therefore singlet-triplet energies are vertical. From cyclooctatetraene onwards, all $C_{(n-2k)}B_{(2k)}H_{(m+2k)}$ structures have more than one isomer, for $1 \leq k < n/2$. Hence, one might inquire the extent of change in energy differences and singlet-triplet energies in different isomers of a given structure. When isomers arise under k {C=C ↔ $B(H_2)B$} substitutions in a given structure, these are labelled as {(**I**), (**II**), ... } and ordered in increasing energy, as displayed in Figures 1–5. All structures and isomers correspond to planar structures, except for cyclooctatetraene C_8H_8 ($k = 0$) and structures $C_{(8-2k)}B_{(2k)}H_{(8+2k)}$ ($k = 1$–4). All structures and isomers presented in this work, as computed with the B3LYP/cc-pVTZ model, correspond to energy minima. B3LYP/cc-pVTZ optimised geometries, in cartesian coordinates (Å), of the systems included in the work are presented in Supplementary Materials.

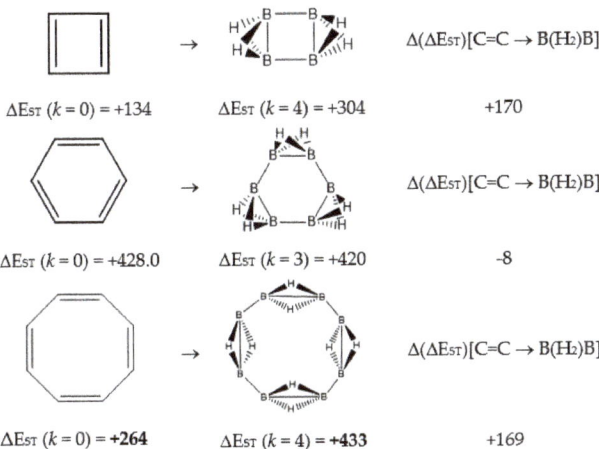

Figure 1. Singlet-triplet gap ΔE_{ST}(kJ·mol^{-1}) for cyclobutadiene, benzene and cyclooctatetraene and the equivalent B_nH_{m+n} structures, and the differences $\Delta(\Delta E_{ST})[C=C \rightarrow B(H_2)B] = \Delta E_{ST}(B_nH_{m+n}) - \Delta E_{ST}(C_nH_m)$.

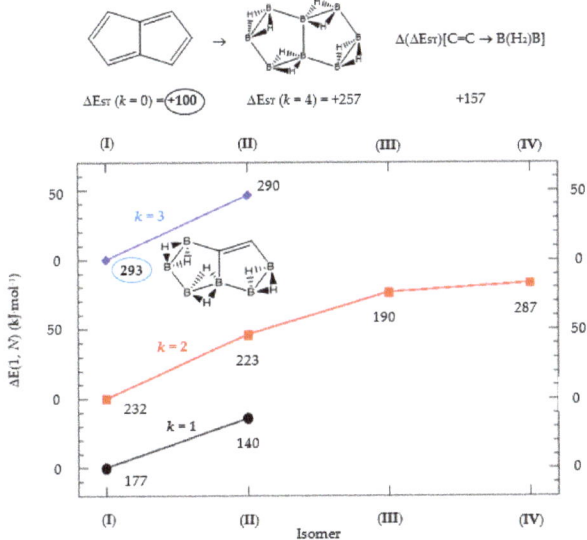

Figure 2. Isomer structure vs. $\Delta E(\mathbf{I}, N_{iso})$ (kJ·mol^{-1}) for pentalene, with $\Delta E(\mathbf{I}, N) = E(N_{iso}) - E(\mathbf{I})$. N_{iso} is the number of isomers for a given k in $C_{(n-2k)}B_{(2k)}H_{(m+2k)}$. In addition, for each structure the singlet-triplet gap is shown (kJ·mol^{-1}). The encircled singlet-triplet energy gaps correspond to the maximum and minimum values.

Table 1. Cyclobutadiene, benzene, cyclooctatetraene and the corresponding hybrids $C_{(n-2k)}B_{(2k)}H_{(m+2k)}$, with formula ($k$), structure, energies (B3LYP/cc-pVTZ, in au), singlet-triplet vertical energy gaps (kJ·mol^{-1}) with the B3LYP and B97D functionals, $\Delta E_{ST,B3LYP}$ and $\Delta E_{ST,B97D}$ (italic) respectively, and point-group symmetry (PGS) in parentheses. For benzene and hybrids we also show the molecular electrostatic potential (MEP) on the 0.001 au electron density isosurface (blue and red colors indicate values $V(r) > +0.015$ au and $V(r) < -0.015$ au, respectively), and the anisotropy of the induced current density (ACID). Different isomers of a given structure are labelled as {(**I**), (**II**), ... } and ordered in increasing energy. All geometries correspond to energy minima. In-plane *exo* hydrogen atoms bound to carbon and boron not shown for clarity. B3LYP/cc-pVTZ computations.

	Cyclobutadiene $C_{(4-2k)}B_{(2k)}H_{(4+2k)}$		
Formula (k)	C_4H_4 (0)	$C_2B_2H_6$ (1)	B_4H_8 (2)
Structure			
Energy	−154.734983	−129.488436	−104.182558
$\Delta E_{ST,B3LYP}$ (PGS)	133.5 (D_{2h})	272.9 (C_{2v})	303.6 (D_{2h})
$\Delta E_{ST,B97D}$	*124.8*	*265.2*	*272.3*

		Benzene $C_{(6-2k)}B_{(2k)}H_{(6+2k)}$		
C_6H_6 (0)	$C_4B_2H_8$ (1)	$C_2B_4H_{10}$ (2)	B_6H_{12} (3)	
−232.333368	−206.929239	−181.596841	−156.324833	
428.0 (D_{6h})	325.9 (C_{2v})	361.2 (C_{2v})	419.8 (D_{3h})	
414.4	*312.9*	*353.1*	*373.2*	

		Cyclooctatetraene $C_{(8-2k)}B_{(2k)}H_{(8+2k)}$		
C_8H_8 (0)	$C_6B_2H_{10}$ (1)	$C_4B_4H_{12}$ (2)	$C_2B_6H_{14}$ (3)	B_8H_{16} (4)
−309.697808	−284.398420	−259.096253	−233.769903	−208.442935
264.4 (D_{2d})	312.2 (C_s)	(**I**) 377.1 (C_{2v})	375.3 (C_s)	433.1 (D_{2d})
234.3	*284.1*	*344.2*	*364.4*	*381.0*
		−259.084995		
		(**II**) 355.7 (C_2)		
		332.9		

Table 2. Pentalene and the corresponding hybrids $C_{(8-2k)}B_{(2k)}H_{(6+2k)}$, with formula ($k$), structure, energies (au), singlet-triplet vertical energy gaps ΔE_{ST} (kJ·mol^{-1}) and point-group symmetry (PGS) in parentheses. Different isomers of a given structure are labelled as {(I), (II), ... } and ordered in increasing energy. All geometries correspond to energy minima. In-plane *exo* hydrogen atoms bound to carbon and boron not shown for clarity. B3LYP/*cc*-pVTZ computations.

Pentalene $C_{(8-2k)}B_{(2k)}H_{(6+2k)}$				
C_8H_6 (0)	$C_6B_2H_8$ (1)	$C_4B_4H_{10}$ (2)	$C_2B_6H_{12}$ (3)	B_8H_{14} (4)
−308.476075	−283.193095	−257.898516	−232.562207	−207.215121
100.1 (C_{2h})	(I) 176.7 (C_s)	(I) 232.3 (C_{2h})	(I) 292.9 (C_s)	257.3 (C_{2h})
	−283.179450	−257.881067	−232.544645	
	(II) 140.2 (C_s)	(II) 222.5 (C_s)	(II) 290.4 (C_s)	
		−257.869529		
		(III) 189.9 (C_s)		
		−257.866865		
		(IV) 286.8 (C_{2h})		

Table 3. a. Benzocyclobutadiene Kekulé structures K_1 and the corresponding hybrid boranes, with formula (k), structure, energies (au), singlet-triplet vertical energy gaps ΔE_{ST} (kJ·mol^{-1}) and point-group symmetry (PGS) in parentheses. **b.** Kekulé structures K_2. **c.** Kekulé structures K_3. Different isomers of a given structure are labelled as {(I), (II), ... } and ordered in increasing energy. B3LYP/cc-pVTZ computations. All geometries correspond to energy minima. In-plane *exo* hydrogen atoms bound to carbon and boron not shown for clarity.

(a)				
Benzocyclobutadiene Kekulé Structure K_1: $C_{(8-2k)}B_{(2k)}H_{(6+2k)}$				
C_8H_6 (0)	$C_6B_2H_8$ (1)	$C_4B_4H_{10}$ (2)	$C_2B_6H_{12}$ (3)	B_8H_{14} (4)
−308.466884	−283.206069	−257.868229	−232.534049	−207.187765
195.3 (C_{2v})	(I) 303.0 (C_{2v})	(I) 225.8 (C_s)	(I) 361.7 (C_{2v})	264.0 (C_{2v})

Table 3. *Cont.*

(a) Benzocyclobutadiene Kekulé Structure K_1: $C_{(8-2k)}B_{(2k)}H_{(6+2k)}$

C_8H_6 (0)	$C_6B_2H_8$ (1)	$C_4B_4H_{10}$ (2)	$C_2B_6H_{12}$ (3)	B_8H_{14} (4)
	−283.141948 (II) 285.1 (C_{2v})	−257.847770 (II) 338.7 (C_{2v})	−232.521784 (II) 308.5 (C_s)	
	−283.125618 (III) 117.7 (C_s)	−257.820144 (III) 265.9 (C_s)	−232.488989 (III) 254.3 (C_{2v})	
		−257.791707 (IV) 108.4 (C_{2v})		

(b) Benzocyclobutadiene Kekulé Structure K_2: $C_{(8-2k)}B_{(2k)}H_{(6+2k)}$

C_6H_6 (0)	$C_6B_2H_8$ (1)	$C_4B_4H_{10}$ (2)	$C_2B_6H_{12}$ (3)	B_6H_{14} (4)
−308.466884 195.3 (C_{2v})	−283.206069 (I) C_{2v} 302.9 = (I) in K_1	−257.880840 (I) C_{2v} 367.5	−232.542344 (I) C_s 342.3	−207.194531 C_{2v} 310.3
	−283.162243 (II) C_{2v} 267.2	−257.869238 (II) C_s 280.1	−232.529224 (II) C_{2v} 323.8	
	−283.161645 (III) C_s 220.2	−257.848002 (III) C_s 280.4	−232.517824 (III) C_{2v} 348.8	
		−257.842989 (IV) C_{2v} 328.5		

Table 3. *Cont.*

(c)

Benzocyclobutadiene Kekulé Structure K_3: $C_{(8-2k)}B_{(2k)}H_{(6+2k)}$

C_8H_6 (0)	$C_6B_2H_8$ (1)	$C_4B_4H_{10}$ (2)	$C_2B_6H_{12}$ (3)	B_6H_{14} (4)
−308.466884 195.3 (C_{2v})	−283.151132 (I) 137.7 (C_s)	−257.848644 (I) 229.1 (C_s)	−232.527701 (I) 210.9 (C_s)	−207.196234 374.0 (C_{2v})
	−283.125618 (II) 117.6 (C_s)	−257.841430 (II) 219.6 (C_{2v})	−232.517894 (II) 237.0 (C_s)	
		−257.838383 (III) 173.8 (C_s)		
		−257.791707 (IV) 108.5 (C_{2v})		

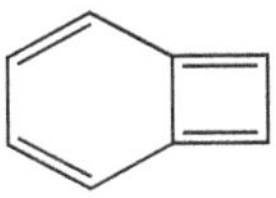

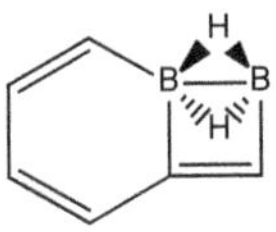

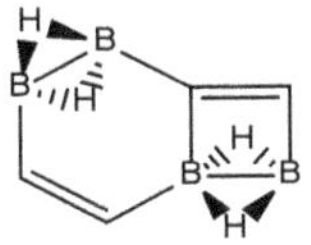

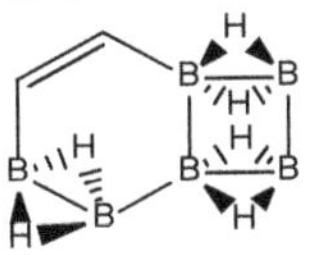

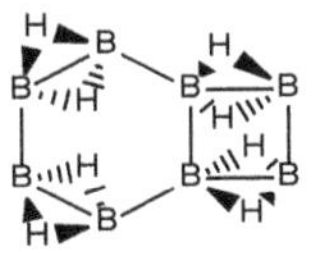

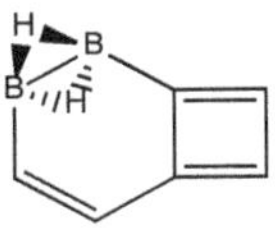

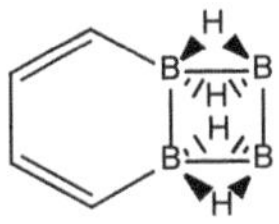

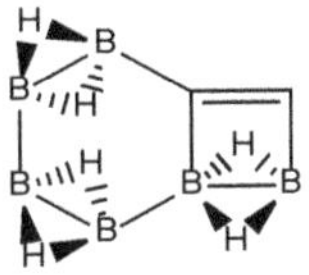

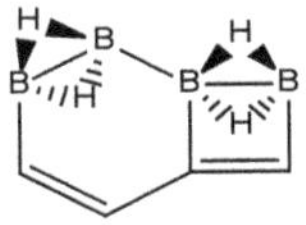

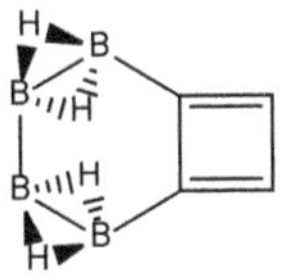

Table 4. a. Naphthalene Kekulé structures K_1 and the corresponding hybrid boranes, with formula (k), structure, energies (au), singlet-triplet vertical energy gaps ΔE_{ST} (kJ·mol^{-1}) and point-group symmetry (PGS) in parentheses. **b.** Kekulé structures K_2. Different isomers of a given structure are labelled as {(I), (II), ... } and ordered in increasing energy. B3LYP/cc-pVTZ computations. All geometries correspond to energy minima. In-plane *exo* hydrogen atoms bound to carbon and boron not shown for clarity.

(a)

Naphthalene Kekulé Structure K_1: $C_{(10-2k)}B_{(2k)}H_{(8+2k)}$

$C_{10}H_8$ (0)	$C_8B_2H_{10}$ (1)	$C_6B_4H_{12}$ (2)	$C_4B_6H_{14}$ (3)	$C_2B_8H_{16}$ (4)	$B_{10}H_{18}$ (5)
−386.025545 304.3 (D_{2h})	−360.701217 (I) 325.9 (C_s)	−335.372607 (I) 414.4 (C_{2v})	−310.030119 (I) 287.8 (C_s)	−284.691550 (I) 316.1 (D_{2h})	−259.335521 313.6 (D_{2h})

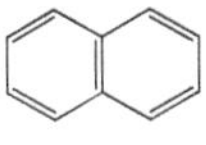

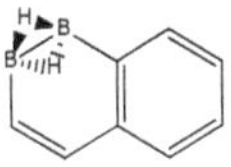

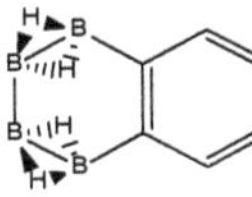

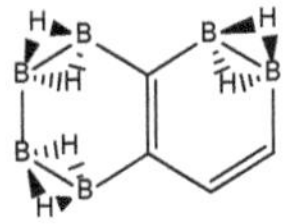

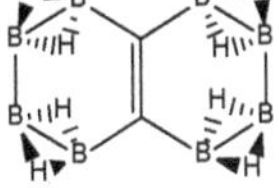

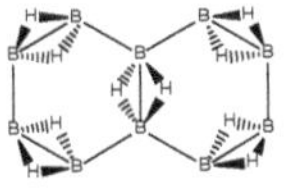

Table 4. *Cont.*

(a)

Naphthalene Kekulé Structure K_1: $C_{(10-2k)}B_{(2k)}H_{(8+2k)}$

$C_{10}H_8$ (0)	$C_8B_2H_{10}$ (1)	$C_6B_4H_{12}$ (2)	$C_4B_6H_{14}$ (3)	$C_2B_8H_{16}$ (4)	$B_{10}H_{18}$ (5)
−360.673209	−335.365370	−310.015638	−284.676167		
(**II**) 340.7 (D_{2h})	(**II**) 266.4 (C_{2h})	(**II**) 393.0 (C_{2v})	(**II**) 353.7 (C_s)		
	−335.364766	−310.013813			
	(**III**) 300.4 (C_{2v})	(**III**) 385.0 (C_{2h})			
	−335.346472	−310.011145			
	(**IV**) 322.7 (C_s)	(**IV**) 327.6 (C_{2v})			

(b)

Naphthalene Kekulé Structure K_2: $C_{(10-2k)}B_{(2k)}H_{(8+2k)}$

$C_8B_2H_{10}$ (1)	$C_6B_4H_{12}$ (2)	$C_4B_6H_{14}$ (3)	$C_2B_8H_{16}$ (4)	$B_{10}H_{18}$ (5)
−360.701217	−335.372607	−310.028387	−284.687476	−259.338265
(**I**) 325.9 (C_s)	(**I**) 414.4 (C_{2v})	(**I**) 297.5 (C_{2v})	(**I**) 331.2 (C_s)	307.0 (C_{2v})
= (**I**) in K_1	= (**I**) in K_1			
−360.678565	−335.362684	−310.025977	−284.679641	
(**II**) 223.1 (C_s)	(**II**) C_s 303.3	(**II**) 301.6 (C_s)	(**II**) C_{2v} 351.0	
−360.675945	−335.356631	−310.014490	−284.675123	
(**III**) C_{2v} 187.8	(**III**) 256.1 (C_s)	(**III**) C_s 251.0	(**III**) 313.2 (C_s)	
	−335.352046	−310.014211		
	(**IV**) 244.0 (C_s)	(**IV**) 333.7 (C_s)		

Table 4. Cont.

(b)

Naphthalene Kekulé Structure K_2: $C_{(10-2k)}B_{(2k)}H_{(8+2k)}$				
$C_8B_2H_{10}$ (1)	$C_6B_4H_{12}$ (2)	$C_4B_6H_{14}$ (3)	$C_2B_8H_{16}$ (4)	$B_{10}H_{18}$ (5)
	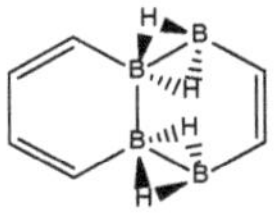 −335.335292 (V) C_{2v} 252.3	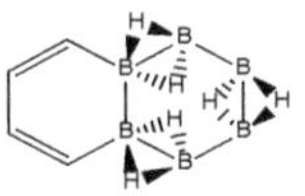 −309.998099 (V) C_{2v} 234.9		

Table 5. Azulene the corresponding hybrid boranes, with formula (k), structure, energies (au), singlet-triplet vertical energy gaps ΔE_{ST} (kJ·mol^{-1}) and point-group symmetry (PGS) in parentheses. Different isomers of a given structure are labelled as {(I), (II), ... } and ordered in increasing energy. B3LYP/cc-pVTZ computations. All geometries correspond to energy minima. In-plane *exo* hydrogen atoms bound to carbon and boron not shown for clarity. All structures and isomers have C_s symmetry except azulene $C_{10}H_8$ and the equivalent boron structure $B_{10}H_{18}$ both with C_{2v} symmetry.

Azulene $C_{(10-2k)}B_{(2k)}H_{(8+2k)}$			
$C_{10}H_8$ (0)	$C_8B_2H_{10}$ (1)	$C_6B_4H_{12}$ (2)	$C_4B_6H_{14}$ (3)
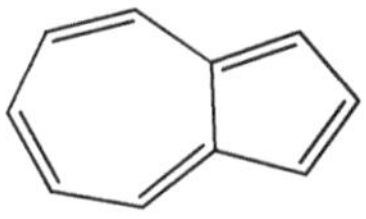 −385.971486 196.3 (C_{2v})	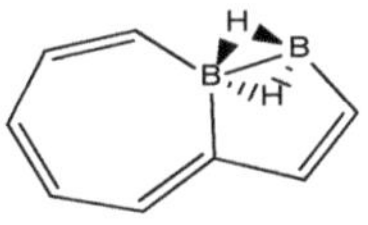 −360.664485 (I) 224.4 (C_s)	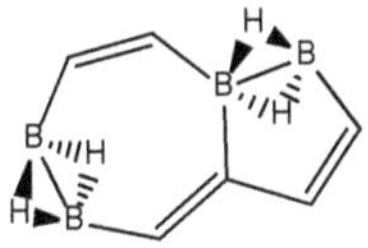 −335.342898 (I) 260.7 (C_s)	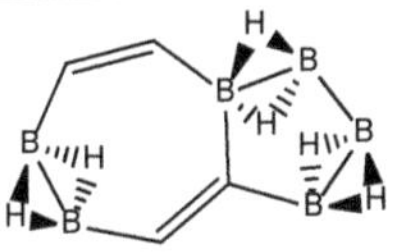 −310.017400 (I) 329.9 (C_s)
	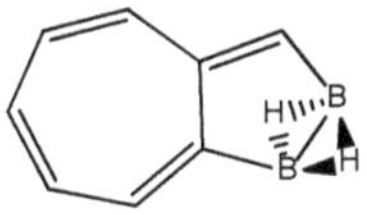 −360.653691 (II) 180.5 (C_s)	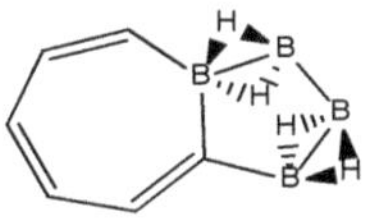 −335.338375 (II) 235.9 (C_s)	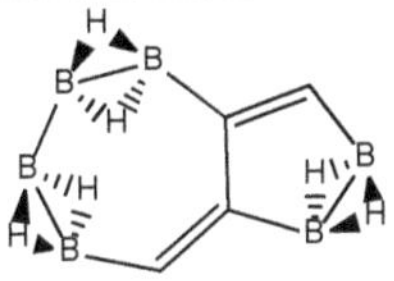 −310.010310 (II) 252.5 (C_s)
	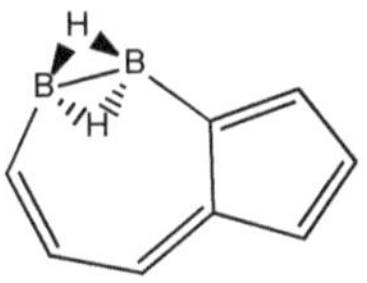 −360.644455 (III) 203.1 (C_s)	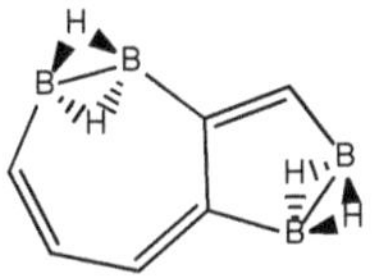 −335.335924 (III) 207.6 (C_s)	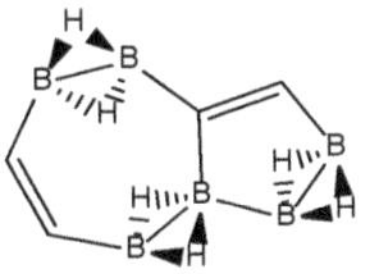 −310.009032 (III) 335.1 (C_s)
	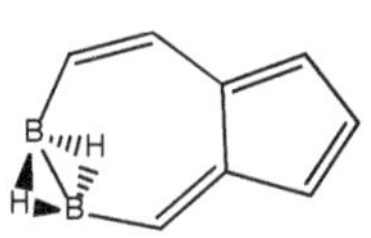 −360.643019 (IV) 183.3 (C_s)	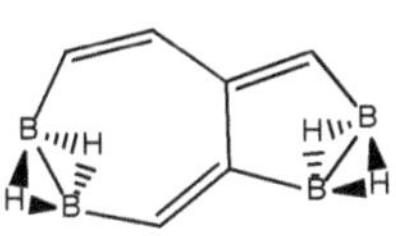 −335.333474 (IV) 239.7 (C_s)	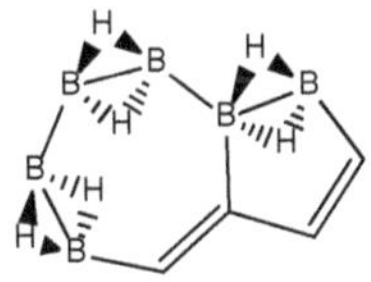 −310.006879 (IV) 248.1 (C_s)

Table 5. *Cont.*

Azulene C$_{(10-2k)}$B$_{(2k)}$H$_{(8+2k)}$			
C$_{10}$H$_8$ (0)	C$_8$B$_2$H$_{10}$ (1)	C$_6$B$_4$H$_{12}$ (2)	C$_4$B$_6$H$_{14}$ (3)
	−360.637439 (**V**) 179.7 (C$_s$)	−335.332218 (**V**) 212.6 (C$_s$)	−310.005650 (**V**) 250.4 (C$_s$)

Azulene C$_{(10-2k)}$B$_{(2k)}$H$_{(8+2k)}$	
C$_6$B$_4$H$_{12}$ (2)	C$_4$B$_6$H$_{14}$ (3)
−335.324248 (**VI**) 219.3 (C$_s$)	−309.999674 (**VI**) 327.6 (C$_s$)
−335.321141 (**VII**) 272.0 (C$_s$)	−309.999405 (**VII**) 319.5 (C$_s$)
−335.320601 (**VIII**) 217.0 (C$_s$)	−309.996626 (**VIII**) 249.8 (C$_s$)
−335.319266 (**IX**) 208.9 (C$_s$)	−309.992937 (**IX**) 210.9 (C$_s$)
−335.313370 (**X**) 163.1 (C$_s$)	−309.989768 (**X**) 262.6 (C$_s$)

Table 5. Cont.

Azulene $C_{(10-2k)}B_{(2k)}H_{(8+2k)}$	
$C_2B_8H_{16}$ (4)	$B_{10}H_{18}$ (5)
−284.679284 (I) 315.4 (C_s)	−259.332171 (I) 293.3 (C_s)
−284.675773 (II) 311.6 (C_s)	
−284.667945 (III) 320.1 (C_s)	
−284.666066 (IV) 317.1 (C_s)	
−284.666006 (V) 313.2 (C_s)	

At this point we should emphasize the geometrical changes that a C_nH_m conjugated hydrocarbon undergoes through k {C=C ↔ B(H$_2$)B} substitutions, with k = {0, 1, 2, ..., $n/2$}, in a given hybrid boron-carbon $C_{(m-2k)}B_{(2k)}H_{(n+2k)}$ structure.

In Table 1 and Figure 1, we gather the structures, energies and singlet-triplet gaps for cyclobutadiene, benzene and cyclooctatetraene and the corresponding $C_{(n-2k)}B_{(2k)}H_{(m+2k)}$ structures. We start off with the series for n = 4 in cyclobutadiene: $C_{(4-2k)}B_{(2k)}H_{(4+2k)}$, with k = {0, 1, 2}. Given the antiaromatic nature of cyclobutadiene, this molecule is very reactive with tendency to dimerize and can be observed by matrix isolation below 35 K [26]. Substitution of one {C=C} moiety by one {B(H$_2$)B} moiety in C_4H_4 (k = 1) leads to a $C_2B_2H_6$ cyclic structure, with a singlet-triplet energy gap 140 kJ·mol^{-1} larger than in C_4H_4. A second {C=C ↔ B(H$_2$)B} substitution in C_4H_4 (k = 2) leads to cyclic tetraborane(8), with a even larger singlet-triplet energy gap, 170 kJ·mol^{-1} higher than in C_4H_4.

The next system to be analyzed is aromatic benzene ($n = 6$) and the boron-carbon hybrids: $C_{(6-2k)}B_{(2k)}H_{(6+2k)}$, with $k = \{0, 1, 2, 3\}$, as gathered in Table 1 and Figure 1. Benzene is a colorless liquid, with a characteristic odor, volatile, very flammable and carcinogenic. This molecule is the paradigm of Hückel theory [1] and the concept of aromaticity itself, with a very interesting debate reaching our days on the role of its correlation energy and the π electron spin-pairing [27]. As opposed to the previous example, the singlet-triplet energy gaps for benzene and cyclic hexaborane(12) B_6H_{12} are quite similar, 8 kJ·mol^{-1} larger in benzene, due to the aromatic nature of the latter. The first {C=C ↔ B(H$_2$)B} substitution in benzene leading to $C_4B_2H_8$ ($k = 1$) decreases the singlet-triplet energy gap by more than 100 kJ·mol^{-1}—due to aromaticity loss—but with further {C=C ↔ B(H$_2$)B} substitutions the singlet-triplet energy gaps increase by 35 kJ·mol^{-1} and 60 kJ·mol^{-1} for $C_2B_4H_{10}$ ($k = 2$) and B_6H_{12} ($k = 3$) respectively. This is a clear indication that breaking the aromaticity of benzene leads, to a first instance, to a less stable structure. However, this stability is increased by along the series for $k = \{2, 3\}$, and three {C=C ↔ B(H$_2$)B} substitutions leads to a structure—D_{3h} cyclic hexaborane(12) B_6H_{12}—with a singlet-triplet energy gap which is only 8 kJ·mol^{-1} lower as compared to benzene. This result is striking and remains at the very origin of the recent proposal we have put forward [24]: To every (poly)cyclic planar conjugated hydrocarbon C_nH_m there corresponds a boron equivalent structure B_nH_{m+n} which is also an energy minimum.

Inclusion of empirical dispersion corrections with the B97D functional [28] shows a systematic decrease of the singlet-triplet energy gaps as compared to the B3LYP results, following a similar tendency, except for the complete borane structures B_nH_{m+n} with considerably lowering of 47 kJ·mol^{-1} for cyclic hexaborane(12), as compared to the B3LYP results. For the cyclooctatetraene series the lowering is also noteworthy when including dispersion corrections in the functional; the structure of the stationary points—energy minima—is maintained with both functionals. As described below in the Discussion section, the bond distances are slightly elongated when dispersion corrections are included, which is in agreement with lower singlet-triplet energy gaps.

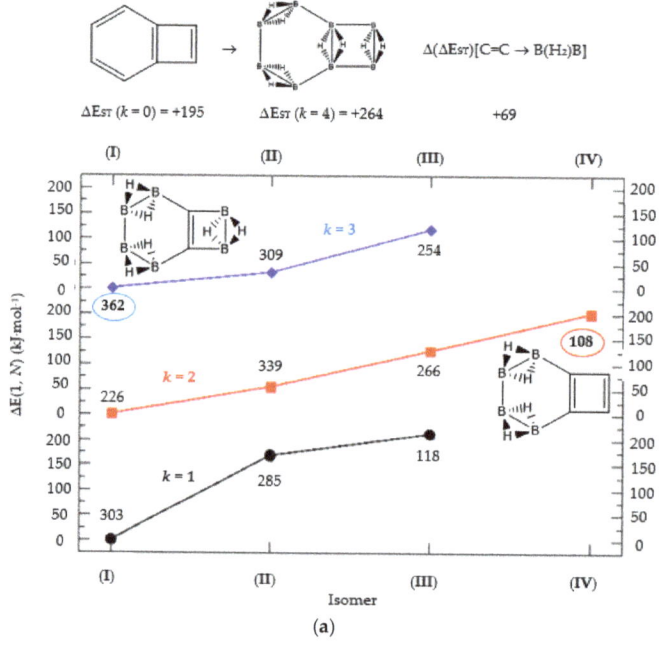

Figure 3. Cont.

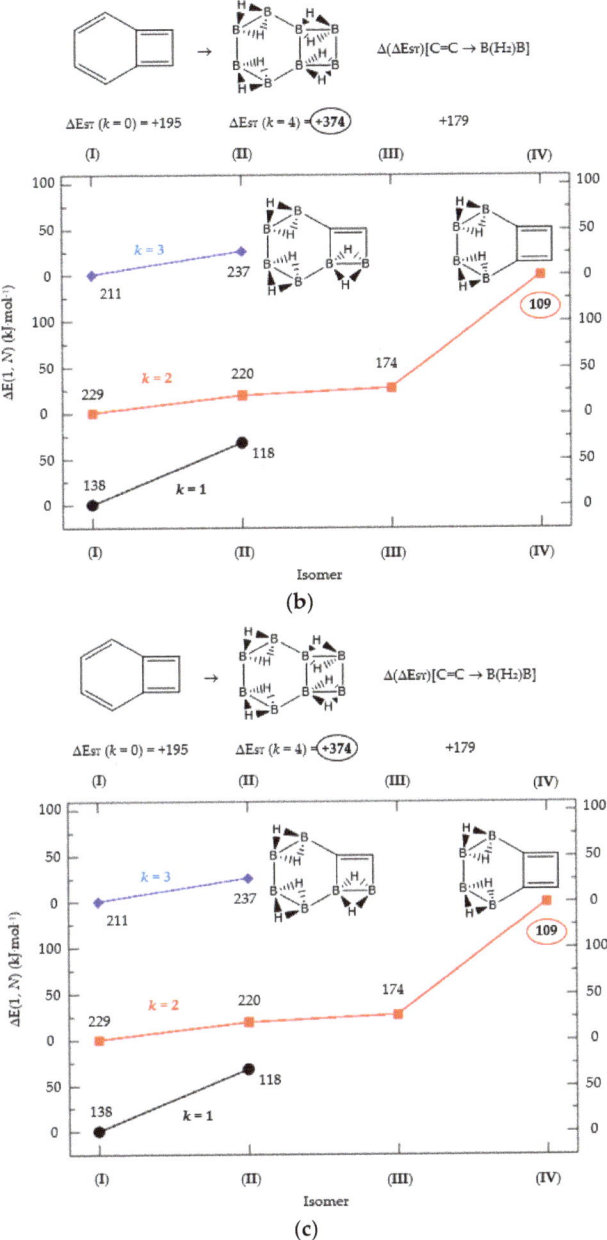

Figure 3. (**a**) Isomer structure vs. $\Delta E(\mathbf{I}, N_{iso})$ (kJ·mol^{-1}) for benzocyclobutadiene Kekulé structure K_1, with $\Delta E(\mathbf{I}, N_{iso}) = E(N_{iso}) - E(\mathbf{I})$. (**b**) Isomer structure vs. $\Delta E(\mathbf{I}, N_{iso})$ (kJ·mol^{-1}) for benzocyclobutadiene Kekulé structure K_2. (**c**) Isomer structure vs. $\Delta E(\mathbf{I}, N_{iso})$ (kJ·mol^{-1}) for benzocyclobutadiene Kekulé structure K_3. N_{iso} is the number of isomers for a given k in $C_{(n-2k)}B_{(2k)}H_{(m+2k)}$. In addition, for each structure the singlet-triplet gap is shown (kJ·mol^{-1}). The encircled singlet-triplet gaps correspond to the maximum and minimum values.

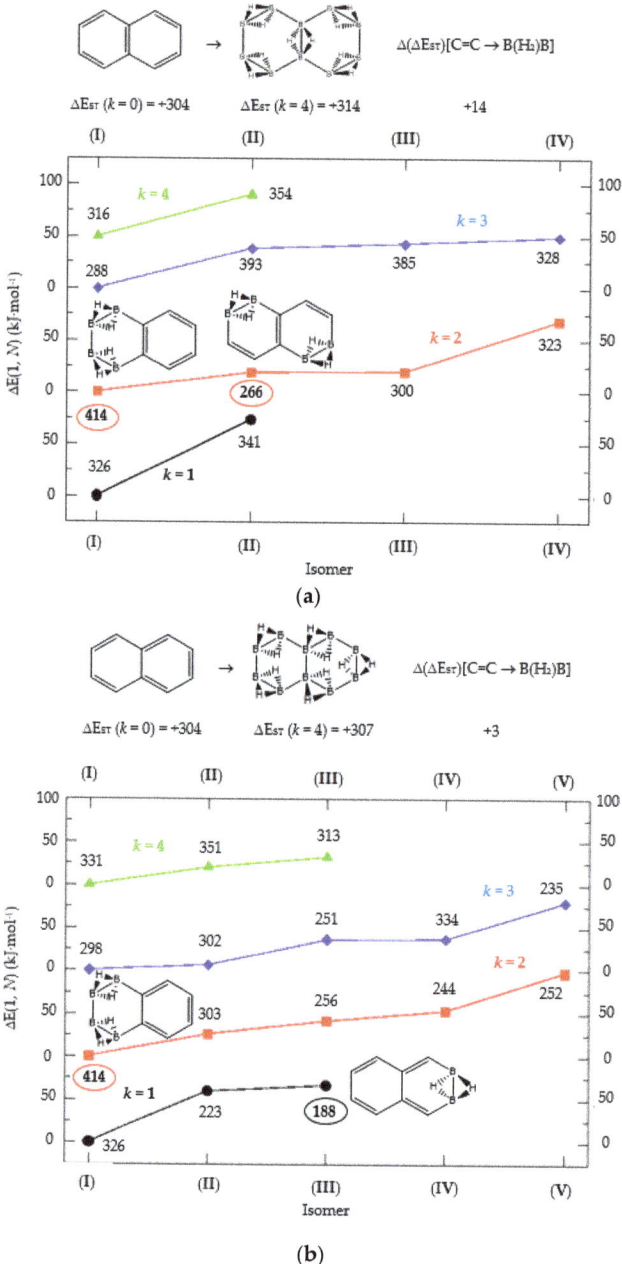

Figure 4. (**a**) Isomer structure vs. $\Delta E(\mathbf{I}, N_{iso})$ (kJ·mol^{-1}) for naphthalene Kekulé structure K_1, with $\Delta E(\mathbf{I}, N_{iso}) = E(N_{iso}) - E(\mathbf{I})$. (**b**) Isomer structure vs. $\Delta E(1, N)$ (kJ·mol^{-1}) for naphthalene Kekulé structure K_2. N_{iso} is the number of isomers for a given k in $C_{(n-2k)}B_{(2k)}H_{(m+2k)}$. In addition, for each structure the singlet-triplet gap is shown (kJ·mol^{-1}). The encircled singlet-triplet gaps correspond to the maximum and minimum values.

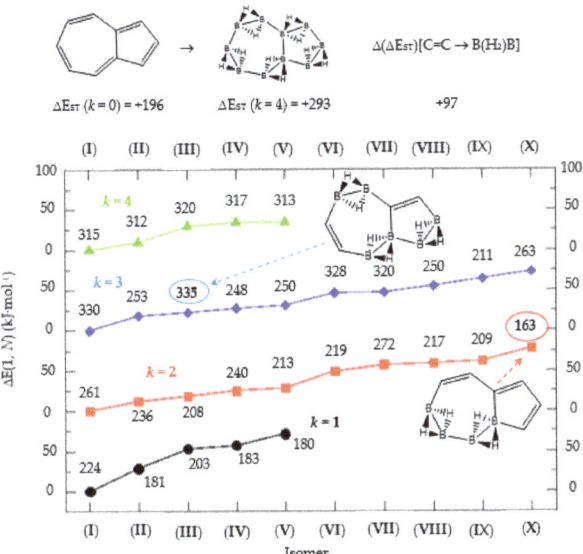

Figure 5. Isomer structure vs. $\Delta E(\mathbf{I}, N_{iso})$ (kJ·mol^{-1}) for azulene, with $\Delta E(\mathbf{I}, N_{iso}) = E(N_{iso}) - E(\mathbf{I})$. N_{iso} is the number of isomers for a given k in $C_{(n-2k)}B_{(2k)}H_{(m+2k)}$. In addition, for each structure the singlet-triplet gap is shown (kJ·mol^{-1}). The encircled singlet-triplet gaps correspond to the maximum and minimum values.

In order to highlight the electronic structure differences in benzene and its borane hybrids we have plotted in Table 1 the molecular electrostatic potential (MEP) for the benzene series $C_{(6-2k)}B_{(2k)}H_{(6+2k)}$, with $k = \{0, 1, 2, 3\}$: The blue and red colors in the MEP indicate negative and positive charge attraction areas: clearly, the benzene π-electron cloud is a positive-charge attraction area [29,30] and the borane substituted areas are negative-charge attraction areas, with a complete blue area above the boron-atoms plane in cyclic hexaborane(12) B_6H_{12}. Just below the MEP in Table 1 we show the anisotropy of the induced current density (ACID) [31] for the same systems, with a clear blocking of the current in the region where a {C=C ↔ B(H$_2$)B} substitution is carried out; therefore π electron delocalization lowers considerably as k increases.

We turn now to an antiaromatic system with $n = 8$ π electrons, cyclooctatetraene and the structural isoelectronic series summarized in Table 1 and Figure 1: $C_{(8-2k)}B_{(2k)}H_{(8+2k)}$, with $k = \{0, 1, 2, 3, 4\}$. Cyclooctetraene is a colorless to light yellow flammable liquid at room temperature and adopts a non-planar D_{2d} structure in the ground state [32,33]. We should emphasize that cyclooctatetraene C_8H_8 ($k = 0$) and the corresponding borane equivalent B_8H_{16} ($k = 4$) are transition state structures in the planar D_{4h} conformation with energy barriers of 44.0 kJ·mol^{-1} [34] and 55.8 kJ·mol^{-1} respectively (B3LYP/cc-pVTZ computations), and thus the barrier for the $D_{2d} \to D_{4h}$ process is larger for the borane compound. Both energy minima correspond to D_{2d} structures, and this is the only series of non-planar hydrocarbon and hybrid boron-carbon structures included in this work. The main purpose of including cyclooctatetraene is due to the non-planarity of the system, which was not considered before [21].

Consecutive {C=C ↔ B(H$_2$)B} substitutions in cyclooctatetraene C_8H_8 ($k = 0$) leads to $C_{(6-2k)}B_{(2k)}H_{(6+2k)}$ structures with a larger singlet-triplet energy gap as k increases from zero to 4, with a final gap which is 170 kJ·mol^{-1} larger as compared to the original hydrocarbon. Cyclooctatetraene is the first instance where we have two different positional isomers in $C_4B_4H_{12}$ ($k = 2$): Isomers **I** and **II**, with lower energy and larger singlet-triplet gap for the more symmetric structure **I** with C_{2v} symmetry. We define N_{iso} as the number of isomers for a given k in $C_{(n-2k)}B_{(2k)}H_{(m+2k)}$. For instance, for azulene and $k = 2$—see Table 5 below, we have a set of ten $C_6B_4H_{12}$ isomers: {**I, II, III**, ... , N_{iso}}, with $N_{iso} = 10$ in Roman numerals, i.e., $N_{iso} = $ **X**.

We consider next the pentalene molecule C_8H_6, a two-pentagon one-side fused conjugated system, also with $n = 8$ π electrons, and therefore antiaromatic. Pentalene dimerizes at 173 K [35], and its derivative, 1,3,5-tri-tert-butylpentalene, has been synthesized [36], as a stabilized planar 8π electron system. In Table 2 and Figure 2 we gather the results for the pentalene series: $C_{(8-2k)}B_{(2k)}H_{(6+2k)}$, with $k = \{0, 1, 2, 3, 4\}$. Again, due to the antiaromatic nature of the hydrocarbon, the singlet-triplet energy gap in the equivalent boron structure ($k = 4$) is 157 kJ·mol^{-1} higher in energy, as in the previous antiaromatic cases for cyclobutadiene and cyclooctatetraene, and the energy gap in pentalene is the lowest along the whole series of structures and isomers derived from $C_{(8-2k)}B_{(2k)}H_{(6+2k)}$ and $1 < k \leq 4$, as the encircled value shown at the top of Figure 2.

The singlet-triplet energy gaps increase along the series for $0 \leq k \leq 3$; the equivalent boron structure B_8H_{14} ($k = 4$) has a lower gap as compared to the previous ($k = 3$) molecule $C_2B_6H_{12}$. There are striking differences in the energy gaps as function of k, the number of {C=C ↔ B(H$_2$)B} substitutions, and also within the set of isomers of a given k, as shown in Figure 2. The largest and lowest singlet-triplet energy gaps in the whole series are shown in Figure 2 as encircled numbers and with the corresponding structure. All structures and isomers for $1 \leq k \leq 4$ have a larger singlet-triplet energy gap as compared to original pentalene ($k = 0$), a clear indication on how unstable this conjugated hydrocarbon is as compared to the boron-carbon hybrid series.

A shown in Table 2 and Figure 2 the isomer energies for a given structure (k) follow the same order as the singlet-triplet gaps, except isomer **IV** for $k = 2$ with an increase of 100 kJ·mol^{-1}. It is striking that isomers with very close electronic energies might have such different singlet-triplet gaps. The energy difference between isomers **III** and **IV** for $k = 2$ is only 7 kJ·mol^{-1}. Hence positional {C=C ↔ B(H$_2$)B} substitutions might have important changes in the interaction of these hybrid boron-carbon isomers with external perturbations, such as photochemical processes.

We follow with another conjugated hydrocarbon with $n = 8$ π electrons: benzocyclobutadiene and the hybrid boron-carbon series $C_{(8-2k)}B_{(2k)}H_{(6+2k)}$, with $k = \{0, 1, 2, 3, 4\}$, similar to pentalene, and displayed in Table 3a–c and Figure 3a–c. Benzocyclobutadiene polymerises readily and reacts as a dienophile in Diels-Alder reactions [37]. As opposed to the previous $n = 8$ systems, in this particular case we need to consider different Kekulé structures, K_1, K_2 and K_3 (Scheme 1) since they give different structural isomers on consecutive k {C=C ↔ B(H$_2$)B} substitutions. The energy gap in the original hydrocarbon is 195 kJ·mol^{-1} (the same for K_1, K_2 and K_3) and we find for the first time that {C=C ↔ B(H$_2$)B} substitutions lead to hybrid systems with lower gaps, e.g., isomer **IV** with $k = 2$ in Table 3a with a gap of 108 kJ·mol^{-1}, the lowest among all as shown in Figure 3a, clearly due to the presence of the strained cyclobutadiene ring in the structure. Only isomers from Kekulé structures K_1 and K_3 keep this tendency as shown in Table 3a–c respectively. There is an exception to this behavior: for the Kekulé structure K_2—Table 3b—the gaps are always larger in the substituted isomers, as compared to benzocyclobutadiene. This is striking and due to the presence of only one double bond C=C in the rectangular cyclobutadiene, hence lowering the strain energy on any further {C=C ↔ B(H$_2$)B} substitution. Another noteworthy result is the singlet-triplet energy gap differences in the boron equivalent structures of K_1, K_2 and K_3, always fairly above benzocyclobutadiene by 70 kJ·mol^{-1}, 115 kJ·mol^{-1} and 180 kJ·mol^{-1} respectively, as shown in Figure 3a–c.

There are many isomers with larger energy gaps as compared to benzocyclobutadiene: The borane structure for planar ($k = 4$) B_8H_{14} in the K_3 configuration—Table 3c and Figure 3c—has the largest gap of all structures and isomers, 374 kJ·mol^{-1}. When strain is released from the rectangular cyclobutadiene moiety in the original hydrocarbon, many isomers have gaps above 300 kJ·mol^{-1}, with the interesting case of isomer **I** of K_2 with $k = 2$—Table 4b—a *cis*-butadiene system fused with two diborane molecules and energy gap 368 kJ·mol^{-1} and the lowest ground state energy. The experimental vertical singlet-triplet gap in butadiene is 310.7 kJ·mol^{-1} [38] (318.9 kJ·mol^{-1} with a B3LYP/cc-pVTZ computation). Therefore, the two diborane molecules fused to the *cis*-butadiene moiety in isomer **I** of $C_4B_4H_{10}$ ($k = 2$) and Kekulé structure K_2 stabilise the system further by 50 kJ·mol^{-1}. We do not intend to give spectroscopic accuracy in these examples, but rather an explanation of the tendency in energy and stability changes upon {C=C ↔ B(H$_2$)B} substitutions in conjugated hydrocarbons. We should emphasize here that all structures and isomers remain planar upon these substitutions in benzocyclobutadiene, checked through frequency computations, thus following the previous examples with the exception of cyclooctatetraene, which is not planar.

Given the topological distribution of single and double bonds in benzocyclobutadiene, only Kekulé structures K_1 and K_3 retain the cyclobutadiene ring in the hybrid boron-carbon series. Hence, for K_1—Table 3a and Figure 3a—one has isomer **III** from $C_6B_2H_8$ ($k = 1$), with an energy gap of 118 kJ·mol^{-1} and isomer **IV** from $C_4B_6H_{10}$ ($k = 2$) with a gap of 108 kJ·mol^{-1} (lowest gap). These isomers are equivalent in K_3—Table 3c and Figure 3c—to isomer **II** from $C_6B_2H_8$ ($k = 1$), and isomer **IV** from $C_4B_6H_{10}$ ($k = 2$). The presence of the cyclobutadiene ring explains why these hybrid boron-carbon isomers have such low singlet-triplet energy gaps.

We turn now to the next aromatic system, naphthalene, with $n = 10$ π electrons, and the series $C_{(10-2k)}B_{(2k)}H_{(8+2k)}$, with $k = \{0, 1, 2, 3, 4\}$. Naphthalene is a white crystalline solid with a characteristic odor and detectable at very low concentrations and reacts more readily than benzene in electrophilic aromatic substitution reactions [37]. The two Kekulé structures K_1 and K_2 shown in Scheme 1 are also necessary in order to include all possible different isomers upon consecutive {C=C ↔ B(H$_2$)B} substitutions. Table 4a,b and Figure 4a,b include the energies and energy gaps for all isomers derived from Kekulé structures K_1 and K_2 in naphthalene, respectively. What differs notably, *viz.*, the energy gaps as compared to antiaromatic systems ($n = 4, 8$)—see above also benzene—is the small difference between the gaps in the original conjugated hydrocarbon $C_{10}H_8$ and the equivalent borane structure $B_{10}H_{18}$, 10 kj·mol^{-1} and 3 kJ·mol^{-1} higher for Kekulé structures K_1 and K_2 respectively. Therefore, from a thermochemical point of view, the synthesis of the planar borane structures derived from benzene and naphthalene, B_6H_{12} and $B_{10}H_{18}$ respectively, should be affordable.

Again, for an aromatic system, consecutive {C=C ↔ B(H$_2$)B} substitutions lead to larger energy gaps compared to naphthalene, with a striking increase for the $k = 2$ isomer **I** in structures K_1 and K_2—both are equivalent—$C_6B_4H_{12}$ ($k = 2$)—Table 4a—which corresponds to a fusion of two diborane molecules to benzene, and giving a further stability to the molecule. Isomers with lower gaps than naphthalene appear both in K_1 and K_2 structures. The two lowest gaps in K_1 correspond to isomers **II** from $C_6B_4H_{12}$ ($k = 2$) and isomer **I** from $C_4B_6H_{14}$ ($k = 3$). The lowest gap in K_2—Table 4b—correspond to the $k = 1$ isomer **III** from structure $C_8B_2H_{10}$; this destabilisation is clearly due to the loss of aromaticity in the whole system on substituting one C=C bond by a B(H$_2$)B moiety on the right ending of the molecule, thus quenching the resonance energy.

We should also emphasize that K_1 set has more isomers than the K_2 set with a gap larger than naphthalene, and all isomers are planar structures corresponding to energy minima for all set of Kekulé structures.

Finally, the last system included in this work, aromatic azulene with $n = 10$ π electrons follows the same series as in naphthalene: $C_{(10-2k)}B_{(2k)}H_{(8+2k)}$, with $k = \{0, 1, 2, 3, 4, 5\}$. Two terpenoids of azulene appear in Nature and offer a rich organic chemistry [39]. In azulene we have one, 5 and 10 isomers for $k = \{0, 5\}$, $k = \{1, 4\}$ and $k = \{2, 3\}$ respectively, as shown in Table 5 and Figure 5. There is only one Kekulé structure leading to different isomers, and the presence of a seven-membered ring fused to a five-membered ring entails a different electronic structure as depicted by the lower energy gap in the original hydrocarbon as compared to benzene and naphthalene.

The azulene equivalent borane structure $B_{10}H_{18}$ has a singlet-triplet energy gap 100 kJ·mol^{-1} above the original system and again there are isomers with lower and higher energy gaps as compared to azulene. The largest and lowest gaps correspond to 335 kJ·mol^{-1} and 163 kJ·mol^{-1} for isomer **III** of structure $C_4B_6H_{14}$ ($k = 3$) and isomer **X** of structure $C_6B_4H_{12}$ ($k = 2$), respectively, as displayed in Figure 5, with the encircled numbers. The general rule is that the average gap increases with k for $1 < k < 4$, namely, the more {C=C ↔ B(H$_2$)B} substitutions, the more stable the system on average.

One {C=C ↔ B(H$_2$)B} substitution in azulene, leading to five C$_8$B$_2$H$_{10}$ isomers (k = 1), shows two isomers (**I, III**) with an energy gap above azulene and three isomers (**II, IV** and **V**) with energy gaps below azulene. The energy gap in isomer **III** is only 4 kJ·mol^{-1} higher than in azulene. Note that the structural difference between isomer **I** and isomer **V** of structure C$_8$B$_2$H$_{10}$ (k = 1) stems from the boron substitution site: pentagon and heptagon respectively, and both on the bridge atomic site. Boron substitution on the pentagon cycle at the bridge position leads to the least reactive species (k = 1) C$_8$B$_2$H$_{10}$ (**I**) as compared to azulene. However, this situation is inverted when the boron substitution site is on the bridge position of the heptagon cycle, leading to the more reactive isomer **V** for C$_8$B$_2$H$_{10}$ (k = 1) as compared to azulene. As commented above, complete boron substitution of azulene (k = 4) leads to a more stable structure, with a singlet-triplet gap 100 kJ·mol^{-1} above azulene. This is remarkable if we take into account that B$_{10}$H$_{18}$ has never been synthesized.

3. Discussion

The experimental CC and BB distances in ethylene and diborane(6) are 1.340 Å and 1.736 Å respectively. These are considerable differences which are due mainly to the presence of two bridge hydrogen atoms in diborane(6), changing the electronic structure of the substituted systems considerably. However, the role of these two bridge hydrogen atoms in the B(H$_2$)B moiety correspond, from a structural and electronic point of view, to the two π electrons in ethylene: Every {C=C ↔ B(H$_2$)B} substitution in the original conjugated C$_n$H$_m$ hydrocarbon leads to a hybrid planar or nonplanar equivalent structure, as we have seen above in the description of the results. The aromaticity and antiaromaticity of a conjugated C$_n$H$_m$ system according to Hückel's 4n + 2 and 4n rule for π electrons, respectively, does not necessarily apply by consecutive {C=C ↔ B(H$_2$)B} substitutions, leading to the hybrid systems C$_{(n-2k)}$B$_{(2k)}$H$_{(m+2k)}$, with k = {0, 1, 2, ... , n/2}. The geometrical parameters for cyclobutadiene, benzene and cyclooctatetraene and the hybrid boron-carbon systems are gathered in Table 6.

According to Table 6, inclusion of empirical dispersion corrections in the B97D functional leads to slightly larger distances in a systematic way as compared to the B3LYP results, without changing the minimum energy nature of the stationary points for the conjugated hydrocarbon structures and the corresponding boron-carbon hybrids.

As regards to the singlet-triplet energy gaps in the same three model systems, we gathered in Table 1 the B3LYP and B97D energy gaps, with a systematic lowering when including dispersion corrections, and following the same trend; since our goal here is a tentative prediction, from a theoretical point of view, of the stabilities in hybrid boron-carbon systems which are experimentally unknown, the qualitative description is valid, without pretending spectroscopic accuracy.

A further assessment on the degree of aromaticity in the hybrid boron-carbon systems included in this work can be tackled with the comparison of the nucleus-independent chemical shifts (NICS) [40], at the centre of ring—NICS(0)—and 1 Å above this point, perpendicular to the ring—NICS(1). NICS is a computational method which provides the extent of absolute magnetic shielding at the centre of a ring. The values are reported with a reversed sign to make them compatible with the chemical shift conventions of NMR spectroscopy, and negative NICS values indicate aromaticity and positive values antiaromaticity. In Table 7 we gather the NICS(0) and NICS(1) values for cyclobutadiene and benzene with the corresponding hybrids.

Cyclobutadiene is the most antiaromatic system, and the NICS values decrease with more {C=C ↔ B(H$_2$)B} substitutions. Clearly, benzene is the most aromatic system considered here and the first hybrid with one {C=C ↔ B(H$_2$)B} substitution retains certain degree of aromaticity. A further {C=C ↔ B(H$_2$)B} substitution leads to a positive NICS(0) and a small negative NICS(1). Complete {C=C ↔ B(H$_2$)B} substitutions in benzene leads to a non-aromatic cyclic hexaborane(12) B$_6$H$_{12}$. The ACID plots (Table 1) of the benzene derivatives show that each {C=C ↔ B(H$_2$)B} substitution produces gaps in the isosurface as indication of reduced aromaticity. In Figure 6 we display the correlation between the singlet-triplet energy gap vs. NICS(0) and NICS(1) from Table 7. The correlations are quite linear except for the two encircled hybrid structures.

Table 6. Geometrical distances (Å) in optimised geometries of $C_{(n-2k)}B_{(2k)}H_{(m+2k)}$ systems for cyclobutadiene, benzene and cyclooctatetraene: d(C-C), d(C=C), d(C-H), d(B-B), $d(B(H_b)_2B)_b$, d(B-H), and $d(H_b-H_b)$. H_b corresponds to a bridge hydrogen atom, i.e., diborane(6) can be drawn as $H_2B(H_b)_2BH_2$. The first and second (italic) value for each distance corresponds respectively to B3LYP/cc-pVTZ and *B97D/cc-pVTZ* computations, the latter with a functional including dispersion corrections.

	d(C-C)	d(C=C)	d(C-H)	d(B-B)	$d(B(H_b)_2B)$	d(B-H)	$d(H_b-H_b)$
Ethylene	—	1.324 *1.332*	1.083 *1.089*	—	—	—	—
Diborane(6)	—	—	—	—	1.758 *1.777*	1.185 *1.195*	1.948 *1.973*
Cyclobutadiene (k)	d(C-C)	d(C=C)	d(C-H)	d(B-B)	$d(B(H_b)_2B)$	d(B-H)	$d(H_b-H_b)$
C_4H_4 (0)	1.575 *1.582*	1.329 *1.337*	1.079 *1.085*	—	—	—	—
$C_2B_2H_6$ (1)	—	1.338 *1.347*	1.083 *1.09*	—	1.741 *1.749*	1.185 *1.194*	1.858 *1.886*
B_4H_8 (2)	—	—	—	1.728 *1.739*	1.76 *1.771*	1.189 *1.199*	1.856 *1.883*
Benzene (k)	d(C-C)	d(C=C)	d(C-H)	d(B-B)	$d(B(H_b)_2B)$	d(B-H)	$d(H_b-H_b)$
C_6H_6 (0)	1.391 *1.399*	1.391 *1.399*	1.082 *1.087*	—	—	—	—
$C_4B_2H_8$ (1)	1.439 *1.442*	1.359 *1.369*	1.085 *1.09*	—	1.799 *1.823*	1.186 *1.196*	1.926 *1.951*
$C_2B_4H_{10}$ (2)	—	1.35 *1.359*	1.088 *1.094*	1.695 *1.708*	1.789 *1.812*	1.188 *1.198*	1.926 *1.952*
B_6H_{12} (3)	—	—	—	1.713 *1.724*	1.793 *1.812*	1.19 *1.2*	1.923 *1.948*
Cyclooctatetraene (k)	d(C-C)	d(C=C)	d(C-H)	d(B-B)	$d(B(H_b)_2B)$	d(B-H)	$d(H_b-H_b)$
C_8H_8 (0)	1.469 *1.471*	1.335 *1.343*	1.087 *1.092*	—	—	—	—
$C_6B_2H_{10}$ (1)	1.47 *1.471*	1.338 *1.347*	1.088 *1.093*	—	1.77 *1.788*	1.191 *1.202*	1.944 *1.972*
$C_4B_4H_{12}$ (2) (I)	—	1.341 *1.349*	1.09 *1.096*	—	1.785 *1.804*	1.19 *1.202*	1.94 *1.966*
$C_4B_4H_{12}$ (2) (II)	1.475 *1.476*	1.34 *1.348*	1.089 *1.095*	1.694 *1.706*	1.776 *1.791*	1.191 *1.202*	1.941 *1.969*
$C_2B_6H_{14}$ (3)	—	1.343 *1.351*	1.091 *1.096*	1.696 *1.706*	1.777 *1.796*	1.191 *1.202*	1.938 *1.964*
B_8H_{16} (4)	—	—	—	1.698 *1.708*	1.786 *1.8*	1.191 *1.202*	1.934 *1.96*

Table 7. NICS(0) and NICS(1), in ppm for cyclobutadiene and benzene and the corresponding hybrids.

Cyclobutadiene	NICS(0)	NICS(1)	Benzene	NICS(0)	NICS(1)
□	26.8	17.4	⬡	−8.2	−10.4
(C₂B₂H₄)	10.2	6.3	(C₄B₂H₄)	−3.6	−5.9
(B₄H₄)	4.7	2.4	(C₂B₄H₈)	0.5	−1.4
			(B₆H₁₂)	2.2	1.0

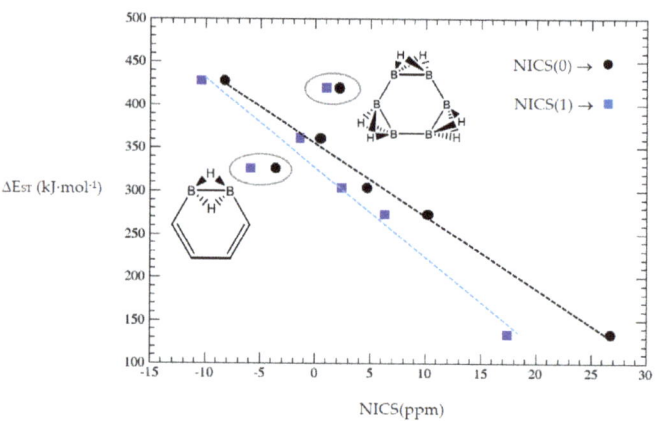

Figure 6. Singlet-triplet energy gaps vs. NICS(0) and NICS(1) for cyclobutadiene and benzene series.

The most noticeable changes upon {C=C ↔ B(H$_2$)B} substitutions for a given conjugated hydrocarbon C$_n$H$_m$, from an energy point of view and singlet-triplet gap, is in the loss of aromaticity or in the inclusion or permanence of very unstable moieties, such as cyclobutadiene. For instance, for antiaromatic cyclobutadiene and cyclooctatetraene—Table 1 and Figure 1—the minimum and maximum singlet-triplet energy gaps correspond respectively to the conjugated hydrocarbon C$_n$H$_m$ and the corresponding borane systems B$_n$H$_{m+n}$ respectively, namely, with all $k = n/2$ {C=C ↔ B(H$_2$)B} substitutions. This is remarkable, since planarity of borane molecules is an exception rather a rule. The loss of aromaticity in benzene upon one {C=C ↔ B(H$_2$)B} substitution leads to the minimum energy gap structure, with the largest gap for benzene itself. Pentalene has the lowest energy gap as shown in Table 2 and Figure 2, as an antiaromatic system, with the larger gap for the $k = 3$ isomer (I), where only one C=C double bond remains in the structure, releasing the constrained hydrocarbon structure due to longer BB and B(H$_b$)$_2$B distances. The particular case of benzocyclobutadiene, also antiaromatic, is remarkable due to the three different Kekulé structures leading to different isomers, Table 3a–c and Figure 3a–c. Thus, for K_1, the minimum singlet-triplet gap comes from the $k = 2$, isomer (IV), which is nothing but a cyclobutadiene molecule with two diborane(6) moieties attached to one of the C=C double bonds forming a six-member cycle: This isomer has a gap which is even lower than in cyclobutadiene itself! On the other hand, and following the cases of cyclobutadiene and cyclooctatetraene, the complete borane structure (k = 4) B_6H_{14} has the largest gap. For Kekulé structure K_2, benzocyclobutadiene itself has the lowest gap and the largest gap corresponds to k = 2 isomer (I), a *cis*-butadiene molecule with two diborane(6) molecules fused and forming a six- and four-member ring on the =CH_2 and =CH- moieties respectively. Hence the conjugation of the two double bonds is maintained even with the inclusion of diborane moieties. As for K_3, the minimum gap corresponds to the k = 2 isomer (IV) which is cyclobutadiene with two fused diborane molecules forming a six-member ring; note that whenever a cyclobutadiene is maintained in the {C=C ↔ B(H_2)B} substitutions the energy gap is very, thus predicting a very reactive system. Again, the maximum energy gap corresponds the completed borane structure with four {C=C ↔ B(H_2)B} substitutions: This is a rule for antiaromatic systems.

We now turn to aromatic naphthalene with the K_1 and K_2 Kekulé structures as displayed in Table 4a–b and Figure 4a,b. Due to the indistinguishable nature of the two valence-bond Kekulé structures in benzene, the k = 2 isomer (I) is equivalent for K_1 and K_2 and corresponds to a benzene molecule with two diborane molecules fused forming an additional six-member ring. If we use a valence-bond or multiconfigurational wave function, these isomers would have a different energy due to the different spin-coupling patterns. In our approximation, this isomer has the largest energy gap due to the aromaticity of benzene, which is only 14 kJ·mol^{-1} higher than in benzene. The lowest energy gaps correspond to structures where the cyclic aromaticity is somehow broken due to {C=C ↔ B(H_2)B} substitutions in concrete places: for K_1 this corresponds to k = 2 isomer (II), with three consecutive C=C bonds passing through the bridge C=C bond but with two B(H_2)B moieties on opposite rings, thus destroying the cyclic aromaticity. As for K_2, the lowest gap corresponds to the k = 1 isomer (III) and similarly the aromaticity is destroyed by one {C=C ↔ B(H_2)B} substitution on the ending edge. We should emphasize that in this latter case the energy gap is much lower as compared to the minimum gap structure for K_1 since the three consecutive alternating C=C bonds gives further stability of the k = 2 isomer (II) structure.

Finally, azulene boron-carbon hybrids give a range of singlet-triplet gaps from 163 kJ·mol^{-1} to 335 kJ·mol^{-1}, corresponding to structures k = 2 isomer (X) and k = 3 isomer (III) respectively, as shown in Table 5 and Figure 5. Due to the larger seven-member ring the structures are less constrained, but the presence of a cyclopentadiene ring lowers the stability. The lowest energy gap structure—k = 2 isomer (X)—consists of a cyclopentadiene ring where one of the connecting atoms is a boron atom; this structure should be very reactive indeed.

4. Computational Methods

The quantum-chemical computations of the structures and isomers included in this work were carried out with the B3LYP/cc-pVTZ model [41–44] and with the scientific software Gaussian16 [45]. All geometries correspond to energy minima, checked through frequency computations, with geometry optimisation thresholds of 0.00045 Hartree/Bohr and 0.00030 Hartree/Bohr for maximum force and root-mean-square (RMS) force respectively, and 0.0018 Bohr and 0.0012 Bohr for maximum displacement and RMS displacement respectively. For the computations of the $D_{2d} \rightarrow D_{4h}$ barriers in cyclooctatetraene C_8H_8 and the boron substituted system B_8H_{16}, the optimized geometries showed zero and one imaginary frequency for the D_{2d} and D_{4h} geometries, respectively. The MEP and ACID plots for benzene and the $C_{(6-2k)}B_{(2k)}H_{(6+2k)}$ series, k = {1–3} (Table 1), were also computed with Gaussian16.

5. Conclusions

The goal of this work was to check the structural stability from any conjugated hydrocarbon C_nH_m through the boron-carbon hybrid series $C_{(n-2k)}B_{(2k)}H_{(m+2k)}$, obtained upon k {C=C ↔ B(H_2)B} substitutions for k = {0, 1, 2, ... , $n/2$}, leading to the structurally equivalent complete borane B_nH_{m+n} structure, for k = $n/2$. We have chosen planar and non-planar conjugated hydrocarbons: cyclobutadiene

(n = 4), benzene (n = 6), cyclooctatetraene (n = 8), pentalene (n = 8), naphthalene (n = 10) and azulene (n = 10). As a result, all hybrid boron-carbon structures appear as energy minima from the quantum-chemical computations and therefore, from a thermodynamic point of view, they should be a synthetic target for experimentalists working on planar boron chemistry. This is unusual if we consider that borane polyhedral molecules are 3D curved structures. In the particular case of cyclooctatetraene C_8H_8, a non-planar structure, we have shown that the one-to-one structural equivalence $C_nH_m \leftrightarrow B_nH_{m+n}$ also holds; the energy barrier for loss of planarity from a planar transition state structure (D_{4h}) to an energy minimum twisted structure (D_{2d}) is larger for the cyclic borane B_8H_{16}, as compared to cyclic octatetraene. Structural isomers appear for a given number k of {C=C $\leftrightarrow$ B(H$_2$)B} substitutions in an isoelectronic series $C_{(n-2k)}B_{(2k)}H_{(m+2k)}$ with k = {0, 1, 2, ... , $n/2$} for all conjugated hydrocarbons, except for cyclobutadiene and benzene, with a maximum number of isomers of 10 for azulene structures with $C_6B_4H_{12}$ (k = 2) and $C_4B_6H_{14}$ (k = 3) formulae. These isomers are ordered in increasing energy {**I, II,** ... , N_{iso}}. Striking energy differences in an isomeric series stem from one {C=C $\leftrightarrow$ B(H$_2$)B} substitution in the cyclobutadiene cycle from benzocyclobutadiene with an energy jump of 125 kJ·mol^{-1}. As regards to the energy order of vertical singlet-triplet energy gaps, they do not necessarily follow the same trend as compared to the energy profile for the isomers of a given k in a $C_{(n-2k)}B_{(2k)}H_{(m+2k)}$ structure.

The vertical singlet-triplet energy gaps in conjugated hydrocarbons C_nH_m vs. B_nH_{m+n} change strikingly in antiaromatic systems, when the number of π electrons is $4n$, with the energy gap always much larger for the borane systems. On the other hand, for conjugated hydrocarbons with ($4n$ + 2) electrons—Hückel rule for aromaticity—the difference in vertical singlet-triplet energy gaps between C_nH_m and B_nH_{m+n} is minor for benzene and naphthalene and larger for azulene. From these results a potential rule arises, which will be thoroughly checked in a future work and holds for the systems included here: the lowest and largest singlet-triplet energy gap for a $C_{(n-2k)}B_{(2k)}H_{(m+2k)}$ series of structures—with n multiple of 4—always corresponds to the k = 0 and k = $n/2$ structure respectively, namely, to the original conjugated hydrocarbon C_nH_m and the complete borane structure B_nH_{m+n}. We should also emphasize that, if aromatic structures—e.g., π sextets in six-member rings—are maintained in a given isoelectronic $C_{(n-2k)}B_{(2k)}H_{(m+2k)}$ series, then the singlet-triplet energy gap is large and similar to the original conjugated hydrocarbon and the structure is predicted to be stable. On the other hand, the presence of an antiaromatic structure—e.g., cyclobutadiene moieties with 4π electrons—in any $C_{(n-2k)}B_{(2k)}H_{(m+2k)}$ system, lowers considerably the energy gap thus predicting an unstable structure. Finally, the existence of similar structures derived from phenanthrene derivatives with one {C=C $\leftrightarrow$ B(H$_2$)B} substitution in the conjugated hydrocarbon [19–23], gives support to the possibility of synthesizing or characterizing the hybrid boron-carbon isoelectronic $C_{(n-2k)}B_{(2k)}H_{(m+2k)}$ structures presented in this work.

Supplementary Materials: B3LYP/cc-pVTZ optimised geometries, in cartesian coordinates (Å), of the systems included in the work.

Author Contributions: Conceptualization, J.M.O.-E., I.A. and J.E., writing—review and editing J.M.O.-E., I.A. and J.E. All authors have read and agreed to the published version of the manuscript.

Funding: This research was funded by Spanish MICINN, grant number CTQ2018-094644-B-C22 and Comunidad de Madrid, grant number P2018/EMT-4329 AIRTEC-CM.

Conflicts of Interest: The authors declare no conflict of interest.

References

1. Hückel, E. Quantentheoretische Beiträge zum Benzolproblem. *Z. Phys.* **1931**, *70*, 204–286. [CrossRef]
2. Pauling, L. *The Nature of the Chemical Bond and the Structure of Molecules and Crystals*, 3rd ed.; Cornell University Press: Ithaca, NY, USA, 1960.
3. Heitler, W.; London, F. Wechselwirkung neutraler Atome und homöopolare Bindung nach der Quantenmechanik. *Z. Phys.* **1927**, *44*, 455–472. [CrossRef]
4. Hund, F. Zur Deutung einiger Erscheinungen in den Molekelspektren. *Z. Phys.* **1926**, *36*, 657. [CrossRef]

5. Mulliken, R.S. Electronic states and band spectrum structure in diatomic Molecules. I. Statement of the postulates. Interpretation of CuH, CH, and Co band-types. *Phys. Rev.* **1926**, *28*, 481. [CrossRef]
6. Woodward, R.B.; Hoffmann, R. Stereochemistry of electrocyclic reactions. *J. Am. Chem. Soc.* **1965**, *87*, 395–397. [CrossRef]
7. Kamtekar, K.T.; Monkman, A.P.; Bryce, M.R. Recent advances in white organic light-emitting materials and devices (WOLEDs). *Adv. Mater.* **2010**, *22*, 572–582. [CrossRef] [PubMed]
8. Shavitt, I. Geometry and singlet-triplet energy gap in methylene: A critical review of experimental and theoretical determinations. *Tetrahedron* **1985**, *41*, 1531–1542. [CrossRef]
9. Chutjian, A.; Hall, R.I.; Trajmar, S. Electron-impact excitation of H_2O and D_2O at various scattering angles and impact energies in the energy-loss range 4.2–12 eV. *J. Chem. Phys.* **1975**, *63*, 892–898. [CrossRef]
10. Stock, A. *The Hydrides of Boron and Silicon*; Cornell University Press: Ithaca, NY, USA, 1933.
11. Brown, H.C. From little acorns to tall oaks from boranes through organoboranes. In *Nobel Lecture, 8 December 1979*; World Scientific Publishing Co.: Singapore, 1993.
12. Marder, T.B.; Lin, Z. (Eds.) *Contemporary Metal Boron Chemistry I.: Borylenes, Boryls, Borane Sigma-Complexes, and Borohydrides*; Springer: Berlin, Germany, 2008.
13. Lipscomb, W.N. *Boron Hydrides*; Dover Publications Inc.: Mineola, NY, USA, 2012.
14. Štíbr, B. Carboranes other than $C_2B_{10}H_{12}$. *Chem. Rev.* **1992**, *92*, 225–250. [CrossRef]
15. Grimes, R.N. *Carboranes*, 3rd ed.; Academic Press: Cambridge, MA, USA, 2016.
16. Saxena, A.K.; Hosmane, N.S. Recent advances in the chemistry of carborane metal complexes incorporating d- and f-block elements. *Chem. Rev.* **1993**, *93*, 1081–1124. [CrossRef]
17. Poater, J.; Viñas, C.; Bennour, I.; Escayola, S.; Solà, M.; Teixidor, F. Too persistent to give up: Aromaticity in boron clusters survives radical structural changes. *J. Am. Chem. Soc.* **2020**, *142*, 9396–9407. [CrossRef] [PubMed]
18. Nishino, H.; Fujita, T.; Cuong, N.T.; Tominaka, S.; Miyauchi, M.; Iimura, S.; Hirata, A.; Umezawa, N.; Okada, S.; Nishibori, E.; et al. Formation and characterization of hydrogen boride sheets derived from MgB_2 by cation exchange. *J. Am. Chem. Soc.* **2017**, *139*, 13761–13769. [CrossRef] [PubMed]
19. Wrackmeyer, B.; Thoma, P.; Kempe, R.; Glatz, G. 9-Borafluorenes—NMR spectroscopy and DFT calculations. Molecular structure of 1,2-(2,2'-diphenylene)-1,2-diethyldiborane. *Collect. Czech. Chem. Commun.* **2010**, *75*, 743–756. [CrossRef]
20. Das, A.; Hübner, A.; Weber, M.; Bolte, M.; Lerner, H.-W.; Wagner, M. 9-H-9-Borafluorene dimethyl sulfide adduct: A product of a unique ring-contraction reaction and a useful hydroboration reagent. *Chem. Commun.* **2011**, *47*, 11339–11341. [CrossRef] [PubMed]
21. Hübner, A.; Qu, Z.-W.; Englert, U.; Bolte, M.; Lerner, H.-W.; Holthausen, M.C.; Wagner, M. Main-chain boron-containing oligophenylenes via ring-opening polymerization of 9-H-9-borafluorene. *J. Am. Chem. Soc.* **2011**, *133*, 4596–4609. [CrossRef] [PubMed]
22. Hübner, A.; Diefenbach, M.; Bolte, M.; Lerner, H.-W.; Holthausen, M.C.; Wagner, M. Confirmation of an early postulate: B-C-B two-electron–three-center bonding in organo(hydro)boranes. *Angew. Chem. Int. Ed.* **2012**, *51*, 12514–12518. [CrossRef] [PubMed]
23. Kaese, T.; Hübner, A.; Bolte, M.; Lerner, H.-W.; Wagner, M. Forming B–B Bonds by the Controlled Reduction of a Tetraaryldiborane(6). *J. Am. Chem. Soc.* **2016**, *138*, 6224–6233. [CrossRef]
24. Oliva-Enrich, J.M.; Kondo, T.; Alkorta, I.; Elguero, J.; Klein, D.J. Concatenation of diborane leads to new planar boron chemistry. *Chem. Phys. Chem.* **2020**. [CrossRef]
25. Rashid, Z.; van Lenthe, J.H. Generation of Kekulé valence structures and the corresponding valence bond wave function. *J. Comput. Chem.* **2010**, *32*, 696–708. [CrossRef]
26. Cram, D.J.; Tanner, M.E.; Thoams, R. The taming of cyclobutadiene. *Angew. Chem. Int. Ed.* **1991**, *30*, 1024–1027. [CrossRef]
27. Liu, Y.; Kilby, P.; Frankcombe, T.J.; Schmidt, T.W. The electronic structure of benzene from a tiling of the correlated 126-dimensional wavefunction. *Nature* **2020**, *11*, 1–5. [CrossRef] [PubMed]
28. Grimme, S. Semiempirical GGA-type density functional constructed with a long-range dispersion correction. *J. Comp. Chem.* **2006**, *27*, 1787–1799. [CrossRef] [PubMed]
29. Ma, J.C.; Dougherty, D.A. The cation–π interaction. *Chem. Rev.* **1997**, *97*, 1303–1324. [CrossRef] [PubMed]
30. Alkorta, I.; Rozas, I.; Elguero, J. Interaction of anions with perfluoro aromatic compounds. *J. Am. Chem. Soc.* **2002**, *124*, 8593–8598. [CrossRef]

31. Geuenich, D.; Hess, K.; Köhler, F.; Herges, R. Anisotropy of the induced current density (ACID), a general method to quantify and visualize electronic delocalization. *Chem. Rev.* **2005**, *105*, 3758–3772. [CrossRef]
32. Claus, K.H.; Krüger, C. Structure of cyclooctatetraene at 129 K. *Acta Cryst. Sect. C* **1988**, *44*, 1632–1634. [CrossRef]
33. Krygowski, T.M.; Pindelska, E.; Cyrański, M.K.; Häfelinger, G. Planarization of 1,3,5,7-cyclooctatetraene as a result of a partial rehybridization at carbon atoms: An MP2/6-31G* and B3LYP/6-311G** study. *Chem. Phys. Lett.* **2002**, *359*, 158–162. [CrossRef]
34. Nishinaga, T.; Ohmae, T.; Iyoda, M. Recent studies on the aromaticity and antiaromaticity of planar cyclooctatetraene. *Symmetry* **2010**, *2*, 76. [CrossRef]
35. Bally, T.; Chai, S.; Neuenschwander, M.; Zhu, Z. Pentalene: Formation, electronic, and vibrational structure. *J. Am. Chem. Soc.* **1997**, *119*, 1869–1875. [CrossRef]
36. Hafner, K.; Süss, H.U. 1,3,5-Tri-tert-butylpentalene. A stabilized planar 8π electron system. *Angew. Chem. Int. Ed. Engl.* **1973**, *12*, 575–577. [CrossRef]
37. Carey, F.A.; Sundberg, R.J. *Advanced Organic Chemistry. Part A: Structure and Mechanisms*, 2nd ed.; Plenum Press: New York, NY, USA, 1984.
38. Mosher, O.A.; Flicker, W.M.; Kuppermann, A. Triplet states in 1,3-butadiene. *Chem. Phys. Lett.* **1973**, *19*, 332–333. [CrossRef]
39. Gordon, M. The Azulenes. *Chem. Rev.* **1952**, *50*, 127–200. [CrossRef]
40. Schleyer, P.V.R.; Maerker, C.; Dransfeld, A.; Jiao, H.; Hommes, J.R.V.E. Nucleus-independent chemical shifts: A simple and efficient aromaticity probe. *J. Am. Chem. Soc.* **1996**, *118*, 6317–6318. [CrossRef] [PubMed]
41. Becke, A.D. Density-functional thermochemistry. III. The role of exact exchange. *J. Chem. Phys.* **1993**, *98*, 5648–5652. [CrossRef]
42. Lee, C.; Yang, W.; Parr, R.G. Development of the Colle-Salvetti correlation-energy formula into a functional of the electron density. *Phys. Rev. B* **1988**, *37*, 785–789. [CrossRef]
43. Vosko, S.H.; Wilk, L.; Nusair, M. Accurate spin-dependent electron liquid correlation energies for local spin density calculations: A critical analysis. *Can. J. Phys.* **1980**, *58*, 1200–1211. [CrossRef]
44. Stephens, P.J.; Devlin, F.J.; Chabalowski, C.F.; Frisch, M.J. Ab initio calculation of vibrational absorption and circular dichroism spectra using density functional force fields. *J. Phys. Chem.* **1994**, *98*, 11623–11627. [CrossRef]
45. Frisch, M.J.; Trucks, G.W.; Schlegel, H.B.; Scuseria, G.E.; Robb, M.A.; Cheeseman, J.R.; Scalmani, G.; Barone, V.; Petersson, G.A.; Nakatsuji, H.; et al. *Gaussian 16*; Revision C.01; Gaussian, Inc.: Wallingford, CT, USA, 2016.

Sample Availability: Samples of the compounds are not available from the authors.

Publisher's Note: MDPI stays neutral with regard to jurisdictional claims in published maps and institutional affiliations.

© 2020 by the authors. Licensee MDPI, Basel, Switzerland. This article is an open access article distributed under the terms and conditions of the Creative Commons Attribution (CC BY) license (http://creativecommons.org/licenses/by/4.0/).

Article

m-Carborane as a Novel Core for Periphery-Decorated Macromolecules

Ines Bennour, Francesc Teixidor, Zsolt Kelemen and Clara Viñas *

Institut de Ciència de Materials de Barcelona (ICMAB-CSIC), Campus UAB, 08193 Barcelona, Spain; bennourines@ymail.com (I.B.); teixidor@icmab.es (F.T.); kelemen.zsolt@mail.bme.hu (Z.K.)
* Correspondence: clara@icmab.es; Tel.: +34-935-801-853; Fax: +34-935-805-729

Academic Editor: Ashok Kakkar
Received: 31 May 2020; Accepted: 16 June 2020; Published: 18 June 2020

Abstract: Closo m-$C_2B_{10}H_{12}$ can perform as a novel core of globular periphery-decorated macromolecules. To do this, a new class of di and tetrabranched m-carborane derivatives has been synthesized by a judicious choice of the synthetic procedure, starting from 9,10-I_2-1,7-$closo$-$C_2B_{10}H_{10}$. The 2a-NPA (sum of the natural charges of the two bonded atoms) value for a bond, which is defined as the sum of the NPA charges of the two bonded atoms, matches the order of electrophilic reaction at the different cluster bonds of the icosahedral o-and m- carboranes that lead to the formation of B-I bonds. As for m-carborane, most of the 2a-NPA values of B-H vertexes are positive, and their functionalization is more challenging. The synthesis and full characterization of dibranched 9,10-R_2-1,7-$closo$-carborane (R = CH_2CHCH_2, $HO(CH_2)_3$, $Cl(CH_2)_3$, $TsO(CH_2)_3$, $C_6H_5COO(CH_2)_3$, $C_6H_5COO(CH_2)_3$, $N_3(CH_2)_3$, CH_3CHCH, and $C_6H_5C_2N_3(CH_2)_3$) compounds as well as the tetrabranched 9,10-R_2-1,7-R_2-$closo$-$C_2B_{10}H_8$ (R = CH_2CHCH_2, $HO(CH_2)_3$) are presented. The X-ray diffraction of 9,10-$(HO(CH_2)_3)_2$-1,7-$closo$-$C_2B_{10}H_{10}$ and 9,10-$(CH_3CHCH)_2$-1,7-$closo$-$C_2B_{10}H_{10}$, as well as their Hirshfeld surface analysis and decomposed fingerprint plots, are described. These new reported tetrabranched m-carborane derivatives provide a sort of novel core for the synthesis of 3D radially grown periphery-decorated macromolecules that are different to the 2D radially grown core of the tetrabranched o-carborane framework.

Keywords: m-carborane; electrophilic substitution; coupling reaction; organic branches; Hirshfeld Study

---

1. Introduction

Icosahedral carborane clusters with empirical formula $C_2B_{10}H_{12}$ can be in three different isomers: 1,2-$closo$-$C_2B_{10}H_{12}$ (o-carborane), 1,7-$closo$-$C_2B_{10}H_{12}$ (m-carborane; **1**), and 1,12-$closo$-$C_2B_{10}H_{12}$ (p-carborane). Figure 1 displays a schematic representation of the isomers with their vertexes numbering. Despite their common icosahedral geometry, they display similarities, but also important differences. Among the similarities are the high stabilities and 3D geometrical properties, their very similar 3D aromatic character [1,2] that leads to display great inertia to keep the original scaffold upon electrophilic substitution, their dual-mode as electron-withdrawing through carbon or electron-donating through boron vertexes [3–5], their molecular volume that is high compared to rotating benzene [6], and high hydrophobicity [7–9]. Among the differences are the dipolar moment and their different reactivity towards boron elimination [8], and the lowest unoccupied molecular orbital (LUMO) geometrical disposition that is responsible for many of the physical properties of the isomers. Among the three of them, the most extensively studied is the o-carborane. Some tips to keep in mind between the three isomers when substitution is sought are: First, the weak acidic C_c-H bond (C_c = cluster C atom) [10] can be deprotonated using a strong base followed by an electrophilic reaction to form the C_c-R bond [8,11]. Second, the B-H hydrogen atoms with the hydridic character on (B(9,12), B(8,10), and B(4,5,7,11)) are subjected to electrophilic substitution to form B-halogen units [8,11] that, if followed by a Kumada cross-coupling reaction, may lead to the introduction of organic moieties to make B-R vertexes [3–8]. This procedure does not proceed equally for all B-Hs, for instance, at the B(3,6) vertices, which are the most electron-deficient vertexes, their functionalization does not take place by using the same process at the other cluster's vertices [8,11]. Substitution on these positions can be achieved via the deboration-capping multistep reaction or via the metallation procedure [11–13]. The diverse regioreactivity of o-carborane, has been exploited and adapted to make o-carborane an exceptional core for developing a large variety of multibranched molecules, globular macromolecules, dendrimers (Figure 2b), and so on [14–18]. Moreover, new versatile synthons have been explored through the multi-functionalization of B and C_c atoms jointly, which make the o-carborane clusters an exciting platform for new materials [6,8,18–25].

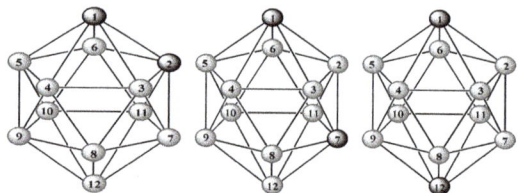

Figure 1. Icosahedral 1,2-*closo*-$C_2B_{10}H_{12}$, 1,7-*closo*-$C_2B_{10}H_{12}$ (**1**) and 1,12-*closo*-$C_2B_{10}H_{12}$ isomers with their vertexes numbering. Dark circles are C_c-H vertexes and grey ones are B-H vertexes.

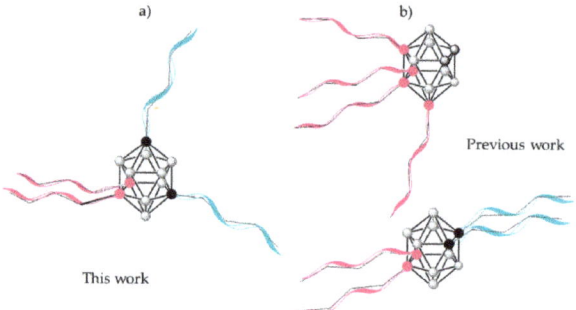

Figure 2. Schematic view of the two types of radially expanded tetrabranched core for constructing dendritic structures. Circles colour: dark grey correspond to C_c-H bonds, black to C atoms, pink to Boron atoms and grey to B-H vertexes in the tetra-branched clusters.

On the other hand, the reactivity of m-carborane is less studied but for the C_c-H vertexes, which are less acidic as compared to the C_c-H vertexes of the o-isomer [10,26]. Using a similar strategy as for o-carborane, a wide variety of 1-R-1,7-*closo*-$C_2B_{10}H_{11}$ and 1,7-R$_2$-1,7-*closo*-$C_2B_{10}H_{10}$ derivatives has been developed [27–35]. To some extent, the current state of knowledge of the m-carborane functionalization through the B-H vertexes is in an odd situation. As compared to the o-carborane, a much-limited number of protocols leading to modify the B-H vertexes in the m-cluster have been reported [8,11]. In this context, very few derivatives of m-carborane with a functional group that is bonded to B(9) or B(9) and B(10) synchronously have been described [27–35]. The Pd-catalyzed cross-coupling reaction of 9-X-1,7-*closo*-$C_2B_{10}H_{11}$ and 9,10-X$_2$-1,7-*closo*-$C_2B_{10}H_{10}$ (X= halogen atom) represents one of these examples of derivatization [34,36,37]. By contrast to the o-carborane, no multibranched m-carborane structures with a general formula 1,7-R$_2$-9,10-R'$_2$-1,7-*closo*-$C_2B_{10}H_8$ have been reported despite the potential of its structure and the relatively high reactivity of the C_c-H bonds that should allow the reaction to a great extent. Notably, as shown in Figure 2, the *m*-carborane core provides a 3D radially growth core while *o*-carborane a 2D one.

Consequently, we became interested in introducing organic branches connected to B(9) and B(10) to prepare a new set of 9,10-R$_2$-1,7-*closo*-C$_2$B$_{10}$H$_{10}$ derivatives. In the second part of this paper, we functionalized the two C$_c$-H of 9,10-(CH$_2$=CHCH$_2$)$_2$-1,7-*closo*-C$_2$B$_{10}$H$_{10}$ to form the quadruped-shaped structure with a general formula 1,7-R$_2$-9,10-R'$_2$-1,7-*closo*-C$_2$B$_{10}$H$_{10}$ which might serve as versatile precursors with free ends for further reaction.

2. Results and Discussion

2.1. Synthesis of di-Branched m-carborane Derivatives at the 9,10 Vertexes

Versatile strategy for the synthesis of the two branches B(9,10) *m*-carborane derivatives (9,10-R$_2$-1,7-*closo*-C$_2$B$_{10}$H$_{10}$) was achieved by using 9,10-I$_2$-1,7-*closo*-C$_2$B$_{10}$H$_{10}$ as the starting compound.

The synthesis of 9,10-I$_2$-1,7-*closo*-C$_2$B$_{10}$H$_{10}$; (**2**) has been reported by using two different methodologies: i) the electrophilic iodination reaction of icosahedral *closo m*-carborane (**1**) by using a molar Equiv. Of iodine: monochloride, which acts as an electrophilic agent, in the presence of catalytic amounts of aluminum chloride, and ii) using iodine as an electrophilic agent in a very acidic media (HNO$_3$:H$_2$SO$_4$, 1:1). The target compound **2** was obtained in 60% and 87% yield, respectively [38,39]. Our study focused on the synthesis of new Boron disubstituted *closo m*-carborane derivatives at the 9 and 10 vertexes began with the synthesis of **2** in 87% yield by combining the two reported methods: an equimolar ratio of *m*-carborane (**1**): iodine in acidic HNO$_3$:H$_2$SO$_4$ (1:1) solution was left under reflux to react for 3 h (Scheme 1).

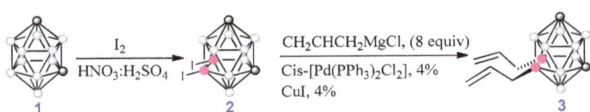

Scheme 1. Synthesis of 9,10-(CH$_2$=CHCH$_2$)$_2$-1,7-*closo*-C$_2$B$_{10}$H$_{10}$ (**3**). Dark circles are C$_c$-H vertexes, pink circles are boron atoms, and grey circles are B-H vertexes.

To produce the B-C bonds on **1**, a useful and general method is the Kumada cross-coupling reaction on B-iodinated *m*-carborane **2** with Grignard reagents in the presence of Pd(II) and Cu(I) catalysts. To achieve the di-branched *m*-carborane derivatives at the 9, 10 vertexes, the cross-coupling reaction on **2** was studied using CH$_2$CHCH$_2$MgCl Grignard derivative in the presence of [PdCl$_2$(PPh$_3$)$_2$] and CuI as catalysts to give the 9,10-(CH$_2$CHCH$_2$)$_2$-1,7-*closo*-C$_2$B$_{10}$H$_{10}$ (**3**) in 95% yield [40].

The terminal olefin groups in **3** are ready for further reactions on them, enabling the *m*-carborane cluster to become the template for a new type of macromolecules having a rigid head and two appended branches. As a first example of these molecules, compound **3** was converted to 9,10-(HOCH$_2$CH$_2$CH$_2$)$_2$-*closo*-1,7-C$_2$B$_{10}$H$_{10}$, (**4**) following the hydroboration/oxidation reaction on **3** by using BH$_3$·THF as hydroboration agent and subsequent oxidation with H$_2$O$_2$ in a basic aqueous solution. After workup, crystalline white pure solid, **4**, was obtained in 93% yield. The ^{1}H NMR spectrum displays a new broad peak at 3.43 ppm, which supports the presence of the O-H group in **4**. Also, the ^{1}H and ^{13}C{^{1}H} NMR spectra revealed that the reaction had proceeded by an anti-Markovnikov addition, therefore having the two hydroxyl groups at terminal positions. No hindered hydroboranes were thus needed for the control of the reaction's regioselectivity.

These terminal alcohol groups anticipate versatile chemistry for radial growth, given the availability of the terminal hydroxyl groups for further elongation of the chains. Moreover, the C$_c$–H vertexes on the rigid *m*-carborane head are ready for derivatization or supramolecular assembly. Then, the *m*-carborane cluster, as *o*-carborane does, provides a singular platform for the construction of highly dense multibranched molecules with a wide range of possibilities. Therefore, derivatives of *m*-carborane with precise patterns of substitution, which are sterically different from the ones of *o*-carborane but complementary, can be prepared by a judicious choice of the synthetic procedure. Consequently, the substitution of the terminal hydroxyl units in **4** by chloro (**5**), ester (**6**), tosyl (**7**) or azide (**8**) groups, which enable the branches to grow by a subsequent coupling reaction with nucleophilic agents, was achieved (Scheme 2).

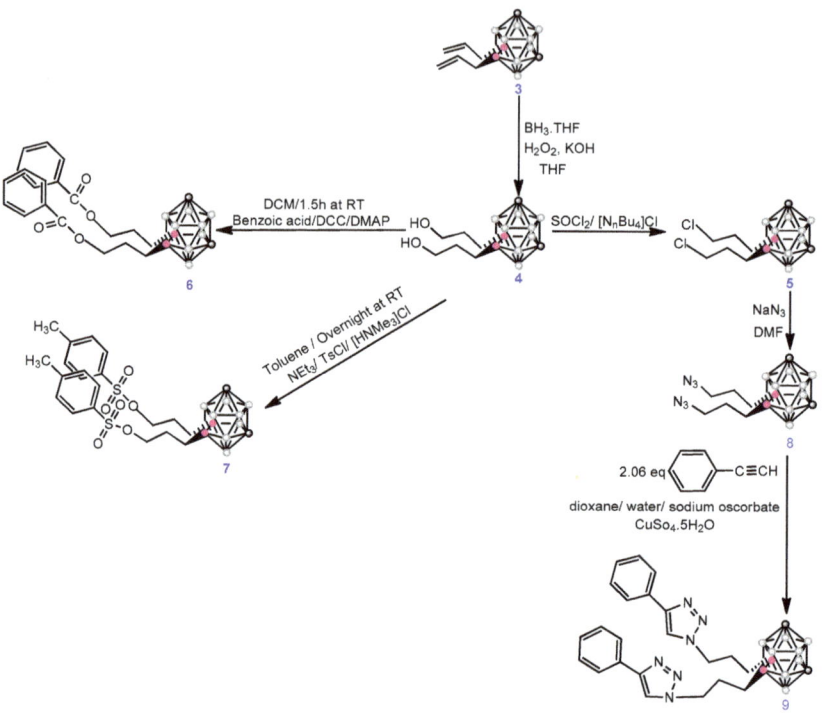

Scheme 2. Derivatization reactions on 9,10-(CH$_2$=CHCH$_2$)$_2$-1,7-*closo*-C$_2$B$_{10}$H$_{10}$ **3**. Dark circles are C$_c$-H vertexes, pink circles are boron atoms, and grey circles are B-H vertexes.

Chlorination in **3** was achieved using SOCl$_2$ and [Nbu$_4$]Cl salt to give 9,10-(ClCH$_2$CHCH$_2$)$_2$-1,7-*closo*-C$_2$B$_{10}$H$_{10}$ (**5**) in 92% yield.

As an example of the esterification of the terminal alcohol groups, compound **6** was obtained in 90% yield by Steglich esterification [41,42] with benzoic acid using *N*,*N*'-dicyclohexylcarbodiimide as a coupling reagent and the *N*,*N*-dimethylaminopyridine as a catalyst.

Furthermore, alcohol groups were converted to tosylate groups by performing the reaction of **3** with tosyl chloride, Net$_3$ as a base, and [HNMe$_3$]Cl as a catalyst [43], obtaining 9,10-(TsOCH$_2$CH$_2$CH$_2$)$_2$-1,7-*closo*-C$_2$B$_{10}$H$_{10}$, **7**, in 85% yield.

Overall, we have succeeded in the preparation of these new *m*-carborane derivatives **5–7** thanks to the primary alcohol groups, which undergo chain extension reactions. Owing to the formation of **5**, a new way of functionalization is opened to prepare the azide derivative **8**, which in turn opens the way to perform the Azide-Alkyne Huisgen Cycloaddition commonly known as the click reaction. Compound **8** was obtained in 81% yield by avigorous stirring of **5**, with excess of NaN$_3$ and [Nbu$_4$]Cl in a mixture of toluene and water, at reflux for 24–48 h. An example of the click reaction on **8** was the compound **9** synthesis in 86% yield by simple reaction with phenylaceylene, sodium ascorbate, and hydrated CuSO$_4$ as a catalyst in a mixture of dioxane/water.

Therefore, di-branched *m*-carborane derivatives (Scheme 2) with precise patterns of substitution are prepared by judicious choice of the synthetic procedure using similar conditions to the preparation of *o*-carborane derivatives present in previous work [14].

2.2. Synthesis of tetra-Branched m-carborane Derivatives at the 1,7,9,10 Vertexes

The above described di-branched 9,10-(CH$_2$=CHCH$_2$)$_2$-1,7-*closo*-C$_2$B$_{10}$H$_{10}$ (**3**) derivative, which still possesses its two C$_c$-H vertexes are ready for derivatization, offers the possibility to obtain globular icosahedral *m*-carborane derivatives with four branches as a new dendritic structure (Figure 2) by the incorporation of functional groups at the two carbon vertexes. Consequently, starting with **3** the four branched 1,7-(CH$_2$=CHCH$_2$)$_2$-9,10-(CH$_2$=CHCH$_2$)$_2$-1,7-*closo*-C$_2$B$_{10}$H$_8$, **10**, is obtained in two steps: i) removing the acidic hydrogen atoms with two equivalents of BuLi and ii) by electrophilic reaction with two equivalents of allylbromide (Scheme 3). From **10**, the tetraalcohol 1,7-(OHCH$_2$CH$_2$CH$_2$)$_2$-9,10-(OHCH$_2$CH$_2$CH$_2$)$_2$-1,7-*closo*-C$_2$B$_{10}$H$_8$, **11**, can be achieved by hydroboration. In the same way, **11** is ready as a core for constructing a tetra-branched *m*-derivatives using the judicious choice of synthetic procedure (Scheme 3).

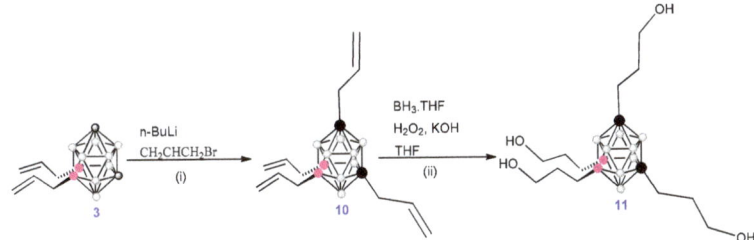

Scheme 3. (i) Deprotonation reaction on C$_c$-H of B(9,10)-disubstituted *m*-carborane derivative **3** with *n*-BuLi followed by nucleophilic substitution with allyl bromide. (ii) Hydroboration/oxidation process on terminal olefinic groups in **10** by using BH$_3$·THF, H$_2$O$_2$ in basic aqueous solution (KOH) to obtain **11**. Dark circles are C$_c$-H vertexes, pink circles are boron atoms, black circles are C$_c$ atoms and grey circles are B-H vertexes.

The synthesis of the tetra-substituted dianionic compound [9,10-({3,3′-Co(8′-(OCH$_2$CH$_2$)$_2$-1′,2′-C$_2$B$_9$H$_{10}$)(1″,2″-C$_2$B$_9$H$_{11}$)}$_2$-1,7-C$_2$B$_{10}$H$_8$)$^{2-}$ was attempted in THF starting with compound **3** in a two steps reaction: i) the deprotonation of C$_c$-H vertexes of **3** using two equivalents of *n*-BuLi to form the intermediate Li$_2$[9,10-(CH$_2$=CHCH$_2$)$_2$-1,7-*closo*-C$_2$B$_{10}$H$_8$]] salt and, ii) the nucleophilic attack of this salt to the dioxanate ring of the zwitterion [3,3′-Co-(8-(CH$_2$CH$_2$O)$_2$-1,2-C$_2$B$_9$H$_{10}$)(1′,2′-C$_2$B$_9$H$_{11}$)] in the same way as Li$_2$[1,7-*closo*-C$_2$B$_{10}$H$_{10}$]] had performed (see Scheme 4) [44]. Nevertheless, unexpectedly, the synthesis of the dianionic compound was not achieved while the isomerization of allyl branches to propenyl ones took place giving the isomer 9,10-(CH$_3$CH=CH)$_2$-1,7-*closo*-C$_2$B$_{10}$H$_{10}$, **12**, in 80% yield (Scheme 4).

The reason for this unexpected reaction can be the comparable acidity of the allyl groups and the C$_c$-H of the *m*-carborane unit, which may allow a deprotonation/protonation isomerization of the allyl group as it is well known for allylbenzenes [45]. The pKa value of the unsubstituted carborane clusters, which are insoluble in water, have been determined by two methods [6,10]. The pKa by using Streitwieser's scale provides the 27.9 value for the isomers *m*-carborane, while the one obtained by polarography is 24 [46]. Both experimental techniques support that unsubstituted *m*-carborane is a very weak Brønsted acid [46]. The allyl isomerization of **3** to propenyl in **12**, which takes place in THF, is supported by the formation of solvent separated ion pairs that prevent the carboranyl anion to act as a nucleophile. To verify this hypothesis, Density-functional theory (DFT) calculations were performed (details in the S.I.). The proton affinity (PA) of the cluster carbon atom is 332.8 kcal/mol (at B3LYP-D3/6-311+G**, PCM=tetrahydrofuran level of theory), while the proton affinity of allylic carbon atom has a somewhat higher value (342.3 kcal/mol). This moderate difference (ΔPA = 9.5 kcal/mol) probably allows for the above-mentioned mechanism. The question arises whether the same process does not occur in the case of the analog *o*-carborane based compounds [44]. It is known that cluster carbon in *m*-carborane is more than 1000 times less acidic than its *orto* isomer [47,48] therefore the difference between the two positions (allylic *vs* carboranyl) is larger as it was verified by our calculations (ΔPA = 18.6 kcal/mol) as well. It should be highlighted that Li⁺ mediated isomerizations on allyl substituents bonded at the C_c vertexes of the *o*-carborane cluster was previously demonstrated as well, as Et₂O does not tend to induce isomerization, whereas THF or DME produces the propenyl isomer [49]. A similar mechanism should be considered as well.

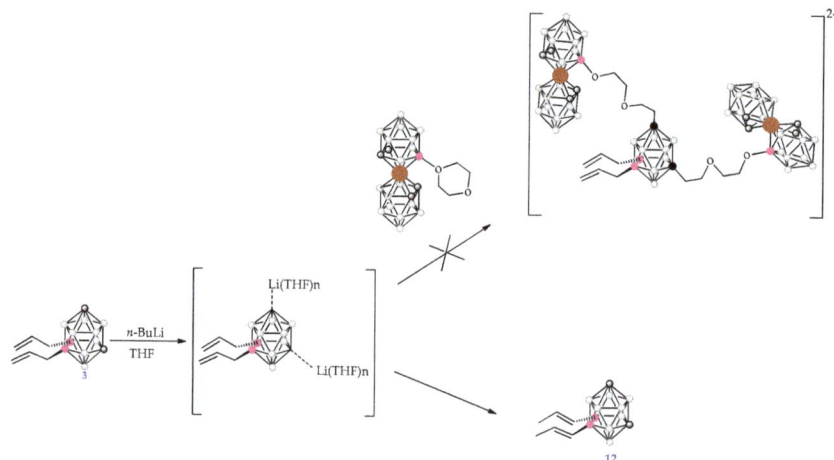

Scheme 4. Top: Designed a synthetic reaction to achieve the dianionic species. Bottom: Achieved reaction was the isomerization of 9,10-(CH₂=CHCH₂)₂-1,7-*closo*-C₂B₁₀H₁₀, to 9,10-(CH₃CH=CH)₂-1,7-*closo*-C₂B₁₀H₁₀. Dark circles are C_c-H vertexes, pink circles are boron atoms, and grey circles are B-H vertexes.

¹H-NMR spectrum of **12** supported the allyl branches isomerization to propenyl ones but, this process was unambiguously proven by X-ray diffraction of **12** from good crystals, which were grown from its acetone solution.

2.3. Characterization of di-Branched m-carborane Derivatives at the 9,10 Vertexes

The electrophilic substitution of the *o*-carborane led to the formation of the tetrasubstituted 8,9,10,12-I₄-1,2-*closo*-C₂B₁₀H₈ compound [5,42,43] in which the B-I vertexes reside at the compacted adjacent positions antipodal to the two cluster carbon C_c atoms. Conversely, the iodination electrophilic substitution takes place only at the B(9) and B(10) vertexes of the *m*-isomer.

We reported that the 2a-NPA value for a bond, defined as the sum of the NPA charges of the two bonded atoms (e.g., B-H or C-H), matches the order of attack on the different cluster' bonds [50]. Calculated NPA charges of the two bonded atoms (2a-NPA, calculated at B3LYP-D3/6-311+G** level of theory) of *ortho*- and *meta-closo*-carborane (present in Table 1) explain the higher accessibility of the B-H vertexes of *o*-carborane cluster to undergo electrophilic reaction, which drive to the formation of B-I bonds. While in the case of *o*-carborane there are four negative 2a-NPA values, in the case of *m*-carborane, there are only two. However, these positions exhibit higher reactivity towards electrophilic agents. Since in the case of *m*-carborane most of the 2a-NPA values of B-H vertexes are positive, the functionalization of this compound is more challenging. Table 1 shows that the electron density at the B-H vertexes of *m*-carborane follow a different trend B(9), B(10) >>> B(5), B(12) > B(4), B(6),

B(8), B(11) B(2), B(3) than o-carborane, which is B(9,12) > B(8,10) > B(4,5,7,11) > B(3,6) [3,51]. Contrary to the o-carborane that contains two positive natural charges; the m-carborane presents four positive natural charges on BH vertex, which explain the difficulty of the substitution of the B-H vertexes. Figure 3 shows that LUMO in o-carborane is located between the C atoms, whereas it is not the case for m-carborane where it is more disperse. Therefore, the carbon cluster position in the carborane has an important role related to the substitution of the B-H vertexes. Using the electrophilic iodination, it is possible to derivatize only B(9) and B(10) because these boron atoms do not have any connection with the C_c in the *meta* isomer. On the contrary, for the o- isomer the same procedure allows the attack to all B-H vertexes except B(3) and B(6) that are adjacent to both carbon clusters [52].

Table 1. Theoretical calculations of natural charges, 2a-NPA charges, and cumulative build-up of the cluster-only total charge (CTC) of *ortho-closo* and *meta-closo* carborane. See Figure 1 for the numbering of the clusters' vertexes.

	o-closo-$C_2B_{10}H_{12}$			m-closo-$C_2B_{10}H_{12}$	
	NPA	2a-NPA		NPA	2a-NPA
C(1)	−0.498	−0.198	C(1)	−0.654	−0.354
C(2)	−0.498	−0.198	B(2)	0.151	0.215
B(3)	0.159	0.213	B(3)	0.151	0.215
B(4)	0.000	0.054	B(4)	0.001	0.070
B(5)	0.000	0.069	B(5)	0.023	0.087
B(6)	0.159	0.213	B(6)	0.001	0.070
B(7)	0.000	0.069	C(7)	−0.654	−0.354
B(8)	−0.165	−0.087	B(8)	0.001	0.070
B(9)	−0.140	−0.067	B(9)	−0.165	−0.087
B(10)	−0.165	−0.087	B(10)	−0.165	−0.087
B(11)	0.000	0.054	B(11)	0.001	0.070
B(12)	−0.140	−0.067	B(12)	0.023	0.087
CTC	−1.288	-	CTC	−1.286	-

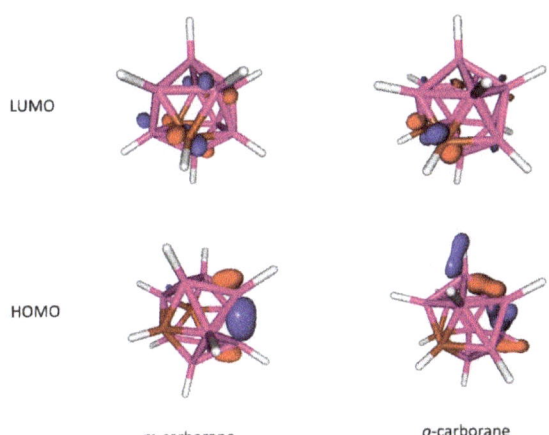

Figure 3. Comparison of the HOMO and LUMO orbitals of the o- and m-carborane.

The starting (**2** and **3**) and new compounds (**4–9**) were fully characterized by ^1H, $^1H\{^{11}B\}$, ^{11}B, $^{11}B\{^1H\}$, $^{13}C\{^1H\}$, and 2D COSY $^{11}B\ \{^1H\}$-$^{11}B\ \{^1H\}$ NMR spectroscopic techniques to be taken as inputs for the discussion of the influence of the substituents at the 9,10 vertexes on the Boron disubstituted *closo* m-carborane derivatives.

The $^{11}B\{^1H\}$ NMR spectrum of the parent cluster **1** displays four signals with intensities 2:2:4:2 from low to high field −5.6, −9.5, −12,0 and −15.4 ppm, which corresponds to a weighted average

^{11}B{^{1}H} NMR chemical shift, <δ(^{11}B)> ≈ −10.9 ppm [53]. Conversely, the ^{11}B {^{1}H} NMR spectrum of **2** displays four signals with intensities 2:4:2:2 from low to high field at −3.0, −10.4, −17.0, and −19.5 ppm, which corresponds to a weighted average <δ(^{11}B)> ≈ −12 ppm. The presence of the two iodo groups bonded to the B(9,10) in **2** produces a <δ(^{11}B)> upfield of −1.1 ppm in the ^{11}B NMR. The upfield resonance of the ^{11}B {^{1}H} NMR spectrum of **2** at −19.5 ppm does not split into a doublet in the ^{11}B-NMR spectrum supporting that it corresponds to the B-I at the 9 and 10 vertexes.

^{11}B {^{1}H}-^{11}B {^{1}H} 2D COSY NMR is of enormous use and potential in polyhedral boron chemistry because it provides a way of rapidly assigning ^{11}B resonances [54,55]. To assign the resonances of compounds **1** and **2** to the different cluster's vertexes by NMR spectroscopy, the two-dimensional ^{11}B {^{1}H}-^{11}B {^{1}H} COSY NMR spectra of compounds **1** and **2** were run (See Supplementary Information). Once the B(9,10) has been unambiguously assigned in compounds **1** and **2**, it is possible to confirm that the substitution of hydrogen by iodo causes significant shielding (−10 ppm) on the boron atoms that support the iodo units. ^{11}B {^{1}H}-^{11}B {^{1}H} 2D COSY NMR spectra of compounds **1** and **2** allow assigning the vertexes' resonances for **1** and **2** (see Supplementary Information). On the other hand, the assignment of the hydrogen atoms to the respective boron cluster vertexes was done by running the selective irradiation ^{1}H{^{11}B} NMR spectra (Table 2) which confirms the presence of four signals in **1** and only three in **2**. Notably, all proton resonances were shifted upfield in the ^{1}H{^{11}B} NMR spectrum which demonstrates the influence of the Iodo groups on all clusters' vertexes.

Table 2. The ^{11}B{^{1}H} and ^{1}H{^{11}B} chemical shifts of icosahedral compounds **1** and **2**. Spectra were recorded in (CD$_3$)$_2$CO. See Figure 1 for vertexes numbering.

	1,7-*closo*-C$_2$B$_{10}$H$_{12}$		9,10-I$_2$-1,7-*closo*-C$_2$B$_{10}$H$_{10}$	
	^{11}B{^{1}H} (ppm)	^{1}H{^{11}B} (ppm)	^{11}B{^{1}H} (ppm)	^{1}H{^{11}B} (ppm)
B(5,12)	−6.6	2.27	−4.1	2.76
B(9,10)	−10.5	2.10	−20.7	-
B(4,6,8,11)	−13.3	2.19	−11.9	2.93
B(5,12)	−17.0	2.64	−18.8	3.15

The ^{11}B{^{1}H} NMR spectrum of **3** displays four signals with intensities 2:2:4:2 from low to high field at 0.6, −5.4, −12.5, and −19.1 ppm, which corresponds to a weighted average <δ(^{11}B)> of ca. −9.8 ppm. The peak at +0.6 ppm does not split into a doublet in the ^{11}B-NMR spectrum, which supports the substitution of an iodo by carbon from the allyl group, which causes a downfield shift on the boron vertexes. On the other hand, the ^{1}H and ^{1}H{^{11}B} NMR spectra are useful to identify the presence of the organic fragments linked to the carborane cluster. Figure 4a shows the presence of three new signals (area ratio 1:2:2, from low to the high field), which are related to the typical resonances of terminal allyl groups. The protons bonded to the boron vertex (H$_d$) appear as a doublet (^1J(H,H) = 7.7 Hz) at 1.78 ppm. There is an overlap of H$_a$ and H$_b$ resonances, which should appear, each one, as a double doublet, but looks like a triplet at 4.88 ppm. H$_c$ is the most complicated proton of the allyl group because of the presence of four different protons at its neighboring carbon atoms. This appears in the range 5.97–5.82 ppm as a multiplet. The coupling constants of H$_c$ with neighbors is shown in Figure 4b.

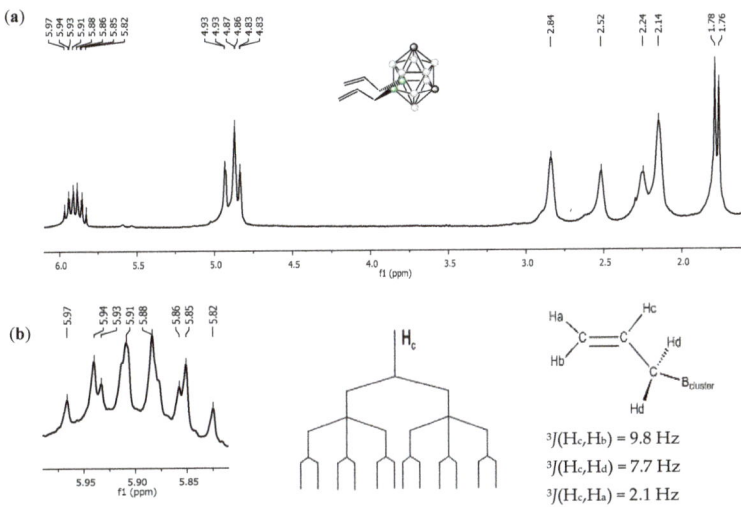

Figure 4. (a) The ^{1}H {^{11}B} spectrum of **3** in CDCl$_3$. (b) H$_c$ allyl resonance as well as the schematic coupling between H$_c$ protons and the protons of the allyl branches with the corresponding coupling constant values.

Compounds **4**, **5**, **6**, and **7** were also characterized by ^{1}H, ^{1}H{^{11}B}, ^{11}B, ^{11}B{^{1}H} and ^{13}C{^{1}H} NMR spectroscopy.

Table 3 lists the ^{11}B{^{1}H} NMR chemical shifts for the B(9,10) disubstituted *m*-carborane derivatives while Table 4 summarizes the ^{1}H, ^{13}C{^{1}H} NMR spectra and the stretching frequency of C$_c$-H in the IR spectra for the B(9,10) disubstituted *m*-carborane derivatives. The presence of organic branches connected to B(9) and B(10) causes a resonance downfield shift about +11 ppm on these boron atoms. Therefore, the average chemical shift value <δ(^{11}B)> = −10.9 ppm of parent **1** is around −9.6 ppm for 9,10-R$_2$-1,7-*closo*-C$_2$B$_{10}$H$_{12}$ derivatives (R=CH$_2$=CH−CH$_2$, HO(CH$_2$)$_3$, Cl(CH$_2$)$_3$, PhCOO(CH$_2$)$_3$, CH$_3$-C$_6$H$_4$-SO$_3$(CH$_2$)$_3$). There is no difference in these two features between the two isomers, *ortho* and *meta*.

Table 3. ^{11}B{^{1}H} NMR chemical shifts (in ppm) of di-branched (**2–9** and **12**) and tetra-branched (**10** and **11**) *m*-carborane derivatives. <δ(^{11}B)> corresponds to the weighted average ^{11}B{^{1}H} NMR spectrum (in ppm). Spectra were recorded in (CD$_3$)$_2$CO and referenced to external BF$_3$·Et$_2$O unless noted otherwise: #(CD$_3$)$_2$SO.

	R	B(9,10)	Δ	B(5,12)	B(4,6,8,11)	B(2,3)	<δ(^{11}B)>
1	H	−9.5	-	−5.6	−11.9	−15.4	−10.9
2	9,10-I$_2$	−19.4	−9.9	−3.0	−10.4	−17.0	−12.0
3	9,10-(CH$_2$=CH−CH$_2$)$_2$	0.6	+10.1	−5.4	−12.5	−19.1	−9.8
4	9,10-(HO(CH$_2$)$_3$)$_2$	1.8	+11.3	−5.3	−12.7	−19.4	−9.7
5	9,10-(Cl(CH$_2$)$_3$)$_2$	1.3	+10.8	−5.2	−12.5	−19.0	−9.6
6	9,10-(PhCOO(CH$_2$)$_3$)$_2$	1.6	+11.1	−5.2	−12.5	−19.0	−9.5
7	9,10-(CH$_3$-C$_6$H$_4$-SO$_3$(CH$_2$)$_3$)$_2$	1.2	+10.7	−5.3	−12.6	−19.1	−9.7
8	9,10-(N$_3$(CH$_2$)$_3$)$_2$	1.4	+10.9	−5.3	−12.6	−19.1	−9.6
9	9,10-(C$_6$H$_5$C$_2$N$_3$(CH$_2$)$_3$)$_2$ #	1.4	+10.9	−5.5	−12.4	−19.0	−9.6
10	1,7,9,10-(CH$_2$=CH−CH$_2$)$_4$	0.8	+10.3	−5.3	−10.2	−15.9	−8.2
11	1,7,9,10-(HO(CH$_2$)$_3$)$_4$	1.4	+10.9	−5.5	−10.5	−15.7	−8.2
12	9,10-(CH$_3$CH=CH)$_2$	−0.5	+9.0	−5.8	−12.5	−19.8	−10.1

Table 4. Chemical shift of ^{1}H and ^{13}C{^{1}H} NMR spectra (in ppm) and stretching frequencies of C$_c$-H (in cm^{-1}) in the IR spectra for the 9,10-R$_2$-1,7-*closo*-C$_2$B$_{10}$H$_{10}$ derivatives. NMR spectra were run in (CD$_3$)$_2$CO unless noted otherwise: * CDCl$_3$ and # (CD$_3$)$_2$SO.

	R	δ^{1}H(C$_c$-H)	Δδ^{1}H	δ^{13}C(C$_c$-H)	Δδ^{13}C(C$_c$-H)	ν(C$_c$-H)
1	H	3.65 2.91 *	- -	56.17 55.23 *	-	-
2	I	4.11	0.46	58.18	+2.01	-
3	CH$_2$=CH–CH$_2$-	3.50	−0.15	52.35*	−2.88*	3062
4	HO(CH$_2$)$_3$-	3.46	−0.19	52.95	−3.22	3038
5	Cl(CH$_2$)$_3$-	3.54	−0.11	53.28	−2.89	3063
6	PhCOO(CH$_2$)$_3$-	3.55	−0.10	53.27	−2.90	3064
7	CH$_3$-C$_6$H$_4$-SO$_3$-(CH$_2$)$_3$-	3.48	−0.17	53.26	−2.91	3064
8	N$_3$(CH$_2$)$_3$-	3.54	−0.11	53.26	−2.91	3065
9	C$_6$H$_5$C$_2$N$_3$(CH$_2$)$_3$-	3.86 #	0.21	Not observed	Not observed	3057
12	CH$_3$CH=CH-	2.82 *	0.09	Not observed	Not observed	3046

Table 3 shows a downfield shift (Δδ = +10.1 ppm) of the B(9,10) resonances of 9,10-(allyl)$_2$-1,7-*closo*-C$_2$B$_{10}$H$_{10}$ (3) *vs* the corresponding B(9,10)-H ones in the parent *m*-carborane. A similar downfield (Δδ = +10.6 ppm) is reported for the B(9,12) vertexes of 9,12-(allyl)$_2$-1,2-*closo*-C$_2$B$_{10}$H$_{10}$ with respect to the B(9,12)-H vertexes of the parent *o*-carborane [14]. The ^{11}B{^{1}H} NMR spectrum provides information on the electron density surrounding B atoms in the cluster vertexes, so it can be concluded that the effect of a B-allyl vertex concerning to the former B-H in the ^{11}B{^{1}H} NMR of both isomers is almost the same, Δδ +10.6 ppm and +10.1 ppm, for *o*- and *m*-, respectively. However, there is a major difference in the chemical shifts of the B-allyl nuclei of the two isomers: δ = +7.75 pm for 9,12-(allyl)$_2$-1,2-*closo*-C$_2$B$_{10}$H$_{10}$ and δ = +0.6 pm for 9,10-(allyl)$_2$-1,7-*closo*-C$_2$B$_{10}$H$_{10}$. We should remember that B-allyl vertexes are located antipodal to the C$_c$ vertexes in the *o*- isomer but antipodal to B vertexes in the *m*-isomer. This fact indicates a quite relevant different electronic surrounding in the B-allyl sites in both isomers, which depends on the atoms' nature at the antipodal vertexes.

Table 4 summarizes the ^{1}H and ^{13}C{^{1}H} NMR spectra and stretching frequencies of C$_c$-H bonds in the IR spectra for the reported 9,10-R$_2$-1,7-*closo*-C$_2$B$_{10}$H$_{10}$ derivatives; the presence of the allyl branches at the B(9,10) vertexes produces an upfield of the carbon and hydrogen atoms resonances of the C$_c$-H concerning to the parent *m*-carborane in their ^{1}H and ^{13}C{^{1}H} NMR spectra.

In Table 5, the comparison of the influence of the substituents at the B(9,12) in the *o*-carborane and the B(9,10) in the *m*-carborane is listed. To notice is that the influence on the chemical shift of the B-halogen (halogen = Cl, Br, I) vertexes in both isomers is the same: iodo is larger than bromo and bromo is larger than chloro. This is due to the i) electronegativity of halogen atoms, which follows the trend Cl > Br > I and ii) π back donation of halogen is I > Br > Cl.

Table 5. ^{1}H- and ^{13}C{^{1}H} NMR chemical shift values (in ppm) of C$_c$-H vertexes for several 9,12-R$_2$-1,2-*closo*-C$_2$B$_{10}$H$_{10}$ and 9,10-R$_2$-1,7-*closo*-C$_2$B$_{10}$H$_{10}$ derivatives. NMR spectra were run in *(CD$_3$)$_2$CO or #CDCl$_3$.

	δ^{1}H(C$_c$-H)	Δδ^{1}H	δ^{13}C(C$_c$-H)	Δδ^{13}C
1,2-*closo*-C$_2$B$_{10}$H$_{12}$	3.56 # 4.40 *	- -	54.46 # 56.20 *	-
9,12-I$_2$-1,2-*closo*-C$_2$B$_{10}$H$_{10}$ [39]	4.00 #	+0.44 #	52.23 #	−2.23 #
9,12-Br$_2$-1,2-*closo*-C$_2$B$_{10}$H$_{10}$ [56]	4.78 *	+0.38 *	48.50 *	−7.70 *
9,12-(CH$_2$=CH–CH$_2$)$_2$-1,2-*closo*-C$_2$B$_{10}$H$_{10}$ [14]	3.42 #	−0.14 #	48.26 #	−6.2 #
9,12-(OHCH$_2$CH$_2$CH$_2$)$_2$-1,2-*closo*-C$_2$B$_{10}$H$_{10}$ [14]	4.30 *	−0.10 *	49.17 *	−7.03 *

Table 5. Cont.

	$\delta^1H(C_c\text{-H})$	$\Delta\delta^1H$	$\delta^{13}C(C_c\text{-H})$	$\Delta\delta^{13}C$
9,10-I$_2$-1,7-*closo*-C$_2$B$_{10}$H$_{10}$ [39]	3.16 # 4.08 *	+0.40 # +0.32 *	- 58.18 *	- +2.01 *
9,10-Cl$_2$-1,7-*closo*-C$_2$B$_{10}$H$_{10}$ [56]	3.76 *	+0.13 *	51.60 *	−4.57 *
9,10-Br$_2$-1,7-*closo*-C$_2$B$_{10}$H$_{10}$ [56]	3.88 *	+0.25 *	54.00 *	−2.17 *
9,10-(CH$_2$=CH–CH$_2$)$_2$-1,7-*closo*-C$_2$B$_{10}$H$_{10}$	2.84 # 3.50 *	−0.07 # −0.13 *	52.35 # -	−2.88 # -
9,10-(OHCH$_2$CH$_2$CH$_2$)$_2$-1,7-*closo*-C$_2$B$_{10}$H$_{10}$	3.49 *	−0.14 *	52.95 *	−3.22 *

2.4. Characterization of Tetrabranched m-carborane Derivatives at the 1,7,9,10 Vertexes

From the analysis of the ^{1}H NMR spectra of **10**, it is seen that the original signal corresponding to the protons linked to the carbon cluster, which appear at 2.83 ppm, vanishes while new signals at 5.58, 5.00 and 2.56 ppm corresponding to the allyl branches on these C$_c$ vertexes are distinguished.

An important influence of the presence of the organic branches linked to the two C$_c$ atoms is observed in the ^{11}B downfield shift of the B(2) and B(3) vertexes that move from −19.1 ppm in **3** to -15.7 ppm in **10**. To notice is the upfield shift of $<\delta(^{11}B)>$ when moving from **2** to **10**: $<\delta(^{11}B)> = -12.0$ ppm in compound **2** (with two B-I and two C$_c$-H vertexes), $<\delta(^{11}B)> = -10.9$ ppm on the parent *m*-carborane, $<\delta(^{11}B)> = -9.8$ ppm in **3** (with two B-allyl and two C$_c$-H vertexes), $<\delta(^{11}B)> = -8.2$ ppm in **10** (with two B-allyl and two C$_c$-allyl vertexes) (see Table 3). Consequently, the incorporation of organic branches at the cluster vertexes produces a downfield of $<\delta(^{11}B)>$ in the ^{11}B NMR while the iodo groups have the opposite effect, supporting that cluster-only total charge is dissimilarly affected by electron-withdrawing substituents than electron-donating ones.

2.5. Structural Description

2.5.1. Crystallographic Studies

A search in the Cambridge Structural Database [57] showed just 3 hits (CUWMUD, TOKCUR and YOZSOV) for 9,10-R$_2$-1,7-*closo*-C$_2$B$_{10}$H$_{10}$ for R = -CCH, -CH$_2$C$_6$H$_4$, and -C$_6$H$_5$, respectively [34,36,37]. In this paper, we contribute with two additional X-ray structures that provide a broader view of the *m*-carborane derivatives.

To get information in such a family of compounds, good crystals of 9,10-(HOCH$_2$CH$_2$CH$_2$)$_2$-1,7-*closo*-C$_2$B$_{10}$H$_{10}$ (**4**) and 9,10-(CH$_3$CH=CH)$_2$-1,7-*closo*-C$_2$B$_{10}$H$_{10}$ (**12**) suitable for X-ray- diffraction were grown from an acetone solution at low temperature. Compound **4** was solved in the triclinic system, with a P 1 space group with four molecules in the asymmetric unit (Z = 4) and all atoms laid on the 1(a) Wyckoff positions. Compound (**12**) also solved in the triclinic system, but in a different space group (P-1) with two molecules in the asymmetric unit (Z = 2) and all atoms laid in 1(i) Wyckoff position. Figure 5 shows the crystal structures of **4** and **12** with the corresponding atom labels. Table 6 displays all crystallographic data and selected bond distances and angles are in the Supplementary Information.

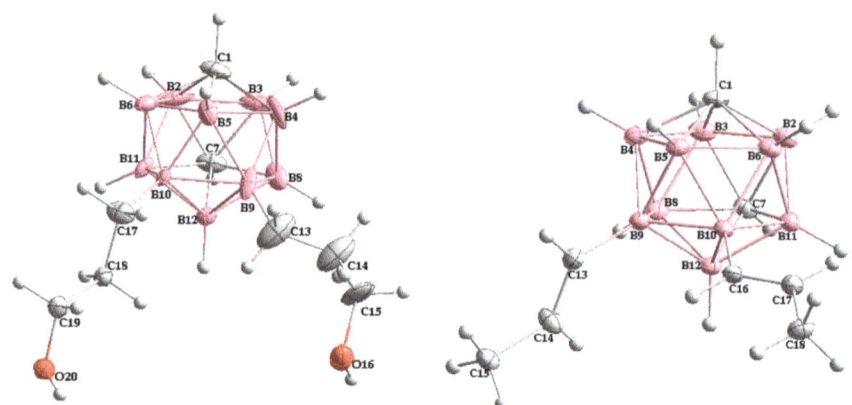

Figure 5. ORTEP presentation of 9,10-(HOCH$_2$CH$_2$CH$_2$)$_2$-1,7-*closo*-C$_2$B$_{10}$H$_{10}$ (**4**) and 9,10-(CH$_3$CH=CH)$_2$-1,7-*closo*-C$_2$B$_{10}$H$_{10}$ (**12**) showing the atom numbering and displacement. Ellipsoids are at 30% and 50% probability level, respectively.

Table 6. Crystal data and structure refinement.

Compound	4	12
Empirical formula	C$_8$H$_{24}$B$_{10}$O$_2$	C$_8$H$_{20}$B$_{10}$
Formula weight(g/mol)	260.37	224.34
Temperature (K)	100(2)	
Wavelength Å	0.71073	
Crystal system	Triclinic	
Space group	P 1	P -1
Unit cell dimensions	a = 10.7502(13)Å b = 11.1738(13)Å c = 14.2934(16)Å α = 73.346(5)° β = 75.220(5)° γ = 72.062(5)°	a = 6.9248(11)Å b = 7.4642(11)Å c = 13.454(2)Å α = 90.768(4)° β = 94.188(4)° γ = 95.488(4)°
Volume Å^3	1538.5(3)	690.23(18)
Z	4	2
ρcal (g/cm^3)	1.124	1.079
Absorption coefficient (mm^{-1})	0.062	0.049
F (000)	552	236
Theta range for data collection	2.74 to 28.30°	2.96 to 27.52°
Index ranges	−14<=h<=14 −14<=k<=14 −19<=l<=19	−9<=h<=8 −9<=k<=9 −17<=l<=17
Refinement method	Full-matrix least-squares on F2	
Final R indices [I > 2σ(I)]	7879 data; I>2σ(I) R1 = 0.1499 wR2 = 0.3573 all data R1 = 0.2389 wR2 = 0.4145	2428 data I>2σ(I) R1 = 0.0530, wR2 = 0.1302 all data R1 = 0.0745 wR2 = 0.1444

Compound **4** is the first example of a B(9,10) disubstituted *closo* 1,7-carborane derivative with a terminal O-H group. Furthermore, compound **12** is the first example of *closo m*-carborane with branches containing double bonds. For this, the behaviour of the two branches in the crystal network has been studied in detail. Exploring the crystal self-assembly, the presence of H⋯H short contacts in the range from 1.207 Å to 2.240 Å for compound **4** and equal to 2.252 Å for compound **12** are noticed, which are presented in Figures 6 and 7, respectively. In carborane chemistry, the dihydrogen H⋯H short contacts are related to the presence of two types of H atoms: the acidic C_c-H and the hydride B-H [58]. The supramolecular structure of **4** has an extensive network of hydrogen bonding due to the presence of the terminal OH groups (Figure 6a,b) with intermolecular distances shorter that sum of Van Der Waals radii ($\sum$vdW) minus 0.8 Å [59] and O-H⋯O angle values of 163.1° and 177.9°. The crystal packing of **4** is also stabilized by O⋯O as shown in Figure 6c. Accordingly, three different types of H⋯H short contacts were observed for **4**: C7-H7⋯H16-O16, O20-H20⋯H16-O16, and B3-H3⋯H14-C14.

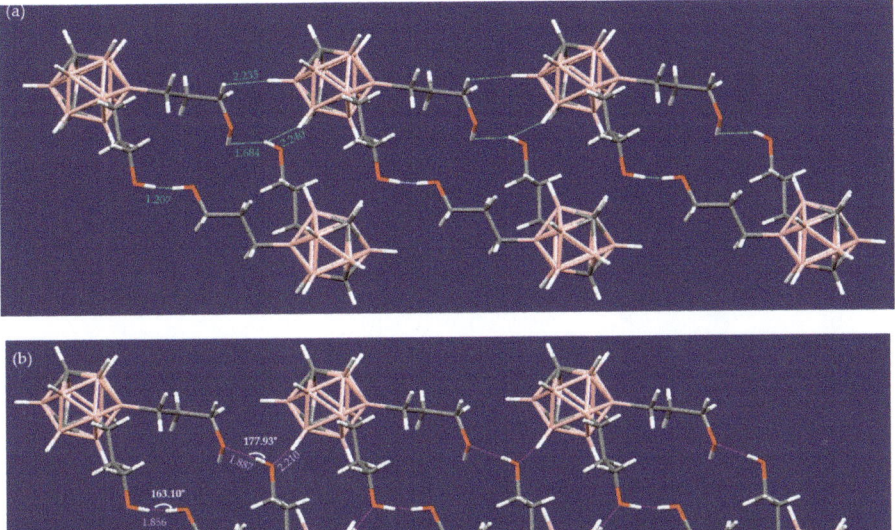

Figure 6. Network presentation of **4** showing all intermolecular contacts as dashed lines: (**a**) H⋯H, (**b**) O⋯H and (**c**) O⋯O (H are omitted for clarity).

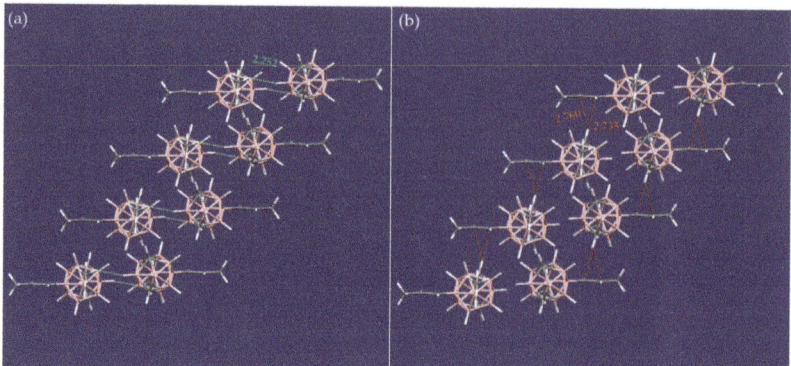

Figure 7. Network presentation of **12** showing all intermolecular contacts as dashed lines: (**a**) C-H ··· H-B and (**b**) C_c-H ··· π interactions.

As expected, the presence of double bonds in **12** has a noticeable role in the stabilization of the supramolecular network (Figure 7). The π electronic effect of the double bond leads to the formation of the π ··· H-C_c contacts (brown dashed lines), which are substantially shorter than 2.90 Å corresponding to the sum of the van der Waals radii ($\sum$vdW) [60]. The layers of **12** are connected into the final 3D structure through the B3-H3 ··· H15A-C15 bonds due to the acceptor character of the hydrogen-bonded to the boron and the donor character of the hydrogen atoms of the -CH_3 group.

2.5.2. Hirshfeld Surface Analysis

The Hirshfeld surface analysis, which is a very valuable method for the analysis of intermolecular contacts that offers a whole-of-the-molecule approach [61], presents three different colours to study the intermolecular interactions in crystal structures. The red colour means the presence of an intermolecular distance shorter that $\sum$vdW, white colour indicates the presence of intermolecular distances close to $\sum$vdW and blue colour designates the contacts longer than $\sum$vdW. Moreover, the shape index on the Hirshfeld surface identify hollows (with shape index < 0) and bumps (with shape index > 0), which are related to the character of each atom; the presence of an acceptor atom is marked by a concavity and the presence of a donor one is marked by a convexity. Therefore, the previous results were corroborated by studying the Hirshfeld surface of both structures using the crystal explorer program [62]. In this respect, Figure 8 presents the d_{norm} of **4** and **12** to visualize the intermolecular interactions and their contribution towards the supramolecular network. The two-dimensional fingerprint plots, which provide information about the percentage of intermolecular contacts present in the Hirshfeld surface is present in the S.I.

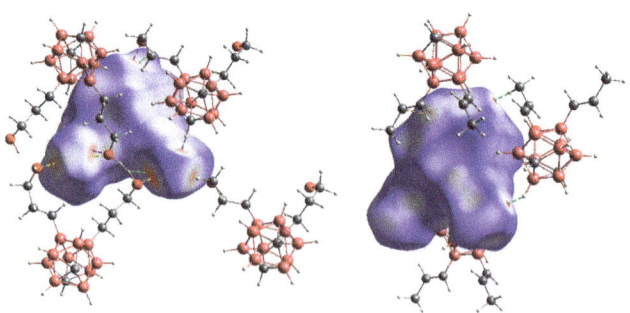

Figure 8. Presentation of close contacts for **4** (on left) and **12** (on right) through the d_{norm}.

The darkest red area in the d_{norm} surface of **4** is observed at the end of the molecule, arising from the O···H short contact as presented in Figure 8 and confirmed in the fingerprint plots (See Supplementary Information). Furthermore, the d_{norm} surface has shown the presence of bright red areas related to the presence of the O···O and H···H short contacts. Because of this packing arrangement, the O atoms at the molecular extremity present an important behaviour on the stability of this molecule by showing close contact values with H atoms of adjacent molecules shorter than $\sum$vdW.

The d_{norm} presentation of compound **12** (Figure 8 and S.I.) shows the presence of dark red points related to the classic H···H bonds. The existence of π···H short contacts is observed as a bright red point. The presence of the π acceptor interactions is indicated by the appearance of red concave triangles surrounded by blue ones in the shape index surface (Figure 9a) [63] while the C_c-H donor is confirmed by the blue convex area (Figure 9b) [64].

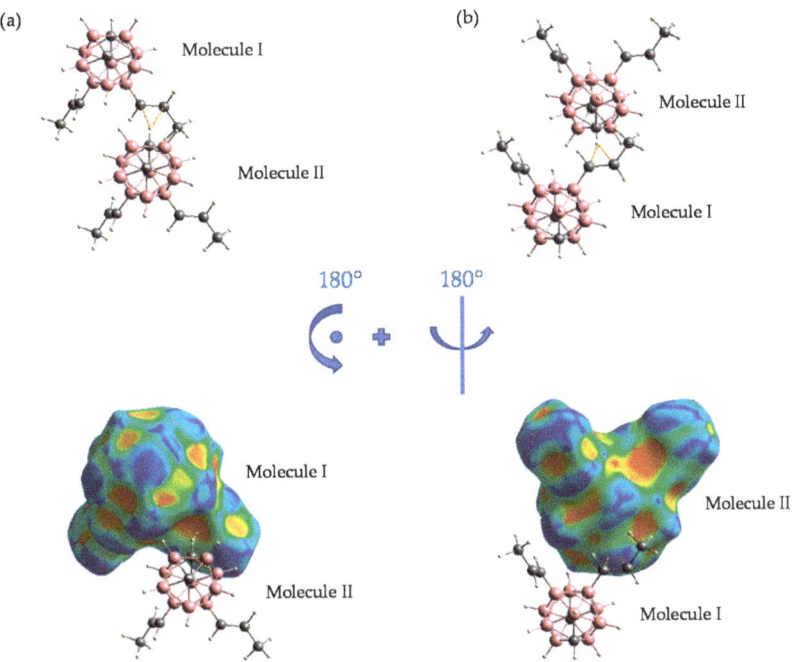

Figure 9. Shape index presentation of **12** showing the red concave and the blue convex areas, which correspond to (**a**) H···π and (**b**) π···H, respectively.

Despite the presence of many strong intermolecular interactions with contacts shorter than the sum of the van der Waals radii minus 0.80 Å, the H···H interactions are the most dominant with 87.8% and 94.2% in compounds **4** and **10** respectively, as shown in the fingerprint plots (See S.I.). The presence of 10.2% contacts O···O in **4** and 5.8% of π···H short contacts in **12** are also relevant.

3. Materials and Methods

3.1. Experimental Section

Materials and instrumentation: All *m*-carborane clusters prepared are air-stable. All manipulations were carried out under nitrogen atmosphere. THF and DMF were distilled from sodium benzophenone before use. Reagents were obtained commercially and used as purchased without purification. 1,7-*closo*-$C_2B_{10}H_{12}$ was obtained from Katchem.

ATR-IR spectra (ν, cm^{-1}) were obtained using the a JASCO FT/IR-4700 spectrometer on a high-resolution (Madrid, Spain). The ^{1}H and ^{1}H{^{11}B} NMR (300.13 MHz), ^{13}C{^{1}H} NMR (75.47 MHz), and ^{11}B and ^{11}B{^{1}H} NMR (96.29 MHz) spectra were recorded on a Bruker ARX300 instrument equipped with the appropriate decoupling accessories (Bruker Biospin, Rheinstetten, Germany)). All NMR spectra were performed in the indicated deuterated solvent at 22 °C. The ^{11}B and ^{11}B{^{1}H} NMR chemical shifts were referenced to external BF$_3$·OEt$_2$, while the ^{1}H, ^{1}H{^{11}B}, and ^{13}C{^{1}H} NMR shifts were referenced to SiMe$_4$. Chemical shifts are reported in units of parts per million downfield from reference, and all coupling constants in Hz.

3.1.1. Synthesis and Characterization of 3

The procedure for the synthesis of 3 was similar to that previously reported [40]. To a stirred solution of 9,10-I$_2$-1,7-*closo*-C$_2$B$_{10}$H$_{10}$ **2**, (300 mg, 1.34 mmol) in THF (15 mL) cooled to 0 °C in an ice-water bath was added, drop wise, a solution of allylmagnesium chloride in THF (6.06 mL, 1 M, 6.06 mmol). After stirring at room temperature for 30 min, [PdCl$_2$(PPh$_3$)$_2$] (21.28 mg, 4% equiv.) and CuI (5.77 mg, 4% equiv.) were added in a single portion, following which the reaction was heated to reflux overnight. The solvent was removed, and 20 mL of diethyl ether were added to the residue. The excess of Grignard reagent was destroyed by slow addition of dilute HCl. The organic layer was separated from the mixture, and the aqueous layer was extracted with diethyl ether (3 × 10 mL). The combined organic phase was dried over MgSO$_4$, filtered and the solvent removed under reduced pressure. The crude product was dissolved in hexane/chloroform mixture (1:1 by volume, ca. 5 mL) and passed rapidly through a bed of silica. The solvent was removed in a vacuum to give 9,10-(CH$_2$=CHCH$_2$)$_2$-1,7-*closo*-C$_2$B$_{10}$H$_{10}$ **3** as a yellowish oil (161.2 mg, 95%). Elemental analysis: calc: %C 42.8, %H 8.7; exp: %C 45.9, %H 8.7. ATR: ν = 3062 (vs, (C$_c$-H and =CH$_2$)), 2972, 2902 (vs, (=CH- and -CH$_2$-)), 2592 (vs, (B-H)), 1634 (vs, (C=C)), 995, 978, 692 (s, (=CH)). ^{13}C{^{1}H} NMR (75.47 MHz, CDCl$_3$) δ: 139.96 (s, *C*H$_2$=CHCH$_2$), 112.09 (s, CH$_2$=*C*HCH$_2$), 52.35 (s, C$_c$-H), 21.67 (m, CH$_2$=CH*C*H$_2$). ^{1}H-NMR (300.13 MHz, CDCl$_3$) δ: 5.89 (m, 2H, CH$_2$=C*H*CH$_2$), 4.88 (m, 4H, C*H*$_2$=CHCH$_2$), 4.94 (s, 2H, C$_c$-*H*), 1.78 (m, 4H, CH$_2$=CHC*H*$_2$). ^{1}H{^{11}B} NMR (300.13 MHz, CDCl$_3$) δ: 5.89 (m, 2H, CH$_2$=C*H*CH$_2$), 4.88 (m, 4H, C*H*$_2$=CHCH$_2$), 4.94 (s, 2H, C$_c$-*H*), 2.50 (s, 2H, B(5,12)-*H*), −2.24 (s, 2H, B(2,3)-*H*), 2.14 (s, 4H, B(4,6,8,11)-*H*), 1.77 (d,^{3}*J*(H,H) = 7.8, 4H, CH$_2$=CHC*H*$_2$). ^{11}B NMR (96.29 MHz, CDCl$_3$) δ: −0.1 (s, 2B, B(9,10)), −6.2 (d, ^{1}*J*(B,H) = 160, 2B, B(5,12)), −13.5 (d, ^{1}*J*(B,H) = 163, 4B, B(4,6,8,11)), 20.3 (d, ^{1}*J*(B,H) = 180, B(2,3)).

3.1.2. Synthesis and Characterization of 4

To a stirred solution of 9,10-(CH$_2$=CHCH$_2$)$_2$-1,7-*closo*-C$_2$B$_{10}$H$_{10}$, **3**, (150 mg, 0.67 mmol) in THF (2 mL) at 0 °C, was added, drop wise, a solution of BH$_3$·THF in THF (1.37 mL, 1 M, 1.37 mmol). The resulting suspension was stirred at 0 °C for 30 min and at room temperature for further 30 min. Then, the reaction mixture was cooled again to 0 °C in an ice-water bath and water (2 mL) was slowly added. When gas evolution had stopped, an aqueous KOH solution (1.68 mL, 3M, 1.94mmol) and subsequently, H$_2$O$_2$ in water (0.20 mL, 35%, 2.30 mmol), were added. Stirring was maintained at room temperature for 1.5 h, after which two liquid phases were observed. The upper organic layer was separated from the mixture and the aqueous layer and washed with THF (3 × 2 mL). The combined organic phase was dried over MgSO$_4$, filtered and the solvent removed in vacuo to give 9,10-(HOCH$_2$CH$_2$CH$_2$)$_2$-1,7-*closo*-C$_2$B$_{10}$H$_{10}$ **4**. Yield: 151 mg (87%). Elemental analysis: calc: %C 36.9, %H 9.22; exp: %C 36.79, %H 8.2. ATR: ν= 3305 (vs, ν$_s$(O-H)), 3038 (vs, C$_c$-H), 2930, 2886,2850, 2823 (vs, C$_{alkyl}$-H), 2585 (s, B-H), 1055, 1005, 978 (s, C-O). ^{13}C{^{1}H} NMR (300.13 MHz, (CD$_3$)$_2$CO) δ: 64.20 (s, HO*C*H$_2$CH$_2$CH$_2$), 52.95 (s, C$_c$), 32.27 (s, HOCH$_2$*C*H$_2$CH$_2$), 10.23 (s, HOCH$_2$CH$_2$*C*H$_2$). ^{1}H-NMR (300.13 MHz, (CD$_3$)$_2$CO) δ: 3.49 (s, 2H, C$_c$-*H*), 3.55 (m, 4H, HOC*H*$_2$CH$_2$CH$_2$), 3.40 (t, 2H, *H*OCH$_2$CH$_2$CH$_2$), 1.60 (m, 4H, HOCH$_2$C*H*$_2$CH$_2$), 0.81 (t, ^{3}*J*(H,H) = 16.7, 4H, HOCH$_2$CH$_2$C*H*$_2$). ^{1}H{^{11}B} NMR (300.13 MHz, (CD$_3$)$_2$CO) δ: 3.49 (s, 2H, C$_c$-*H*), 3.55 (m, 4H, HOC*H*$_2$CH$_2$CH$_2$), 3.40 (t, 2H, *H*OCH$_2$CH$_2$CH$_2$), 2.48,2.20,2.08 (s, 8H, B-*H*), 1.60 (m, 4H, HOCH$_2$C*H*$_2$CH$_2$), 0.81 (t, ^{3}*J*(H,H) = 16.7, 4H, HOCH$_2$CH$_2$C*H*$_2$). ^{11}B NMR (96.29 MHz, (CD$_3$)$_2$CO) δ: 1.9 (s, 2B, B(9,10)), −5.3 (d, ^{1}*J*(B,H) = 157, 2B,

B(5,12)), −12.7 (d, $^1J(B,H) = 160$, 4B, B(4,6,8,11)), −19.3 (d, $^1J(B,H) = 178$, 2B, B(2,3)). Colourless good crystals suitable for X-ray diffraction were grown in acetone.

3.1.3. Synthesis and Characterization of **5**

To a stirred solution of 9,10-(HOCH$_2$CH$_2$CH$_2$)$_2$-1,7-*closo*-C$_2$B$_{10}$H$_{10}$, **4**, (300 mg, 1.14 mmol) and [NBu$_4$]Cl (132.59 mg, 0.478mmol) in dry THF (10mL) at 0 °C, was added SOCl$_2$ dropwise (0.52 mL, 7.076 mmol). The resulting solution was stirred at 0 °C for 1 h and at room temperature overnight. The solvent was removed under reduced pressure, and 8 mL of diethyl ether were added. A solution of Na$_2$CO$_3$ (8 mL, 2 M) was slowly added with stirring. The mixture was thoroughly shaken, and the two layers separated. The aqueous layer was extracted with diethyl ether (3 × 5 mL). Then, the combined organic phase was separated and a solution of HCl (8 mL, 0.1 M) was added, the mixture was thoroughly shaken again. The upper organic layer was separated from the mixture, and the aqueous layer was washed with diethyl ether (3 × 5 mL). Finally, the combined organic phase was dried over MgSO$_4$, filtered and the solvent removed in vacuo to give 9,10-(ClCH$_2$CH$_2$CH$_2$)$_2$-1,7-*closo*-C$_2$B$_{10}$H$_{10}$, **5**. Yield: 310 mg (92%). Elemental analysis: calc: %C 32.32, %H 7.40; exp: %C 33.04, %H 7.60. ATR: ν= 3053 (s, C$_c$-H), 2989, 2972, 2902 (s, C$_{alkyl}$-H), 2626, 2587 (s, B-H), 1310, 1279 (s, CH$_2$-Cl), 728, 647 (s, C-Cl). ^{13}C{^{1}H} NMR (300.13 MHz, (CD$_3$)$_2$CO) δ: 53.28 (s, C$_c$), 47.14 (s, ClCH$_2$CH$_2$CH$_2$), 32.98 (s, ClCH$_2$CH$_2$CH$_2$), 11.47 (s, ClCH$_2$CH$_2$CH$_2$). ^{1}H-NMR (300.13 MHz, (CD$_3$)$_2$CO) δ: 3.53 (s, 2H, C$_c$-*H*), 3.61 (t, 4H, ClC*H*$_2$CH$_2$CH$_2$), 1.87 (m, 4H, ClCH$_2$C*H*$_2$CH$_2$), 0.94 (t, 4H, ClCH$_2$CH$_2$C*H*$_2$). ^{1}H{^{11}B} NMR (300.13 MHz, (CD$_3$)$_2$CO) δ: 3.53 (s, 2H, C$_c$-*H*), 3.61 (t, 4H, ClC*H*$_2$CH$_2$CH$_2$), 2.51, 2.23, 2.13 (s, 8H, B-*H*), 1.87 (m,4H, ClCH$_2$C*H*$_2$CH$_2$), 0.78 (t, 4H, ClCH$_2$CH$_2$C*H*$_2$). ^{11}B NMR (96.29 MHz, (CD$_3$)$_2$CO) δ: 1.3 (s, 2B, B(9,10)), −5.3 (d, $^1J(B,H) = 155$, 2B, B(5,12)), −12.6 (d, $^1J(B,H) = 159$, 4B, B(4,6,8,11)), −19.0 (d, $^1J(B,H) = 179$, 2B, B(2,3)).

3.1.4. Synthesis and Characterization of **6**

To a stirred solution of 9,10-(HOCH$_2$CH$_2$CH$_2$)$_2$-1,7-*closo*-C$_2$B$_{10}$H$_{10}$ (94 mg, 0.361 mmol), **4**, 4-N,N-dimethylaminopyridine (97.17 mg, 0.795 mmol), N,N'-dicyclohexylcarbodiimide (164.11 mg, 0.795 mmol) and benzoic acid (97.17 mg, 0.795 mmol) in dry dichloromethane (10 mL). The resulting solution was stirred at room temperature for 1 h. The white precipitate (dicyclohexylurea) is filtered and then an extraction using 10 mL of HCl (1 M) was done. The aqueous layer was extracted with CH$_2$Cl$_2$ (3 × 10 mL). The combined organic phase was dried over MgSO$_4$, filtered, and the solvent removed under reduced pressure to give 152 mg (90 %) of 9,10-(C$_6$H$_5$COOCH$_2$CH$_2$CH$_2$)$_2$-1,7-*closo*-C$_2$B$_{10}$H$_{10}$, **6**. Elemental analysis: calc: %C 56.41, %H 6.83; exp: %C 55.72, %H 6.81. ATR: ν= 3064 (vs, C$_c$-H), 2988, 2971, 2904 (vs, C$_{alkyl}$-H), 2594, 2565 (s, B-H), 1711 (s, ν$_s$(C=O)), 1599 (s, C-O). ^{13}C{^{1}H} NMR (300.13 MHz, (CD$_3$)$_2$CO) δ: 165.87 (s, C$_6$H$_5$COOCH$_2$CH$_2$CH$_2$), 132.82 (s, C$_{aryl}$), 130.70 (s, C$_{aryl}$), 129.22 (s, C$_{aryl}$), 128.45 (s, C$_{aryl}$), 66.71(s, C$_6$H$_5$COOCH$_2$CH$_2$CH$_2$), 53.27 (s, C$_c$), 25.37 (s, C$_6$H$_5$COOCH$_2$CH$_2$CH$_2$) 10.06 (s, C$_6$H$_5$COOCH$_2$CH$_2$CH$_2$). ^{1}H-NMR (300.13 MHz, (CD$_3$)$_2$CO) δ: 8.03 (d, $^1J(H,H) = 8.1$, 4H, H$_{aryl}$), 7.65 (t, 2H, H$_{aryl}$), 7.53 (m, 4H, $_{aryl}$), 4.30 (t, 4H, $^3J(H,H) = 6.8$, C$_6$H$_5$COOC*H*$_2$CH$_2$CH$_2$), 3.55 (s, 2H, C$_c$-*H*), 1.87 (m, 4H, C$_6$H$_5$COOCH$_2$C*H*$_2$CH$_2$), 0.98 (m, 4H, C$_6$H$_5$COOCH$_2$CH$_2$C*H*$_2$). ^{1}H{^{11}B} NMR (300.13 MHz, (CD$_3$)$_2$CO) δ: 8.03 (d, $^1J(H,H)= 8.1$, 4H, H$_{aryl}$), 7.65 (t, 2H, H$_{aryl}$), 7.51 (m, 4H, H$_{aryl}$), 4.30 (t, $^3J(H,H) = 6.8$, 4H, C$_6$H$_5$COOC*H*$_2$CH$_2$CH$_2$), 3.55 (s, 2H, C$_c$-*H*), 2.52, 2.27, 2.16 (s, 8H, B-H), 1.87 (m, 4H, C$_6$H$_5$COOCH$_2$C*H*$_2$CH$_2$), 0.98 (m, 4H, C$_6$H$_5$COOCH$_2$CH$_2$C*H*$_2$). ^{11}B NMR (96.29 MHz, (CD$_3$)$_2$CO) δ: 1.6 (s, 2B, B(9,10)), −5.2 (d, $^1J(B,H) = 149$, 2B, B(5,12)), −12.5 (d, $^1J(B,H) = 158$, 4B, B(4,6,8,11)), −19.1 (d, $^1J(B,H) = 175$, 2B, B(2,3)).

3.1.5. Synthesis and Characterization of **7**

To a mixture of 9,10-(HOCH$_2$CH$_2$CH$_2$)$_2$-1,7-*closo*-C$_2$B$_{10}$H$_{10}$, **4**, (163 mg, 0.627 mmol) and [HNMe$_3$]Cl (12.76 mg, 0.13 mmol) in 5 mL of dry toluene 1.5 mL of Triethylamine was added. In second flask, the p. toluensulfonyl chloride (363.34 mg, 1.905 mmol) was dissolved in THF, then converted to the first flask at 0°C. The solvent was evaporated and an extraction using the diethyl ether and water. The organic part was dried over MgSO$_4$, filtered, and the solvent removed under reduced pressure to give 9,10-(CH$_3$-C$_6$H$_4$-SO$_3$-CH$_2$CH$_2$CH$_2$)$_2$-1,7-closo-C$_2$B$_{10}$H$_{10}$, **7**, (302 mg, 85%). ATR: ν = 3064 (s, (C$_c$-H)), 2971, 2955, 2895 (s, (C$_{alkyl}$-H)), 2596, 2565 (s, (B-H)), 1349 (s, (S=O)), 1189, 1175 (s, (S-O)), 1097, 981, 954, 918 (s, (C-O)). ^{13}C{^{1}H} NMR (300.13 MHz, (CD$_3$)$_2$CO) δ: 144.77 (s, CH$_3$-C$_6$H$_4$), 133.80 (C$_6$H$_4$-S), 129.94 (s, C$_{aryl}$), 127.76 (s, C$_{aryl}$), 72.67 (s, TsOCH$_2$CH$_2$CH$_2$), 53.26 (s, C$_c$), 37.28 (s, TsOCH$_2$CH$_2$CH$_2$), 20.62 (s, CH$_3$-C$_6$H$_5$), 9.78 (br s,TsOCH$_2$CH$_2$CH$_2$). ^{1}H-NMR (300.13 MHz, (CD$_3$)$_2$CO) δ: 7.82 (d, 4H, ^1J(H,H) = 7.2, H$_{aryl}$), 7.49 (d, 4H,^1J(H,H) = 8.0, H$_{aryl}$), 4.05 (t, ^3J(H,H) = 6.6, 4H, TsO-CH$_2$), 3.45 (br s, 2H, C$_c$-H), 2.47 (s, 6H, CH$_3$C$_6$H$_4$), 1.75 (m, 4H, TsOCH$_2$CH$_2$CH$_2$), 0.77 (m, 4H, TsOCH$_2$CH$_2$CH$_2$). ^{1}H{^{11}B} NMR (300.13 MHz, (CD$_3$)$_2$CO) δ: 7.82 (d, 4H, ^1J(H,H) = 7.2, H$_{aryl}$), 7.49 (d, 4H,^1J(H,H) = 8.0, H$_{aryl}$), 4.05 (t, ^3J(H,H)= 6.6, 4H, TsO-CH$_2$), 3.45 (br s, 2H, C$_c$-H), 2.47 (s, 6H, CH$_3$C$_6$H$_4$), 3.04, 2.03, 2.00 (br s, 8H, BH),1.75 (m, 4H, TsOCH$_2$CH$_2$CH$_2$), 0.77 (m, 4H, TsOCH$_2$CH$_2$CH$_2$). ^{11}B NMR (96.29 MHz, (CD$_3$)$_2$CO) δ: 3.4 (s, 2B, B(9,10)), −5.4 (d, ^1J(B,H) = 153, 2B, B(5,12)), −12.7 (d, ^1J(B,H) = 155, 4B, B(4,6,8,11)), −19.2 (d, ^1J(B,H) = 177, 2B, B(2,3)).

3.1.6. Synthesis and Characterization of **8**

To a stirred solution of previously dried 9,10-(ClCH$_2$CH$_2$CH$_2$)$_2$-1,7-closo-C$_2$B$_{10}$H$_{10}$, **5**, (95 mg, 0.319 mmol) in DMF (10 mL), NaN$_3$ (314.41 mg, 4.83 mmol) was added. At room temperature, the mixture was stirred for 24 h. Then, the solvent was evaporated under vacuum and an extraction with a mixture C$_6$H$_5$CH$_3$-H$_2$O was done. After washing it several times with H$_2$O, the collected organic layer was dried over MgSO$_4$, filtered, and the solvent removed under reduced pressure to give 9,10-(N$_3$CH$_2$CH$_2$CH$_2$)$_2$-1,7-closo-C$_2$B$_{10}$H$_{10}$, **8**, (80.75 mg, 81%). ATR: ν= 3076 (C$_c$-H), 2929, 2890 (C$_{alkyl}$-H), 2591 (B-H), 2095 (C-N). ^{13}C{^{1}H} NMR (300.13 MHz, (CD$_3$)$_2$CO) δ: 53.42 (s, C-N$_3$), 53.26 (C$_c$-H), 11.13 (m, CH$_2$). ^{1}H-NMR (300.13 MHz, (CD$_3$)$_2$CO) δ: 3.84 (s, 2H, C$_c$-H), 3.34 (t, ^3J(H,H) = 6.90, 4H, N$_3$CH$_2$CH$_2$CH$_2$), 1.69 (m, N$_3$CH$_2$CH$_2$CH$_2$), 0.89 (m, 4H, N$_3$CH$_2$CH$_2$CH$_2$). ^{1}H{^{11}B} NMR (300.13 MHz, (CD$_3$)$_2$CO) δ: 3.84 (s, 2H, C$_c$-H), 3.34 (t, ^3J(H,H) = 6.90, 4H, N$_3$CH$_2$CH$_2$CH$_2$), 2.52, 2.23, 2.13 (br s, BH), 1.69 (m, N$_3$CH$_2$CH$_2$CH$_2$), 0.89 (m, 4H, N$_3$CH$_2$CH$_2$CH$_2$). ^{11}B NMR (96.29 MHz, (CD$_3$)$_2$CO) δ: 1.4 (s, 2B, B(9,10)), −5.3 (d, ^1J(B,H)= 157, 2B, B(5,12)), −12.6 (d, ^1J(B,H) = 160, 4B, B(4,6,8,11)), −19.1 (d, ^1J(B,H) = 180, 2B, B(2,3)).

3.1.7. Synthesis and Characterization of **9**

To a solution of 9,10-(N$_3$CH$_2$CH$_2$CH$_2$)$_2$-1,7-closo-C$_2$B$_{10}$H$_{10}$, **8**, (27 mg, 0.086 mmol) in a mixture of dioxane (2 mL) and distilled H$_2$O (2 mL), phenylacetylene (0.04 mL, 0.36 mmol), sodium ascorbate (17.037 mg, 0.086 mmol) and CuSO$_4$.5H$_2$O (21.47 mg, 0.086 mmol) were added in this order. After 20 min, a yellow solid started to be formed. The reaction was stopped after 3 h, when the walls of the small flask were full yellow solid and the solution was green. Then, the yellow solid was separated and very well dried under vacuum to give 9,10-(C$_6$H$_5$C$_2$N$_3$CH$_2$CH$_2$CH$_2$)$_2$-1,7-closo-C$_2$B$_{10}$H$_{10}$, **9**, (76 mg, 86%). Elemental analysis: calc C% 56.23, H% 6.29; exp: %C 56.69, %H 6.88. ATR: ν= 3057 (C$_c$-H), 2924, 2892, 2852, 2826 (C$_{alkyl}$-H, CH$_2$N), 2588 (B-H). ^{13}C{^{1}H} NMR (300.13 MHz, (CD$_3$)$_2$SO) δ: 146.69 (C$_6$H$_5$-C-N=N-N), 131.35 (C$_6$H$_5$-C=C-N), 129.34 (C$_6$H$_5$), 128.22, 125.57 (C$_6$H$_5$), 121.68 (C$_6$H$_5$), 54.15 (C$_c$-H), 52.25 (N-CH$_2$CH$_2$CH$_2$), 30.68 (N-CH$_2$CH$_2$CH$_2$), 11.4 (NCH$_2$CH$_2$CH$_2$). ^{1}H-NMR (300.13 MHz, (CD$_3$)$_2$SO) δ: 8.56 (s, 2H, C=CH-N), 7.85 (d, 4H,^1J = 7.7, C$_6$H$_5$), 7.44 (t, 4H, ^3J = 7.5, C$_6$H$_5$), 7.32 (t, 2H,^3J = 7.1, C$_6$H$_5$), 3.86 (s, 2H, C$_c$-H), 4.34 (m, 4H, NCH$_2$CH$_2$CH$_2$), 1.86 (m, 4H, NCH$_2$CH$_2$CH$_2$), 0.69 (m, 4H, NCH$_2$CH$_2$CH$_2$). ^{1}H{^{11}B} NMR (300.13 MHz, (CD$_3$)$_2$SO) δ: 8.56 (s, 2H, C=CH-N), 7.85 (d, 4H,^1J = 7.7, C$_6$H$_5$), 7.44 (t, 4H, ^3J = 7.5, C$_6$H$_5$), 7.32 (t, 2H,^3J = 7.1, C$_6$H$_5$), 3.86 (s, 2H, C$_c$-H), 4.34 (m, 4H, NCH$_2$CH$_2$CH$_2$), 2.38, 2.10, 2.02 (br s, B-H), 1.86 (m, 4H, NCH$_2$CH$_2$CH$_2$), 0.69 (m, 4H, NCH$_2$CH$_2$CH$_2$). ^{11}B and ^{11}B{^{1}H} NMR (96.29 MHz, (CD$_3$)$_2$SO) δ: 1.4 (s, 2B, B(9,10)-C), −5.5 (2B, B(5,12)), −13.2 (4B, B(4,6,8,11)), −19.9 (2B, B(2,3)).

3.1.8. Synthesis and Characterization of 10

To a stirred solution of 9,10-(CH$_2$=CHCH$_2$)$_2$-1,7-closo-C$_2$B$_{10}$H$_{10}$ (150 mg, 0.67 mmol) in dry THF (10 mL) at 0 °C were added dropwise n-BuLi in hexane (0.98 mL, 1.5 M, 1.47 mmol), the resulting solution was stirred at 0 °C for 1 h. Then the mixture was cooled at −78°C to add dropwise a solution of CH$_2$=CHCH$_2$-Br in dry THF (1.54 mL, 1 M, 1.54 mmol), and allowed to stir overnight at room temperature. Afterwards, the solvent was removed and 10 mL of diethyl ether and 10 mL of HCl (0.25M) were added to the residue. The organic layer was separated from the mixture, and the aqueous layer was extracted with diethyl ether (3 × 10mL). The combined organic phase was dried over MgSO$_4$, filtered, and the solvent removed under reduced pressure to give 71% of 9,10-(CH$_2$=CHCH$_2$)$_2$-1,7-(CH$_2$=CHCH$_2$)$_2$-closo-C$_2$B$_{10}$H$_8$ (145 mg). ^{13}C{^{1}H} NMR (300.13 MHz, (CD$_3$)$_2$CO) δ: 140.17 (s, CH$_2$=CH-CH$_2$-B(9,10)), 133.96 (s, CH$_2$=CHCH$_2$-C(1,2)), 117.98 (s, CH$_2$=CH-CH$_2$-B(9,10)), 111.96 (s, CH$_2$=CH-CH$_2$-C(1,2)), 71.8 (s, C$_c$), 40.97 (s, CH$_2$=CH-CH$_2$-C(1,2)), 21.48 (m, CH$_2$=CH-CH$_2$- B(9,10)). ^{1}H-NMR (300.13 MHz, (CD$_3$)$_2$CO) δ: 5.80 (m, 2H, CH$_2$=CH-CH$_2$-B(9,10)), 5.60 (m, 2H, CH$_2$=CH-CH$_2$-C(1,2)), 4.90 (m, 4H, CH$_2$=CH-CH$_2$-C(1,2)), 4.86 (m, 4H, CH$_2$=CH-CH$_2$-B(9,10)), 2.57 (d, ^{3}J(H-H)=7.3 Hz, 4H, CH$_2$=CH-CH$_2$-C(1,2)), 1.71 (br s, 4H, CH$_2$=CH-CH$_2$-B(9,10)). ^{1}H{^{11}B} NMR (300.13 MHz, (CD$_3$)$_2$CO) δ: 5.80 (m, 2H, CH$_2$=CH-CH$_2$-B(9,10)), 5.60(m, 2H, CH$_2$=CH-CH$_2$-C(1,7)), 4.90 (m, 4H, CH$_2$=CHCH$_2$-C(1,7)), 4.86 (m, 4H, CH$_2$=CHCH$_2$-B(9,10)), 2.57 (d, ^{3}J(H,H) = 7.3 Hz, 4H, CH$_2$=CHCH$_2$-C(1,7)), 2.17 (brs, 2H, B(5,12)-H), 2.12 (brs, 4H, B(4,6,8,11)-H), 1.87 (brs, 2H, B(2,3)-H), 1.71 (brs, 4H, CH$_2$=CH-CH$_2$-B(9,10)). ^{11}B-NMR (96.29 MHz, (CD$_3$)$_2$CO) δ: 0.8 (s, 2B, B(9,10)), −5.4 (d, ^{1}J(B,H) = 154, 2B, B(5,12)), −10.2 (d, ^{1}J(B,H) = 159, 4B, B(4,6,8,11)), −15.9 (d, ^{1}J(B,H) =174, 2B, B(2,3)).

3.1.9. Synthesis and Characterization of 11

To a stirred solution of 9,10-(CH$_2$=CHCH$_2$)$_2$-1,7-(CH$_2$=CHCH$_2$)$_2$-closo-C$_2$B$_{10}$H$_8$, 3, (130 mg, 0.24 mmol) in THF (3.5 mL) at 0 °C, was added, drop wise, a solution of BH$_3$·THF in THF (1.765 mL, 1M, 1.765 mmol). The resulting suspension was stirred at 0 °C for 30 min and at room temperature for further 30 min. Then, the reaction mixture was cooled again to 0 °C in an ice-water bath and water (1mL) was slowly added. When gas evolution had stopped, an aqueous KOH solution (0.583 mL, 3 M, 1.75 mmol) and subsequently, H$_2$O$_2$ in water (0.3 mL, 35%), were added. Stirring was maintained at room temperature for 1.5 h, after which two liquid phases were observed. The upper organic layer was separated from the mixture and the aqueous layer. Then, it was washed with THF (3 × 2 mL). The combined organic phase was dried over MgSO$_4$, filtered and the solvent removed in vacuo to give 9,10-(HOCH$_2$CH$_2$CH$_2$)$_2$-1,7-(HOCH$_2$CH$_2$CH$_2$)$_2$-closo-C$_2$B$_{10}$H$_8$.Yield: 119mg (74%). Elemental analysis: calc: C% 44.7, H% 9.6; exp: %C 45.1, %H 9.8. ^{13}C{^{1}H} NMR (300.13 MHz, (CD$_3$)$_2$CO) δ: 66.04 (s, HOCH$_2$CH$_2$CH$_2$-C(1,7)), 64.18 (s, HOCH$_2$CH$_2$CH$_2$-C(1,7), HOCH$_2$CH$_2$CH$_2$-B(9,10)), 60.52 (s, HOCH$_2$CH$_2$CH$_2$-C(1,7)), 33.38 (s, HOCH$_2$CH$_2$CH$_2$-C(1,7), 33.19 (s, HOCH$_2$CH$_2$CH$_2$-B(9,10)), 23.83 (s, HOCH$_2$CH$_2$CH$_2$B(9,10)). ^{1}H{^{11}B} NMR (300.13 MHz, (CD$_3$)$_2$CO) δ: 3.67–3.51 (m, 12H, HOCH$_2$CH$_2$CH$_2$), 1.59 (m, 8H, HOCH$_2$CH$_2$CH$_2$), 0.99–0.81 (m, 12H, HOCH$_2$CH$_2$CH$_2$). ^{11}B-NMR (96.29 MHz, (CD$_3$)$_2$CO) δ: 1.2 (s, 2B, B(9,10)), −5.6 (d,^{1}J(B,H) = 147, 2B, B(5,12)), −10.6 (d, ^{1}J(B,H) = 150, 4B, B(4,6,8,11)), −15.8 (d, ^{1}J(B,H) = 175, 2B, B(2,3)).

3.1.10. Synthesis and Characterization of 12

To a stirred solution of 9,10-(CH$_2$=CHCH$_2$)$_2$-1,7-closo-C$_2$B$_{10}$H$_{10}$, 3, (20 mg, 0.09 mmol) in THF (3 mL) at 0 °C, was added, drop wise, a solution of BuLi (0.18 mmol, 1.6 M, 0.12 mL). The resulting suspension was stirred at 0 °C for 30 min and at room temperature for further 30 min. In another flask and under nitrogen, 8-{3,3'-Co(8-C$_4$H$_8$O$_2$-1,2-C$_2$B$_9$H$_{10}$)(1',2'-C$_2$B$_9$H$_{11}$) (0.18 mmol, 77 mg) was dissolved in 8 mL of THF. Then, the new solution was transferred to the suspension mixture and the reaction mixture stirred 2 h under reflux at inert atmosphere. The solvent was evaporated, and an extraction took place. The organic layer was evaporated to give a mixture of white and orange compounds. The 8-{3,3'-Co(8-C$_4$H$_8$O$_2$-1,2-C$_2$B$_9$H$_{10}$)(1',2'-C$_2$B$_9$H$_{11}$) was recuperated and the new isomer 9,10-(CH$_3$CH=CH)$_2$-1,7-*closo*-C$_2$B$_{10}$H$_{10}$, 12, was obtained with 80% of yield (16 mg). ATR: ν = 3046 (C$_c$-H), 2998–2849 (m, υs(CH$_3$, =CH)), 2623, 2592 (vs, υs(B-H)), 1634, 1442 (vs, υ$_s$(C=C)), 978 (s, υ$_{as}$(-CH=CH-)). ^{1}H-NMR (300.13 MHz, CDCl$_3$) δ: 5.90 (m, 2H, CH$_3$C*H*=CH), 5.55 (m, 2H, CH$_3$CH=C*H*), 2.82 (s, 2H, C$_c$-*H*), 1.78 (dd, ^3J(H$_a$,H$_b$) = 6.3, ^4J(H$_a$,H$_c$) = 1.6, 6H, C*H$_3$*CH=CH). ^{1}H{^{11}B} NMR (300.13 MHz, CDCl$_3$) δ: 5.90 (m, 2H, CH$_3$C*H*=CH), 5.57 (m, 2H, CH$_3$CH=C*H*), 2.82 (s, 2H, C$_c$-*H*), 2.50 (s, 2H, B(5,12)-*H*), 2.34 (s, 2H, B(2,3)-*H*), 2.19 (s, 4H, B(4,6,8,11)-*H*), 1.78 (d, 6H, C*H$_3$*CH=CH). ^{11}B-NMR (96.29 MHz, (CD$_3$)$_2$CO) δ: −0.5 (s, 2B, B(9,10)), −5.8 (d,^1J(B,H) = 159, 2B, B(5,12)), −12.5 (d, ^1J(B,H) = 163, 4B, B(4,6,8,11)), −19.8 (d, ^1J(B,H) = 180, B(2,3)). Good crystals suitable for X-ray diffraction were grown in acetone.

*3.2. X-ray Structure Determinations of 9,10-(HOCH$_2$CH$_2$CH$_2$)$_2$-1,7-closo-C$_2$B$_{10}$H$_{10}$, 4
and 9,10-(CH$_3$CH=CH)$_2$-1,7-closo-C$_2$B$_{10}$H$_{10}$, 12*

Single-crystal data collections for **4** and **12** were performed with an Bruker D8 QUEST ECO three-circle diffractometer system equipped with a Ceramic x-ray tube (Mo Kα, λ = 0.71076 Å) and a doubly curved silicon crystal Bruker Triumph monochromator (Bruker, Karlsruhe, Germany). The structures were solved by direct methods and refined on F^2 by the SHELXL97 program [65]. The non-hydrogen atoms were refined with anisotropic displacement parameters. The hydrogen atoms were treated as riding atoms using the SHELXL97 default parameters. The crystallographic, structure refinement, and bond parameters for **4** and **10** are reported in CIF-files deposited at CCDC with the reference numbers CCDC 2004945 and 2004946. These data can be obtained free of charge via www.ccdc.cam.ac.uk/conts/retrieving.html (or from the Cambridge Crystallographic Data Centre, 12 Union Road, Cambridge CB2 1EZ, U.K.; fax: +44-1223-336033; or e-mail: deposit@ccdc.cam.ac.uk).

3.3. Hirshfeld Surface Analysis

The Hirshfeld surface analyses were run using the CIF format by the CrystalExplorer program [62]. Hirshfeld surface analysis help to recognize the strong and weak intermolecular interactions area and the nature of these interactions from the electron distribution. The d$_{norm}$ (normalized contact distance) is given by the Equation (1):

$$d_{norm} = \frac{d_i + r_i^{vdw}}{r_i^{vdw}} + \frac{d_e + r_e^{vdw}}{r_e^{vdw}} \qquad (1)$$

where d$_i$ is from the Hirshfeld surface to the nearest atom outside-external, d$_e$ from the Hirshfeld surface to the nearest internal atom, and r^{vdw} is the Van Der Walls radii of the atom

4. Conclusions

All allyl di and tetrabranched derivatives of the *m*-carborane framework have been synthesized. The starting 9,10-(allyl)$_2$-1,7-*closo*-carborane compound was made by Kumada cross-coupling reaction on 9,10-I$_2$-1,7-*closo*-carborane with allyl Grignard reagent in the presence of Pd(II) and Cu(I) as catalysts. These olefin groups have led to a variety of functional groups, alcohol, chloro, tosyl, and azide that have permitted to produce esters and 1,2,3-triazoles by the azide-alkyne cycloaddition, as examples of reactions that show the wide possibilities of this globular icosahedral *m*-carborane to act as a novel core for periphery-decorated macromolecules. Importantly, the four branches in the tetrabranched *m*-carborane derivatives are located in two perpendicular planes and are coplanar in the *o*-carborane isomer. This difference provides novel cores for 3D and 2D radially grown periphery-decorated macromolecules, respectively. Unexpectedly, the isomerization of B-allyl to B-propenyl vertexes in 9,10-(allyl)$_2$-1,7-*closo*-C$_2$B$_{10}$H$_{10}$ was observed in THF. DFT calculation studies conclude that the comparable acidity of the allyl groups and the C$_c$-H of the *m*-carborane unit allows a deprotonation/protonation isomerization of the allyl group as it is well known for allylbenzenes. X-ray crystal structures of 9,10-(OHCH$_2$CH$_2$CH$_2$)$_2$-1,7-*closo*-C$_2$B$_{10}$H$_{10}$ and 9,10-(CH$_3$CHCH)$_2$-1,7-*closo*-C$_2$B$_{10}$H$_{10}$ compounds show an extensive network of hydrogen bonding and π···H-C$_c$ contacts, respectively, due to the presence of alcohol and olefin groups that have been analyzed by Hirshfeld surfaces and decomposed fingerprint plots.

Supplementary Materials: The following are available online, Spectroscopic characterization of compounds 1–12. Figures S1–S78 with IR and NMR spectra and crystal packing of 4 and 12. Table S1–S4 containing the bond lengths and bond angles of crystals 4 and 12; S5-S11 with XYZ coordinates and total energies of the investigated systems.

Author Contributions: Conceptualization, C.V.; methodology, C.V., and I.B.; computational Studies, Z.K.; writing—original draft preparation, I.B., and C.V.; writing—review and editing, I.B., C.V., F.T., and Z.K.; supervision, C.V.; project administration, C.V.; funding acquisition, C.V. All authors have read and agreed to the published version of the manuscript.

Funding: This research was funded by the Spanish MINECO, grant number CTQ2016-75150-R, and Generalitat de Catalunya, grant number 2017 SGR 1720. C.V. and Z.K. thanks European Union's Horizon 2020 Marie Skłodowska-Curie grant agreement MSCA-IF-2016-751587.

Acknowledgments: Dedicated to Todd Marder, who significantly contributed to the Boron chemistry, on his 65th birthday.

Conflicts of Interest: The authors declare no conflict of interest.

References

1. Poater, J.; Solà, M.; Viñas, C.; Teixidor, F. π Aromaticity and Three-Dimensional Aromaticity: Two sides of the Same Coin? *Angew. Chem. Int. Ed.* **2014**, *53*, 12191–12195. [CrossRef] [PubMed]
2. Poater, J.; Viñas, C.; Bennour, I.; Escayola, S.; Solà, M.; Teixidor, F. Too Persistent to Give Up: Aromaticity in Boron Clusters Survives Radical Structural Changes. *J. Am. Chem. Soc.* **2020**, *142*, 9396–9407. [CrossRef] [PubMed]
3. Teixidor, F.; Barbera, G.; Vaca, A.; Kivekäs, R.; Sillanpää, R.; Oliva, J.; Viñas, C. Are Methyl Groups Electron-Donating or Electron-Withdrawing in Boron Clusters? Permethylation of *o*-Carborane. *J. Am. Chem. Soc.* **2005**, *127*, 10158–10159. [CrossRef] [PubMed]
4. Teixidor, F.; Núñez, R.; Viñas, C.; Sillanpää, R.; Kivekäs, R. A discrete P···I-I center dot center dot center dot P assembly: The large influence of weak interactions on the P-31 NMR spectra of phosphane-diiodine complexes. *Angew. Chem. Int. Ed.* **2000**, *39*, 4290–4292. [CrossRef]
5. Spokoyny, A.M.; Machan, C.W.; Clingerman, D.J.; Rosen, M.S.; Wiester, M.J.; Kennedy, R.D.; Stern, C.L.; Sarjeant, A.A.; Mirkin, C.A. A coordination chemistry dichotomy for icosahedral carborane-based ligands. *Nat. Chem.* **2011**, *3*, 590–596. [CrossRef] [PubMed]
6. Scholz, M.; Hey-Hawkins, E. Carbaboranes as Pharmacophores: Properties, Synthesis, and Application Strategies. *Chem. Rev.* **2011**, *111*, 7035–7062. [CrossRef]
7. Plesek, J. Potential applications of the boron cluster compounds. *Chem. Rev.* **1992**, *92*, 269–278. [CrossRef]
8. Grimes, R.N. *Carboranes*, 3rd ed.; Elsevier Inc.: New York, NY, USA, 2016.
9. Issa, F.; Kassiou, M.; Rendina, L.M. Boron in Drug Discovery: Carboranes as Unique Pharmacophores in Biologically Active Compounds. *Chem. Rev.* **2011**, *111*, 5701–5722. [CrossRef]
10. Hermansson, K.; Wojcik, M.; Sjoberg, S. *o*-, *m*-, and *p*-carboranes and their anions: Ab initio calculations of structures, electron affinities, and acidities. *Inorg. Chem.* **1999**, *38*, 6039–6048. [CrossRef]
11. Olid, D.; Núñez, R.; Viñas, C.; Teixidor, F. Methods to produce B-C, B-P, B-N and B-S bonds in boron clusters. *Chem Soc Rev.* **2013**, *42*, 3318–3336. [CrossRef]
12. Quan, Y.; Xie, Z.W. Controlled functionalization of *o*-carborane via transition metal catalyzed B-H activation. *Chem. Rev.* **2019**, *48*, 3660–3673. [CrossRef] [PubMed]
13. Zhang, X.; Zheng, H.; Li, J.; Xu, F.; Zhao, J.; Yan, H. Selective Catalytic B–H Arylation of *o*-Carboranyl Aldehydes by a Transient Directing Strategy. *J. Am. Chem. Soc.* **2017**, *139*, 14511–14517. [CrossRef] [PubMed]
14. Teixidor, F.; Sillanpää, R.; Pepiol, A.; Lupu, M.; Viñas, C. Synthesis of Globular Precursors. *Chem. Eur. J.* **2015**, *21*, 12778–12786. [CrossRef] [PubMed]
15. Teixidor, F.; Pepiol, A.; Viñas, C. Synthesis of Periphery-Decorated and Core-Initiated Borane Polyanionic Macromolecules. *Chem. Eur. J.* **2015**, *21*, 10650–10653. [CrossRef]

16. Kelemen, Z.; Pepiol, A.; Lupu, M.; Sillanpää, R.; Hänninen, M.M.; Teixidor, F.; Viñas, C. Icosahedral carboranes as scaffolds for congested regioselective polyaryl compounds: the distinct distance tuning of C–C and its antipodal B–B. *Chem. Commun.* **2019**, *55*, 8927–8930. [CrossRef]
17. Janczak, S.; Olejniczak, A.; Balabańska, S.; Chmielewski, M.K.; Lupu, M.; Viñas, C.; Lesnikowski, Z.J. Boron Clusters as a Platform for New Materials: Synthesis of Functionalized *o*-Carborane ($C_2B_{10}H_{12}$) Derivatives Incorporating DNA Fragments. *Chem. Eur. J.* **2015**, *21*, 15118–15122. [CrossRef]
18. Kaniowski, D.; Ebenryter-Olbinska, K.; Kulik, K.; Janczak, S.; Maciaszek, A.; Bednarska-Szczepaniak, K.; Nawrot, B.; Lesnikowski, Z. Boron clusters as a platform for new materials: composites of nucleic acids and oligofunctionalized carboranes (C2B10H12) and their assembly into functional nanoparticles. *Nanoscale* **2020**, *12*, 103–114. [CrossRef]
19. Ochi, J.; Tanaka, K.; Chujo, Y. Recent Progress in the Development of Solid-State Luminescent *o*-Carboranes with Stimuli Responsivity. *Angew. Chem. Int. Ed.* **2020**. [CrossRef]
20. Wang, S.; Blaha, C.; Santos, R.; Huynh, T.; Hayes, T.R.; Beckford-Vera, D.R.; Blecha, J.E.; Hong, A.S.; Fogarty, M.; Hope, T.A.; et al. Synthesis and Initial Biological Evaluation of Boron-Containing Prostate-Specific Membrane Antigen Ligands for Treatment of Prostate Cancer Using Boron Neutron Capture Therapy. *Mol. Pharmaceutics.* **2019**, *16*, 3831–3841. [CrossRef]
21. Hosmane, N.S. Boron Science. In *New Technologies and Applications*; CRC Press: Boca Raton, FL, USA, 2012.
22. Hosmane, N.S.; Eagling, R. *Handbook of Boron Chemistry in Organometallics, Catalysis, Materials and Medicine*; World Science Publishers: Hackensack, NJ, USA, 2018.
23. Hey-Hawkins, E.; Viñas Teixidor, C. *Boron-Based Compounds: Potential and Emerging applications in Medicine*; John Wiley & Sons Ltd: Chichester, UK, 2018.
24. Fisher, S.P.; Tomich, A.W.; Lovera, S.O.; Kleinsasser, J.F.; Guo, J.; Asay, M.J.; Nelson, H.M.; Lavallo, V. Nonclassical Applications of *closo*-Carborane Anions: From Main Group Chemistry and Catalysis to Energy Storage. *Chem. Rev.* **2019**, *119*, 8262–8290. [CrossRef]
25. Barth, R.F.; Vicente, M.G.H.; Harling, O.K.; Kiger, W.S.; Riley, K.J.; Binns, P.J.; Wagner, F.M.; Suzuki, M.; Aihara, T.; Kato, I.; et al. Current status of boron neutron capture therapy of high grade gliomas and recurrent head and neck cancer. *Radiat. Oncol.* **2012**, *7*, 146. [CrossRef] [PubMed]
26. Kashin, A.N.; Butin, K.P.; Stanko, V.I.; Beletskaya, I.P. Acidity of *ortho*-, *meta*-, and *para*-barenes. *Russ. Chem. Bull.* **1969**, *18*, 1775–1777. [CrossRef]
27. Gan, L.; Fonquernie, P.G.; Light, M.E.; Norjmaa, G.; Ujaque, G.; Choquesillo-Lazarte, D.; Fraile, J.; Teixidor, F.; Viñas, C.; Planas, J.G. A Reversible Phase Transition of 2D Coordination Layers by B–H···Cu(II) Interactions in a Coordination Polymer. *Molecules* **2019**, *24*, 3204. [CrossRef]
28. Cabrera-González, J.; Chaari, M.; Teixidor, F.; Viñas, C.; Núñez, R. Blue Emitting Star-Shaped and Octasilsesquioxane-Based Polyanions Bearing Boron Clusters. Photophysical and Thermal Properties. *Molecules* **2020**, *25*, 1210. [CrossRef]
29. Mori, S.; Takagaki, R.; Fujii, S.; Urushibara, K.; Tanatani, A.; Kagechika, H. Novel Non-steroidal Progesterone Receptor Ligands Based on *m*-Carborane Containing a Secondary Alcohol: Effect of Chirality on Ligand Activity. *Chem. Pharm. Bull.* **2017**, *65*, 1051–1057. [CrossRef] [PubMed]
30. Eleazer, B.J.; Smith, M.D.; Popov, A.A.; Peryshkov, D.V. Expansion of the (BB) Ru metallacycle with coinage metal cations: Formation of B-M-Ru-B (M = Cu, Ag, Au) dimetalacyclodiboryls. *Chem. Sci.* **2018**, *9*, 2601–2608. [CrossRef]
31. Eleazer, B.J.; Smith, M.D.; Popov, A.A. Peryshkov, Dmitry V. Rapid reversible borane to boryl hydride exchange by metal shuttling on the carborane cluster surface. *Chem. Sci.* **2017**, *9*, 2601–2608. [CrossRef]
32. Eleazer, B.J.; Smith, M.D.; Peryshkov, D.V. Metal- and Ligand-Centered Reactivity of meta-Carboranyl-Backbone Pincer Complexes of Rhodium. *Organometallics* **2016**, *35*, 106–112. [CrossRef]
33. Dziedzic, R.M.; Martin, J.L.; Axtell, J.C.; Saleh, L.M.A.; Ong, T.-C.; Yang, Y.-F.; Messina, M.S.; Rheingold, A.L.; Houk, K.N.; Spokoyny, A.M. Cage-Walking: Vertex Differentiation by Palladium-Catalyzed Isomerization of B(9)-Bromo-*meta*-Carborane. *J. Am. Chem. Soc.* **2017**, *139*, 7729–7732. [CrossRef]
34. Himmelspach, A.; Finze, M. Dicarba-*closo*-dodecaboranes with One and Two Ethynyl Groups Bonded to Boron. *Eur. J. Inorg. Chem.* **2010**, 2012–2024. [CrossRef]
35. Ohta, K.; Endo, Y. Chemistry of boron clusters, carboranes synthesis, structure and application for molecular construction. *J. Synth. Org. Chem. Jpn.* **2007**, *65*, 320–333. [CrossRef]

36. Bayer, M.J.; Herzog, A.; Diaz, M.; Harakas, G.A.; Lee, H.; Knobler, C.B.; Hawthorne, M.F. The Synthesis of Carboracycles Derived from B,B_Bis(aryl) Derivatives of Icosahedral *ortho*-Carborane. *Chem. Eur. J.* **2003**, *9*, 2732–2744. [CrossRef] [PubMed]
37. Jiang, W.; Chizhevsky, I.T.; Mortimer, M.D.; Chen, W.; Knobler, C.B.; Johnson, S.E.; Gomez, F.A.; Hawthorne, M.F. Carboracycles: Macrocyclic Compounds Composed of Carborane Icosahedra Linked by Organic Bridging Groups. *Inorg. Chem.* **1996**, *35*, 5417–5426. [CrossRef] [PubMed]
38. Zheng, Z.; Jiang, W.; Zinn, A.A.; Knobler, C.B.; Hawthorne, M.F. Facile Electrophilic Iodination of Icosahedral Carboranes. Synthesis of Carborane Derivatives with Boron-Carbon Bonds via the Palladium-Catalyzed Reaction of Diiodocarboranes with Grignard Reagents. *Inorg. Chem.* **1995**, *34*, 2095–2100. [CrossRef]
39. Fox, M.A. Icosahedral Carborane Derivatives. Ph.D. Thesis, Durham University, Durham, UK, 1991. Available online: https://www.dur.ac.uk/chemistry/fox.group/publications/phd.thesis/ (accessed on 17 June 2020).
40. Ol'shevskaya, V.A.; Makarenkov, A.V.; Kononova, E.G.; Peregudov, A.S.; Lyssenko, K.A.; Kalinin, V.N. An efficient synthesis of carboranyl tetrazoles via alkylation of 5-R-1H-tetrazoles with allylcarboranes. *Polyhedron* **2016**, *115*, 128–136. [CrossRef]
41. Neises, B.; Steglich, W. Simple Method for the Esterification of Carboxylic Acids. *Angew. Chem. Int. Ed.* **1978**, *17*, 522–524. [CrossRef]
42. Neises, B.; Steglich, W. Esterification of carboxylic acids with dicyclohexylcarbodiimide/4-dimethylaminopyridine: tert-butyl ethyl fumarate. *Org. Synth.* **1985**, *63*, 183.
43. Whitaker, D.T.; Whitaker, K.S.; Johnson, C.R.; Haas, J. *p-Toluenesulfonyl Chloride In Encyclopedia of Reagents for Organic Synthesis*; John Wiley: New York, NY, USA, 2006.
44. Farràs, P.; Cioran, A.M.; Šícha, V.; Teixidor, F.; Štíbr, B.; Grüner, B.; Viñas, C. Toward the Synthesis of High Boron Content Polyanionic Multicluster Macromolecules. *Inorg. Chem.* **2009**, *48*, 8210–8219. [CrossRef]
45. Hassam, M.; Taher, A.; Arnott, G.E.; Green, I.R.; van Otterlo, W.A.L. Isomerization of Allylbenzenes. *Chem. Rev.* **2015**, *115*, 5462–5569. [CrossRef]
46. Reutov, O.A.; Beletskaya, I.P.; Butkin, K.P. *CH-Acids*; Pergamon Press: Oxford, UK, 1978; pp. 13–14, 29, 34, 123–124.
47. Shatenshtein, A.I.; Zakharkin, L.I.; Petrov, E.S.; Yakovleva, E.A.; Yakushin, F.S.; Matic-Vukmirovic, Z.B.; Isaeva, G.G.; Kalinin, V.N. The equilibrium and kinetic acidities of isomeric carborane methines. *J. Organomet. Chem.* **1970**, *23*, 313–322. [CrossRef]
48. Petrov, E.S.; Yakovleva, E.A.; Isaeva, G.G.; Kalinin, V.N.; Zakharkin, L.I.; Shatenshtein, A.I. Thermodynamic and kinetic acidity of the CH bonds of certain ortho- and meta-barenes. *Russ. Chem. Bull.* **1969**, *18*, 1576–1582. [CrossRef]
49. Popescu, A.-R.; Musteti, A.D.; Ferrer-Ugalde, A.; Viñas, C.; Núñez, R.; Teixidor, F. Influential Role of Ethereal Solvent on Organolithium Compounds: The Case of Carboranyllithium. *Chem. Eur. J.* **2012**, *18*, 3174–3184. [CrossRef]
50. Farràs, P.; Viñas, C.; Teixidor, F. Preferential chlorination vertices in cobaltabisdicarbollide anions. Substitution rate correlation with site charges computed by the two atoms natural population analysis method (2a-NPA). *J. Organomet. Chem.* **2013**, *747*, 119–125.
51. Potenza, J.A.; Lipscomb, W.M.; Vickers, G.D.; Schroeder, H. Order of Electrophilic Substitution in 1,2-Dicarbaclovododecaborane(12) and Nuclear Magnetic Resonance Assignment. *J. Am. Chem. Soc.* **1966**, *88*, 628–629. [CrossRef]
52. Barberà, G.; Vaca, A.; Teixidor, F.; Sillanpää, R.; Kivekäs, R.; Viñas, C. From Mono- to Poly-Substituted Frameworks: A Way of Tuning the Acidic Character of Cc-H in *o*-Carborane Derivatives. *Chem. Eur. J.* **2009**, *15*, 9755–9763.
53. Cowie, J.; Reid, B.D.; Watmough, J.M.S.; Welch, A.J. Steric effects in heteroboranes. Part 7: The synthesis and characterisation of arene-ruthenium complexes of C-substituted carbaboranes. Molecular structures of 1-Ph-3-(mes)-3,1,2-*closo*-RuC$_2$B$_9$H$_{10}$ (mes = C$_6$H$_3$-1,3,5) and 1-Ph-2-Me-3-(p-cym)-3,1,2-*closo*-RuC$_2$B$_9$H$_9$ (p-cym = C$_6$H$_4$Me-1-iPr-4), the latter showing an incipient deformation. *J. Organomet. Chem.* **1994**, *481*, 283–293.
54. Venable, T.L.; Hutton, W.C.; Grimes, R.N. Atom connectivities in polyhedral boranes elucidated via two-dimensional J-correlated boron-11-boron-11 FT NMR: a general method. *J. Am. Chem. Soc.* **1982**, *104*, 4716–4717. [CrossRef]

55. Venable, T.L.; Hutton, W.C.; Grimes, R.N. Two-dimensional boron-11-boron-11 nuclear magnetic resonance spectroscopy as a probe of polyhedral structure: Application to boron hydrides, carboranes, metallaboranes, and metallacarboranes. *J. Am. Chem. Soc.* **1984**, *106*, 29–37. [CrossRef]
56. Todd, L.J.; Siedle, A.R.; Bodner, G.M.; Kahl, S.B.; Hickey, J.P. An NMR Study of Icosahedral Heteroatom Borane Derivatives. *J. Magn. Reson.* **1976**, *23*, 301–311. [CrossRef]
57. Bruno, J.; Cole, J.C.; Edgington, P.R.; Kessler, M.; Macrae, C.F.; McCabe, P.; Pearson, J.; Taylor, R. New software for searching the Cambridge Structural Database and visualizing crystal structures. *Acta Crystallogr.* **2002**, *B58*, 389–397. [CrossRef]
58. Fox, M.A.; Hughes, A.K. Cage C-H···X interactions in solid-state structures of icosahedral carboranes. *Coord. Chem. Rev.* **2004**, *248*, 457–476. [CrossRef]
59. Pauling, L. *The Nature of the Chemical Bond*, 3rd ed.; Cornell University Press: Ithaca, NY, USA, 1960.
60. Nishio, M. CH/π hydrogen bonds in crystals. *Cryst. Eng. Comm.* **2004**, *6*, 130–158. [CrossRef]
61. Turner, M.J.; McKinnon, J.J.; Wolff, S.K.; Grimwood, D.J.; Spackman, P.R.; Jayatilaka, D.; Spackman, M.A. *CrystalExplorer17*; The University of Western Australia: Perth, Australia, 2017.
62. Wolff, S.K.; Grimwood, D.J.; McKinnon, J.J.; Jayatilaka, D.; Spackman, M.A. *Crystal Explorer 3.0*; University of Western Australia: Perth, Australia, 2007.
63. Pangajavalli, S.; Ranjithkumar, R.; Ramaswamy, S. Structural, Hirshfeld, Spectroscopic, Quantum Chemical and Molecular docking Studies on 6b′,7′,8′,9′-Tetrahydro-2H,6′H-spiro[acenaphthylene-1,11′-chromeno[3,4-a]pyrrolizine]-2,6′(6a′H,11a′H)-dione. *J. Mol. Struct.* **2020**, *1209*, 127921. [CrossRef]
64. Bojarska, J.; Remko, M.; Fruziński, A.; Maniukiewicz, W. The experimental and theoretical landscape of a new antiplatelet drug ticagrelor: Insight into supramolecular architecture directed by C-H ... F, π...π and C-H...π interactions. *J. Mol. Struct.* **2018**, *1154*, 290–300. [CrossRef]
65. Sheldrick, G.M. A short history of SHELX. *Acta Cryst.* **2008**, *A64*, 112–122. [CrossRef]

Sample Availability: Samples of the compounds are available from the authors.

© 2020 by the authors. Licensee MDPI, Basel, Switzerland. This article is an open access article distributed under the terms and conditions of the Creative Commons Attribution (CC BY) license (http://creativecommons.org/licenses/by/4.0/).

Article

Ruthenacarborane–Phenanthroline Derivatives as Potential Metallodrugs

Martin Kellert [1], Imola Sárosi [1], Rajathees Rajaratnam [2], Eric Meggers [2], Peter Lönnecke [1] and Evamarie Hey-Hawkins [1],*

[1] Institute of Inorganic Chemistry, Faculty of Chemistry and Mineralogy, Leipzig University, Johannisallee 29, 04103 Leipzig, Germany; martin.kellert@uni-leipzig.de (M.K.); sarosi.imola@gmail.com (I.S.); loennecke@uni-leipzig.de (P.L.)
[2] Fachbereich Chemie, Philipps-Universität Marburg, Hans-Meerwein Straße 4, 35043 Marburg, Germany; Rajathees@Rajaratnam.de (R.R.); meggers@chemie.uni-marburg.de (E.M.)
* Correspondence: hey@uni-leipzig.de; Tel.: +49-341-97-36151

Academic Editors: Ashok Kakkar and Stéphane Bellemin-Laponnaz
Received: 10 April 2020; Accepted: 12 May 2020; Published: 15 May 2020

Abstract: Ruthenium-based complexes have received much interest as potential metallodrugs. In this work, four Ru^{II} complexes bearing a dicarbollide moiety, a carbonyl ligand, and a phenanthroline-based ligand were synthesized and characterized, including single crystal diffraction analysis of compounds **2**, **4**, and **5** and an observed side product **SP1**. Complexes **2–5** are air and moisture stable under ambient conditions. They show excellent solubility in organic solvents, but low solubility in water.

Keywords: dicarbollide; ruthenium; metallodrug; kinase inhibitor

1. Introduction

Metal-containing compounds are of increasing interest for applications in medicinal chemistry due to their diverse coordination geometries, unusual reactivities, and useful physicochemical properties [1–3]. Ruthenium shows a low general toxicity [4] and is an excellent metal for this approach. Ru^{II} is usually coordinated in an octahedral or pseudo-octahedral half-sandwich fashion and forms quite stable coordinative or covalent Ru–ligand bonds, which affects cellular metabolism. Furthermore, the reactivity of ruthenium ions is well-known; thus, reactions are predictable and facilitate drug design [5–8]. To date, predominantly ruthenium half-sandwich complexes have been developed as potential anti-cancer agents, antiproliferative drugs, antibiotics, and immunosuppressants [9–14].

Meggers et al., recently reported an interesting class of organometallic protein kinase inhibitors which were inspired by the alkaloid staurosporine (Figure 1) [4,15,16]. For example, the ruthenium half-sandwich complex DW12 is a potent inhibitor of glycogen synthase kinase 3 (GSK-3), whereas staurosporine constitutes a very unselective inhibitor of a large number of protein kinases [5,16–18]. In staurosporine, the important groups for the interaction with protein kinase are the lactam unit, the indolocarbazole heterocycle, and the carbohydrate moiety. In DW12, the ruthenium serves as a purely structural center and enables a geometry, which cannot be easily achieved with purely organic molecules. Thus, the NH group of the maleimide moiety forms a hydrogen bond with the amide carbonyl of Asp133 in the adenosine triphosphate (ATP) binding site of GSK-3. Additionally, one carbonyl group of the maleimide moiety interacts with the NH group of Val135, and the second carbonyl of the same moiety forms a water-mediated hydrogen bond with Asp200. The indole OH group interacts with the carbonyl amide group of Val135. One special feature of DW12 is the presence of the CO ligand, which exhibits a clearly reduced dipolar character due to the interaction with the transition metal ruthenium. Displaying this behavior, the CO ligand is able to undergo an unusual interaction mode with the glycine-rich loop of the ATP binding site. In that way, DW12 achieves a geometry, which seems to be optimal for the ATP binding site of GSK-3, rendering DW12 a more potent inhibitor for GSK-3 than staurosporine.

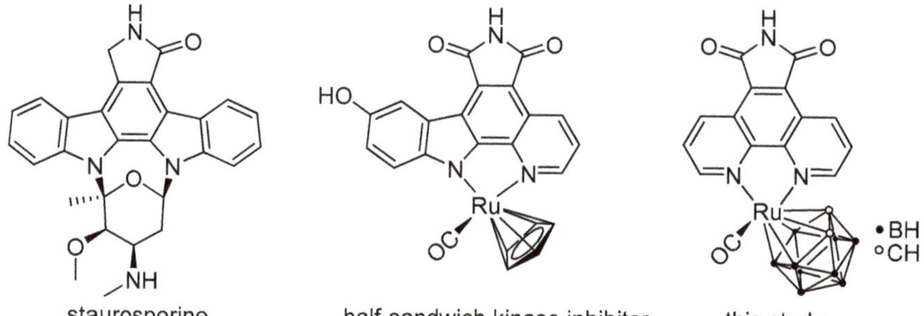

Figure 1. Staurosporine, a ruthenium half-sandwich staurosporine mimic (DW12), and the ruthenacarborane complex synthesized and investigated in this study [17].

In order to design novel potential drugs, bioisosteric replacement has become a wide-spread approach [11,14,19–23]. Thus, the development of drugs in which the carborane moiety mimics a phenyl group or is the pharmacophore itself is actively studied [23–29]. Applications of such carborane-containing drugs are for example cancer therapeutics and enzyme inhibitors [11,14,19,20,22,23,30–37]. Due to its isolobal relationship with the cyclopentadienyl ligand (Cp⁻) the dicarbollide anion (*nido*-carborate(2−), $C_2B_9H_{11}^{2-}$, Cb^{2-}) is a suitable replacement as ligand for transition metals [38,39]. However, distinct activities and reactivities of the respective metal complexes, bearing the Cb^{2-} or the Cp⁻, are observable and caused by the different charge, size, symmetry, and hybridization of the orbitals of the respective ligands [40]. Furthermore, the dicarbollide moiety is highly hydrophobic and could enhance the transport of the corresponding metallodrug across cellular membranes [19,20,22]. Additionally, these clusters are metabolically stable, which renders them robust compounds in biological media [11,23]. Furthermore, the possible regioselective introduction of specific substituents at either the carbon or the boron vertices of the cluster enables customization of the structure and, therefore, the activity of the metallodrug [19,21].

Former studies have shown the importance of the heteroaromatic bidentate pyridocarbazole and CO ligand in DW12 and related complexes for mimicking staurosporine binding in the ATP binding site [5,17,18,41]. Therefore, the presence of a carbonyl ligand and an aromatic moiety are very important features.

In this work, we report the combination of Cb^{2-} with Ru^{II}(CO)–phenanthroline derivatives as mimics for DW12. To our knowledge, this approach to combine the scaffold of an active ruthenium-based protein kinase inhibitor with a dicarbollide moiety has not been pursued before. In DW12, the Cp⁻ ring points away from the ATP binding site towards the aqueous solvent. Thus, there should be sufficient space available in this part of the active site to accommodate larger moieties [42]. Replacing the Cp⁻ ligand with a much bulkier, hydrophobic Cb^{2-} ligand would allow additional van der Waals interactions to be formed with this part of the active site and to profit from the hydrophobic effect which often increases the potency of enzyme inhibitors. Due to the replacement of the cyclopentadienyl ligand with a dicarbollide ligand, the anionic N,N ligand in DW12 must be substituted with a neutral one. As the maleimide moiety in the N,N ligand in DW12 is involved in various hydrogen bond interactions and thus plays an essential role in binding to the ATP binding site of GSK-3, the design of the novel dicarbollide-containing complex should also employ this motif. Therefore, the neutral 5*H*-pyrrolo[3,4-f][1,10]phenanthroline-5,7(6*H*)-dione, which is similar to the anionic 9-hydroxy-5*H*-12λ²-pyrido[2,3-*a*]pyrrolo[3,4-*c*]carbazole-5,7(6*H*)-dione ligand (pyridocarbazole) in DW12, was used as ligand. In combination with the much more hydrophobic and bulkier Cb^{2-} ligand, this should result in increased activity of the respective complex.

2. Results and Discussion

2.1. Ligand and Precursor Syntheses

The precursor 7,8-dicarba-*nido*-undecaborane(13) (**L1**) for the dicarbollide moiety was synthesized according to the literature (Scheme 1) [43]. Details about the synthetic procedure are given in the electronic supplementary information.

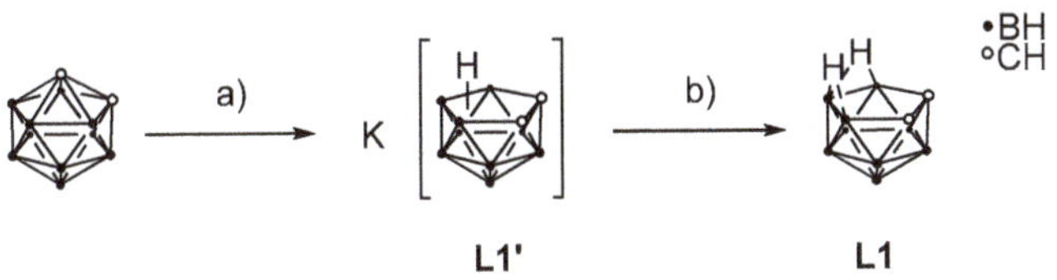

Scheme 1. Two-step synthesis of 7,8-dicarba-*nido*-undecaborane(13) (**L1**). a) MeOH, KOH, 80 °C, 18 h; b) C_6H_6, H_3PO_4, rt, 17 h. Yield over two steps: 70% [43].

The phenanthroline derivatives (Figure 2) that were employed in complexation reactions were prepared according to the literature. Even though 5-nitro-1,10-phenanthroline (**L2**) can be synthesized in excellent yields, it was obtained commercially because of the harsh conditions employed in the synthesis [44]. 1,10-Phenanthroline-5,6-dione (**L3**) was prepared according to the literature [45,46], and 1,10-phenanthrolinopyrrole (**L4**) was formed in a Barton-Zard reaction from 5-nitro-1,10-phenanthroline (**L2**) and ethyl isocyanoacetate under basic conditions followed by hydrolysis of the ester **L4'** (Scheme 2) [47–54] The respective synthetic procedure for **L4** is given in the electronic supplementary information.

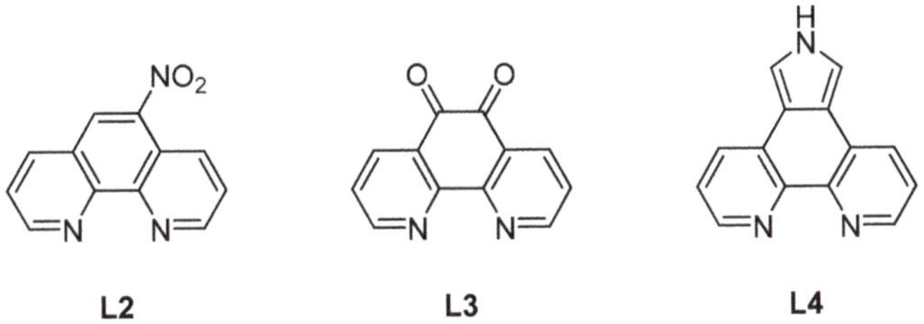

Figure 2. Phenanthroline-based ligands **L2–L4**.

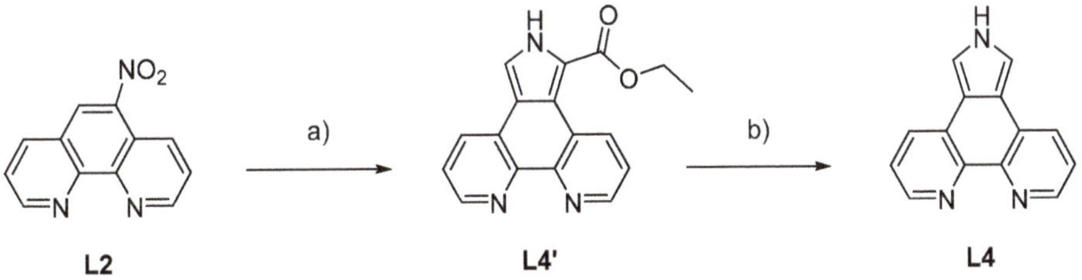

Scheme 2. Preparation of **L4** starting from 5-nitro-1,10-phenanthroline (**L2**) with 1,10-phenanthrolinopyrrole ethyl ester (**L4'**) as intermediate. a) THF, DBU, ethyl isocyanoacetate, rt, 20 h; b) EtOH, 0.2 M NaOH, 110 °C, 8 h. Yield over two steps: 54%; DBU: 1,8-diazabicyclo[5.4.0]undec-7-ene [49].

The precursor molecule for further reactions, [3-$(CO)_3$-*closo*-3,1,2-Ru$C_2B_9H_{11}$] (**1**), was prepared in a redox reaction from triruthenium dodecacarbonyl and **L1** (Scheme 3) [55–57]. The respective synthetic procedure of **1** is given in the electronic supplementary information. Studies showed that the dicarbollide ligand is a much stronger ligand than the Cp ligand [11,39].

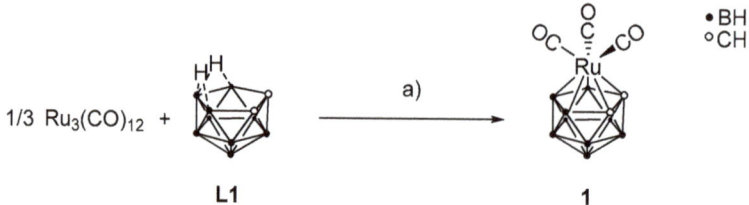

Scheme 3. Preparation of **1** with triruthenium dodecacarbonyl and **L1**. a) C_6H_6, 90 °C, 7 h, 60% [55].

2.2. Synthesis of Ruthenium(II) Complexes

The complexes containing phenanthroline derivatives as ligands were prepared via a ligand exchange reaction following a procedure of Jellis and co-workers for other chelating N-donor ligands (Scheme 4) [58]. In this type of reaction, two carbonyl ligands are oxidized to CO_2 with stoichiometric amounts of trimethylamine N-oxide [59] to facilitate coordination of one bidentate phenanthroline derivative. For complex **2**, both enantiomers *R* and *S* were obtained (Ru is the chiral center). No further investigations were carried out to determine the ratio of the two enantiomers, which is assumed to be close to 1:1.

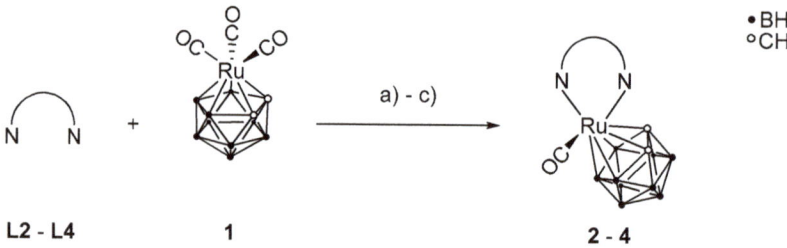

Scheme 4. Preparation of compounds **2**–**4** from the phenanthroline ligands **L2**–**L4** and the ruthenium(II) complex **1**. a) MeCN, trimethylamine N-oxide, rt, 17 h, 13% **2**; b) MeCN, trimethylamine N-oxide, rt, 84 h, 73% **3**; c) MeCN, trimethylamine N-oxide, rt, 48 h, 46% **4**.

Complexes **2**–**4** were characterized via NMR, IR, MS, and, in the case of compounds **2** and **4**, via single crystal X-ray diffraction. In the ^{1}H-NMR spectrum, compound **3** exhibits three doublets of doublets in the aromatic region as expected for the dione ligand **L3**. In the infrared spectrum only one absorption band for a C≡O stretching vibration was observed for **2**–**4** for the single remaining CO ligand. Additionally, in **2**, two absorption bands for the symmetric and asymmetric stretching vibration of the NO_2 group were observed. In the negative ESI-MS spectrum, compound **3** is observable with one additional bromide ion ([M + Br]$^-$, *m/z* = 552). No suitable single crystals could be obtained for complex **3**. The bifunctional ligand **L3** can exhibit several different binding modes: η^2-coordination with both nitrogen or both oxygen atoms. In **3**, coordination via the nitrogen atoms is assumed, as Ru^{2+} is a soft Lewis acid [60,61]. It is also possible for **L3** to act as a bridging ligand to form complexes with different or additional metal centers. In our case the formation of chain-like oligomers formed by linked complex fragments of **3** are not very likely due to the chosen synthetic procedure.

Crystals suitable for single crystal X-ray crystallography were obtained for compounds **2** and **4**. Compound **2** crystallizes from a mixture of dichloromethane and *n*-hexane as orange-red, plate-like crystals in the monoclinic space group $P2_1/n$ with one additional DCM molecule in the asymmetric unit. The solved structure shows a wR_2 value of 9.9% (R_1: 17.8%), which is caused by the low quality of the crystals. Nonetheless, the identity of compound **2** is unambiguous (Figure 3, left). Complex **4** crystallizes as yellow plates from a dichloromethane/*n*-hexane mixture with two independent molecules in the asymmetric unit (Figure 3, right, only one molecule is shown).

Bond lengths and bond angles of **2** and **4** are given in Table 1.

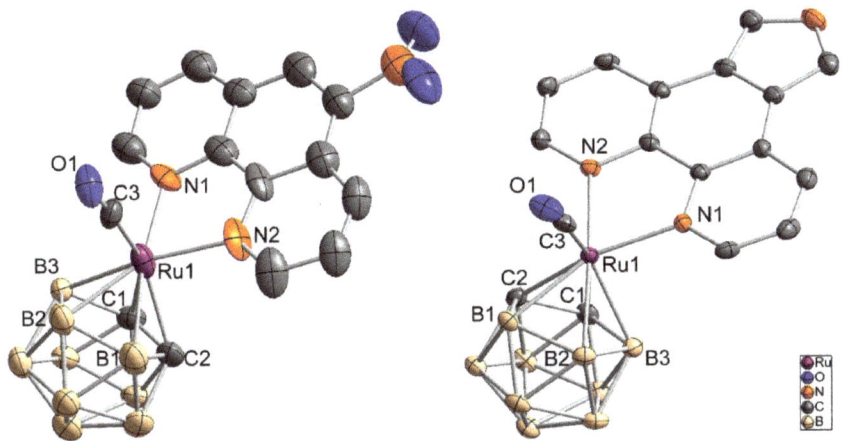

Figure 3. Molecule structure of [3-(CO)-3,3-(**L2**-κ²*N*,*N*)-*closo*-3,1,2-RuC₂B₉H₁₁] (**2**) (**left**, only the *S* enantiomer is shown) and [3-(CO)-3,3-(**L4**-κ²*N*,*N*)-*closo*-3,1,2-RuC₂B₉H₁₁] (**4**) (**right**) as an ellipsoid-stick model with thermal ellipsoids at 50% probability level. Solvent molecules and hydrogen atoms are omitted for clarity.

Table 1. Selected bond lengths (in pm) and angles (in °) in **2**, **4**, and **5** in comparison with [3-CO-3,3-(bipy-κ²*N*,*N*)-*closo*-3,1,2-RuC₂B₉H₁₁] (**I**) (bipy = 2,2′-bipyridine) [58].

Atom Group	2	4 [a]	5	I
Ru(1)–C(1)	226.5(9)	222.3(4) [225.0(4)]	221.3(2)	217.4(4)
Ru(1)–C(2)	222.7(9)	217.6(4) [223.4(49]	222.7(2)	222.4(4)
Ru(1)–B(1)	220(1)	222.4(4) [220.1(4)]	220.6(2)	227.9(4)
Ru(1)–B(2)	219(1))	227.3(4) [220.5(4)]	224.0(2)	227.7(4)
Ru(1)–B(3)	221.5(9)	223.4(4) [223.6(4)]	224.0(2)	220.6(5)
Ru(1)–C(3)	185(1)	185.6(4) [183.5(4)]	184.6(2)	186.6(4)
Ru(1)–N(1)	212.9(8)	210.5(3) [211.6(3)]	212.2(1)	209.3(3)
Ru(1)–N(2)	213.6(7)	212.4(3) [212.2(3)]	211.9(1)	213.5(3)
N(1)···N(2)	261(1)	262.5(5) [262.3(4)]	262.8(2)	261.0(5)
C(3)–O(1)	115(1)	114.7(4) [114.6(4)]	115.0(2)	115.4(4)
N(1)–Ru(1)–C(3)	90.9(4)	91.5(1) [94.9(2)]	93.9(1)	90.1(2)
N(2)–Ru(1)–C(3)	91.0(4)	95.5(1) [94.0(1)]	92.2(1)	92.0(1)
N(1)–Ru(1)–N(2)	75.5(3)	76.7(1) [76.5(1)]	76.6(1)	76.3(1)
Ru(1)–C(3)–O(1)	176.4(9)	173.4(3) [174.1(4)]	173.8(2)	175.3(4)

[a] values of the second independent molecule of **4** are given in [ ].

After the successful preparation of **4**, a selective oxidation of positions 5 and 7 of the phenanthrolinopyrrole moiety was carried out to prepare [3-(CO)-3,3-{1′,10′-NC₅H₃(C(CO)(NH)(CO)C)NC₅H₃-κ²*N*,*N*}-*closo*-3,1,2-RuC₂B₉H₁₁] (**5**) (Scheme 5). For this reaction, excess of *meta*-chloroperoxybenzoic acid (*m*-CPBA) was used. For stronger oxidizing agents, a polymerization of the pyrrole moiety is observed, and milder oxidizing agents lead to lactam scaffolds only and not to the desired maleimide groups [62,63]. Monitoring the reaction using thin layer chromatography showed that it is necessary to add the oxidizing agent successively in small portions during the reaction period. It was observed that only the *ortho* positions of the pyrrole moiety were oxidized, whereas the rest of the complex was unaffected.

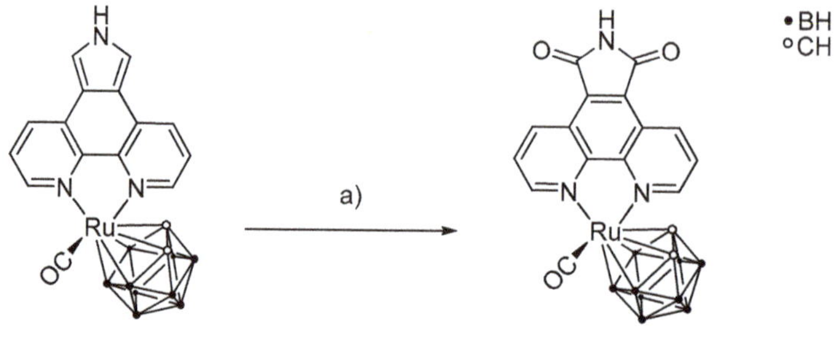

Scheme 5. Synthesis of [3-(CO)-3,3-{1′,10′-NC$_5$H$_3$(C(CO)(NH)(CO)C)NC$_5$H$_3$-κ^{2}N,N}-*closo*-3,1,2-RuC$_2$B$_9$H$_{11}$] (5) from 4 and *meta*-chloroperoxybenzoic acid (*m*-CPBA). a) MeCN, *m*-CPBA, 82 °C, 84 h, 10%.

After column chromatography and recovery of some starting material, the desired complex 5 was isolated in 10% yield as a red, crystalline powder and characterized. In the ^{1}H-NMR spectrum, a low-field shift for the cluster CH and the aromatic CH groups in comparison to 4 was observed. The multiplicity pattern of the aromatic protons indicates an *ortho* substitution of the pyrrole moiety. Two signals are observed in the negative-mode mass spectrum, namely [M − H$^+$]$^-$ and [M − CO − H$^+$]$^-$. The IR spectrum shows the presence of only one CO ligand as well as for the previous complexes. Compound 5 crystallizes from acetonitrile as deep red prisms in the triclinic space group $P\bar{1}$ with three acetonitrile molecules in the asymmetric unit (Figure 4). Bond lengths and bond angles of 5 are given in Table 1.

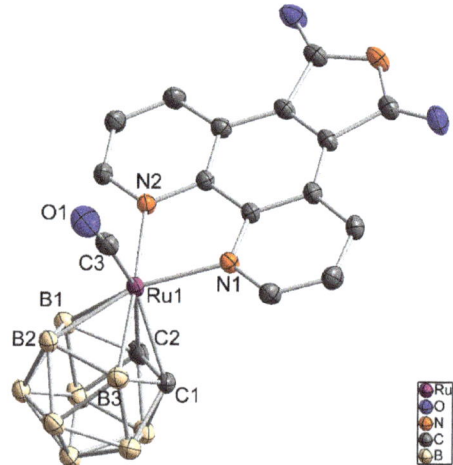

Figure 4. Molecule structure of [3-(CO)-3,3-{1′,10′-NC$_5$H$_3$(C(CO)(NH)(CO)C)NC$_5$H$_3$-κ^{2}N,N}-*closo*-3,1,2-RuC$_2$B$_9$H$_{11}$] (5) as an ellipsoid-stick model with thermal ellipsoids at 50% probability level. Solvent molecules and hydrogen atoms are omitted for clarity.

2.3. Biological Studies

Since the target molecule 5 was designed as a mimic of (*R*)-DW12, protein kinase Pim1 inhibition studies were carried out but were inconclusive, most likely due to a low solubility of 5 in buffer solutions caused by the high hydrophobicity of the dicarbollide moiety. Additional information about the protein kinase inhibition studies is given in the electronic supplementary information.

As mentioned in the introduction, carborane or metallacarborane derivatives are being studied as drugs or enzyme inhibitors [11,14,19,21,23,27,28,30,35]. Due to the presence of the hydrophobic carborane moiety, lack of solubility in aqueous media is often observed, hampering biological studies. Improved water-solubility was achieved by employing charged species with enhanced solubility [11,14,31,35] or incorporating hydrophilic side chains [23]. A novel approach, developed by Hey-Hawkins et al., overcomes the problematic solubility behavior of metallacarborane complexes by employing bovine serum albumin (BSA) [37]; however, it was also observed that BSA can influence the activity of the solubilized drug in specific enzyme inhibition assays.

3. Materials and Methods

All syntheses were carried out using the Schlenk technique and nitrogen as inert gas. Triruthenium dodecacarbonyl, 1,2-dicarba-*closo*-dodecaborane(12), 5-nitro-1,10-phenanthroline, ethyl isocyanoacetate, 1,8-diazabicyclo[5.4.0]undec-7-ene, trimethylamine N-oxide and *meta*-chloroperoxybenzoic acid are commercially available. Trimethylamine N-oxide was dried and purified by sublimation; *meta*-chloroperoxybenzoic acid was purified by washing with phosphate buffer and drying under reduced pressure. 7,8-Dicarba-*nido*-undecaborane(13) (**L1**) [43], 1,10-phenanthrolinopyrrole (**L2**) [49], [3-(CO)$_3$-*closo*-3,1,2-RuC$_2$B$_9$H$_{11}$] (**1**) [55], and 1,10-phenanthrolino-5,6-dione (**L5**) [45,46] were synthesized according to the literature. The oxidation-sensitive compound 7,8-dicarba-*nido*-undecaborane(13) (**L1**) was stored under nitrogen at −55 °C. Ethyl isocyanoacetate was stored under nitrogen atmosphere and *meta*-chloroperoxybenzoic acid under normal atmosphere at 4 °C. All other used chemicals were stored under nitrogen atmosphere at ambient temperatures. All solvents, except acetonitrile, benzene, cyclohexane, and methanol, which were dried over calcium hydride, sodium, or calcium oxide, respectively, were taken from the solvent purification system MB SPS-800 (by MBraun). Ethanol was used as a mixture with water. Petrol ether (40–60 °C) for column chromatography was used as provided.

NMR spectra were measured with an ADVANCE DRX 400 spectrometer from Bruker. The spectrometer frequency for ^{1}H and ^{11}B are 400.13 MHz and 128.38 Hz, respectively. As an internal standard, tetramethylsilane was used for ^{1}H-NMR and the Ξ scale was used for ^{11}B-NMR spectroscopy. Data were interpreted with MestReNova [64]. The numbering scheme of all isolated compounds is given in the electronic supplementary information. Positive- or negative-mode low-resolution electrospray ionization mass spectra (ESI-MS) were recorded with an ESI ESQUIRE 3000 PLUS spectrometer with an IonTrap analyzer from Bruker Daltonics. For these measurements, dichloromethane, acetonitrile, methanol, or a mixture of these solvents were used. Infrared spectra were recorded with a Spectrum 2000 IR spectrometer from PerkinElmer in the range of 400 to 4000 cm^{-1}. All samples were prepared as KBr pellets. The determination of the single crystal structures was carried out with a Gemini-S diffractometer from Oxford Diffraction using MoK$_\alpha$ radiation (λ = 71.073 pm). The visualization of the structures was carried out with Diamond [65]. Additional crystallographic data are given in the electronic supplementary information. For column chromatography, silica gel 60 Å from the company Acros was used. The particle size was in the range of 0.035 to 0.070 mm. The solvents for semi-inert chromatography were obtained from the solvent purification device MB SPS-800. Thin layer chromatography (TLC) was used to monitor reaction processes using glass plates coated with silica gel 60 F$_{254}$ from Merck. Carborane-containing spots were stained with a 5% solution of palladium(II) chloride in methanol.

[3-(CO)-3,3-(L2-κ^{2}N,N)-closo-3,1,2-RuC$_2$B$_9$H$_{11}$] (**2**): 0.10 g (0.32 mmol, 1.00 eq.) **1** were placed in a 250 mL round bottom flask and dissolved in 25 mL acetonitrile. With stirring, a solution of 0.05 g (0.64 mmol, 2.00 eq.) trimethylamine N-oxide in 10 mL acetonitrile was added. The mixture was stirred for 10 min. Subsequently, 0.07 g (0.31 mmol, 0.97 eq.) 5-nitro-1,10-phenanthroline (**L2**), dissolved in 20 mL acetonitrile, were added dropwise to the mixture. The reaction mixture was stirred for

17 h at rt. After the reaction was finished (monitored using TLC), the mixture was filtered and the filtrate was dried under reduced pressure. The product was purified using column chromatography (dichloromethane; R_f = 0.60). **2** (0.02 g, 0.04 mmol, 13%) was obtained as an orange-red crystalline solid. ^{1}H-NMR (400 MHz, CD$_3$CN): δ = 1.00–2.80 (br, 9 H, 9× BH), 3.39 (s, br, 2 H, 2× CH^1), 8.05 (m, 2 H, CH^3, CH^7), 8.88 (d, $^3J_{HH}$ = 8.2 Hz, 1 H, CH^6), 9.10 (s, 1 H, CH^5), 9.23 (d, $^3J_{HH}$ = 8.6 Hz, 1 H, CH^4), 9.49 (d, $^3J_{HH}$ = 5.2 Hz, 1 H, CH^8), 9.53 ppm (d, $^3J_{HH}$ = 5.3 Hz, 1 H, CH^2); ^{11}B{^{1}H}-NMR (128 MHz, CD$_3$CN): δ = −22.2 (s, 1 B, BH), −21.4 (s, 2 B, BH), −9.8 (s, 2 B, BH), −8.9 (s, 2 B, BH), −6.8 (s, 1 B, BH), −2.0 ppm (s, 1 B, BH); ^{11}B-NMR (128 MHz, CD$_3$CN): δ = −21.8 (m, br, 3 B, BH), −9.3 (m, br, 4 B, BH), −6.8 (d, $^1J_{BH}$ = 139 Hz, 1 B, BH), −2.0 ppm (d, $^1J_{BH}$ = 135 Hz, 1 B, BH); IR (KBr): $\tilde{\nu}$ = 2524 (s, νBH-sp^3), 1970 (s, νCO-sp), 1535 (m, ν$_{asym.}$NO$_2$), 1514 (m, νCN-sp^2), 1342 cm^{-1} (m, ν$_{sym.}$NO$_2$); MS (ESI, neg.): found: m/z (%): 549 (100) [M + NO$_3$]$^-$; calcd: m/z: 549 [M + NO$_3$]$^-$.

*[3-(CO)-3,3-(**L3**-κ^{2}N,N)-closo-3,1,2-RuC$_2$B$_9$H$_{11}$] (**3**)*: 0.10 g (0.32 mmol, 1.00 eq.) **1** were placed in a 250 mL round bottom flask and dissolved in 25 mL acetonitrile. Subsequently, 0.05 g (0.64 mmol, 2.00 eq.) trimethylamine N-oxide, dissolved in 10 mL acetonitrile, were added. The mixture was stirred for 10 min at rt. Then, 0.07 g (0.33 mmol, 1.03 eq.) 1,10-phenanthroline-5,6-dione (**L3**), dissolved in 15 mL acetonitrile, were added dropwise to the mixture. The reaction mixture was stirred for 84 h at rt. After the reaction was complete (monitored using TLC), the resulting precipitate was filtered off and the filtrate was concentrated under reduced pressure. The product was purified using column chromatography (dichloromethane; R_f = 0.17). Yield of **3**: 0.11 g (0.23 mmol, 73%), orange solid. ^{1}H-NMR (400 MHz, CD$_3$CN): δ = 1.00–3.90 (br, 9 H, 9 × BH), 3.37 (s, br, 2 H, 2× CH^1), 7.82 (dd, $^3J_{HH}$ = 7.9 Hz, $^3J_{HH}$ = 5.6 Hz, 2 H, 2× CH^3), 8.62 (dd, $^3J_{HH}$ = 7.9 Hz, $^4J_{HH}$ = 1.4 Hz, 2 H, 2× CH^4), 9.26 ppm (dd, $^3J_{HH}$ = 5.6 Hz, $^4J_{HH}$ = 1.4 Hz, 2 H, 2× CH^2); ^{11}B{^{1}H}-NMR (128 MHz, CD$_3$CN): δ = −21.9 (s, br, 3 B, BH), −9.6 (s, br, 4 B, BH), −7.8 (s, 1 B, BH), −1.4 ppm (s, 1 B, BH); ^{11}B-NMR (128 MHz, CD$_3$CN): δ = −21.9 (d, $^1J_{BH}$ = 143 Hz, 3 B, BH), −9.6 (d, $^1J_{BH}$ = 142 Hz, 4 B, BH), −7.8 (d, $^1J_{BH}$ = 144 Hz, 1 B, BH), −1.4 ppm (d, $^1J_{BH}$ = 151 Hz, 1 B, BH); IR (KBr): $\tilde{\nu}$ = 2507 (m, νBH-sp^3), 1984 (m, νCO-sp), 1702 (m, νCO-sp^2), 1691 (m, νCN-sp^2), 802 cm^{-1} (s, 1,2,3-trisubstituted aromatic ring); MS (ESI, neg.): found: m/z (%): 552 (67) [M + Br]$^-$; calcd: m/z: 552 [M + Br]$^-$.

*[3-(CO)-3,3-(**L4**-κ^{2}N,N)-closo-3,1,2-RuC$_2$B$_9$H$_{11}$] (**4**)*: 0.20 g (0.63 mmol, 1.00 eq.) **1** were placed in a 250 mL round bottom flask and dissolved in 25 mL acetonitrile. To this solution, 0.10 g (1.26 mmol, 2.00 eq.) trimethylamine N-oxide, dissolved in 10 mL acetonitrile, were added dropwise. The mixture was stirred for 10 min at rt. Then, 0.28 g (1.28 mmol, 2.03 eq.) 1,10-phenanthrolinopyrrole (**L4**) were added in one portion and the reaction mixture was stirred for 48 h at room temperature. After the reaction was completed (monitored using TLC), the resulting precipitate was filtered off and the filtrate was concentrated under reduced pressure. The product was purified using column chromatography (dichloromethane; R_f = 0.33). Yield of **4**: 0.14 g (0.29 mmol, 46%), yellow crystalline solid. ^{1}H-NMR (400 MHz, CD$_3$CN): δ = 0.70–2.80 (br, 9 H, 9× BH), 3.28 (s, br, 2 H, 2× CH^1), 7.71 (dd, $^3J_{HH}$ = 8.1 Hz, $^3J_{HH}$ = 5.4 Hz, 2 H, 2× CH^3), 7.92 (d, $^3J_{HH}$ = 2.8 Hz, 2 H, 2× CH^5), 8.67 (d, $^3J_{HH}$ = 7.7 Hz, 2 H, 2× CH^4), 9.03 (d, $^3J_{HH}$ = 5.1 Hz, 2 H, 2× CH^2), 10.70 ppm (s, br, 1 H, NH^6); ^{11}B{^{1}H}-NMR (128 MHz, CD$_3$CN): δ = −21.7 (s, br, 3 B, BH), −10.1 (s, 2 B, BH), −9.0 (s, 2 B, BH), −7.2 (s, br, 1 B, BH), −2.7 ppm (s, br, 1 B, BH); ^{11}B-NMR (128 MHz, CD$_3$CN): δ = −21.8 (m, 3 B, BH), −8.8 (m, br, 5 B, BH), −2.7 ppm (d, $^1J_{BH}$ = 139 Hz, 1 B, BH); IR (KBr): $\tilde{\nu}$ = 2523 (m, νBH-sp^3), 1958(s, νCO-sp), 1639 (w, νCN-sp^2), 1600 (w, νCC-sp^2), 803 cm^{-1} (m, 1,2,3-trisubstituted aromatic ring); MS (ESI, neg.): found: m/z (%): 480 (100) [M − H]$^-$; calcd: m/z: 480 [M − H]$^-$.

*[3-(CO)-3,3-{1′,10′-NC$_5$H$_3$(C(CO)(NH)(CO)C)NC$_5$H$_3$-κ^{2}N,N}-closo-3,1,2-RuC$_2$B$_9$H$_{11}$] (**5**)*: 0.31 g (0.65 mmol, 1.00 eq.) **4** were placed in a 250 mL round bottom flask and dissolved in 30 mL acetonitrile. Subsequently, 0.45 g (2.61 mmol, 4.01 eq.) m-CPBA, dissolved in 10 mL acetonitrile, were added under stirring at rt. The reaction mixture was heated under reflux for 84 h. During this time, two additional portions of 0.28 g (1.62 mmol, 2.49 eq.) m-CPBA, dissolved in 10 mL acetonitrile, were added after 24 h and 48 h, respectively. After completion of the reaction (monitored using

TLC), the mixture was cooled to rt; the resulting precipitate was filtered off and the filtrate was concentrated under reduced pressure. The product was purified using column chromatography (dichloromethane/acetonitrile, 10:1, (v/v); R_f = 0.52). Yield of **5**: 0.03 g (0.06 mmol, 9%, corrected, after recovery of starting material **4**: 10%), deep red crystalline powder. In addition, 0.03 g (0.06 mmol) **4** were recovered. ^{1}H-NMR (400 MHz, acetone-d_6): δ = 0.82–3.01 (br, 9 H, 9× BH), 3.53 (s, br, 2 H, 2× CH^1), 8.31 (dd, $^3J_{HH}$ = 8.3 Hz, $^3J_{HH}$ = 5.2 Hz, 2 H, 2× CH^3), 9.59 (dd, $^3J_{HH}$ = 8.3 Hz, $^3J_{HH}$ = 1.4 Hz, 2 H, 2× CH^4), 9.73 (dd, $^3J_{HH}$ = 5.3 Hz, $^3J_{HH}$ = 1.4 Hz, 2 H, 2× CH^2), 10.74 ppm (s, br, 1 H, NH^5); ^{11}B{^{1}H}-NMR (128 MHz, acetone-d_6): δ = −21.7 (s, br, 3 B, BH), −9.3 (s, 2 B, BH), −7.5 (s, 3 B, BH), −1.3 ppm (s, 1 B, BH); ^{11}B-NMR (128 MHz, acetone-d_6): δ = −21.7 (d, $^1J_{BH}$ = 157 Hz, 3 B, BH), −9.2 (d, $^1J_{BH}$ = 162 Hz, 2 B, BH), −7.5 (d, $^1J_{BH}$ = 150 Hz, 3 B, BH), −1.3 ppm (d, $^1J_{BH}$ = 142 Hz, 1 B, BH); IR (KBr): $\tilde{v}$ = 2531 (s, νBH-sp^3), 1956 (s, νCO-sp), 1727 (s, νCO-sp^2), 1695 cm^{-1} (m, νCN-sp^2); MS (ESI, neg.): found: m/z (%): 510 (100) [M − H]$^-$, 482 (26) [M − CO − H]$^-$; calcd: m/z: 510 [M − H]$^-$, 482 [M − CO − H]$^-$.

4. Conclusions

Four air stable ruthenium(II) half-sandwich complexes, **2-5**, which contain a dicarbollide moiety, a carbonyl ligand, and a phenanthroline derivative, were prepared in moderate to good yields and fully characterized. The complexes were designed to mimic the overall shape and structure of the alkaloid staurosporine and the ruthenium half-sandwich complex DW12. Initial inhibition experiments with the ruthenium(II) complexes **4** and **5** against the protein kinase Pim1 were not conclusive, most likely due to a low solubility of **4** and **5** in buffer solutions caused by the high hydrophobicity of the dicarbollide moiety. Thus, although a high hydrophobicity is beneficial for the inhibition of a biological target molecule, a sufficient solubility in aqueous buffer solution must be warranted, which will need to be addressed in future work.

Supplementary Materials: The electronic supplementary information is available online including the numbering scheme of all isolated compounds, additional synthetic procedures for **L1**, **L4**, and **1**, the isolation of **SP1** and its spectroscopic data, crystallographic information of compounds **2**, **SP1**, **4**, and **5**, and information about the protein kinase inhibition studies.

Author Contributions: Conceptualization, M.K. and E.H.-H.; data curation, M.K., R.R., and P.L.; formal analysis, M.K. and P.L.; funding acquisition, E.M. and E.H.-H.; investigation, M.K.; methodology, I.S. and R.R.; project administration, M.K. and E.H.-H.; resources, E.M. and E.H.-H.; supervision, I.S., E.M., and E.H.-H.; validation, M.K. and I.S.; visualization, M.K.; writing—original draft, M.K.; writing—review and editing, E.M., P.L., and E.H.-H. All authors have read and agreed to the published version of the manuscript.

Funding: Funding by the "Europäischer Fonds für regionale Entwicklung (EFRE)", the Free State of Saxony (ESF) and the Graduate School "Leipzig School of Natural Sciences—Building with Molecules and Nano-objects" (BuildMoNa) is gratefully acknowledged.

Acknowledgments: We thank Ramona Oehme, Susann Billig and Claudia Birkemeyer for measuring the mass spectra, Manuela Roßberg and Gunther Wünsche for elemental analysis, and Stefanie Märcker-Recklies for recording the infrared spectra.

Conflicts of Interest: The authors declare no conflicts of interest.

References

1. Thompson, K.H.; Orvig, C. Metal Complexes in Medicinal Chemistry: New Vistas and Challenges in Drug Design. *Dalton Trans.* **2006**, 761–764. [CrossRef]
2. Colotti, G.; Ilari, A.; Boffi, A.; Morea, V. Metals and Metal Derivatives in Medicine. *Mini Rev. Med. Chem.* **2013**, *13*, 211–221. [CrossRef]
3. Cohen, S.M. New Approaches for Medicinal Applications of Bioinorganic Chemistry. *Curr. Opin. Chem. Biol.* **2007**, *11*, 115–120. [CrossRef]
4. Meggers, E.; Atilla-Gokcumen, G.E.; Bregman, H.; Maksimoska, J.; Mulcahy, S.P.; Pagano, N.; Williams, D.S. Exploring Chemical Space with Organometallics: Ruthenium Complexes as Protein Kinase Inhibitors. *Synlett* **2007**, 1177–1189. [CrossRef]

5. Bregman, H.; Williams, D.S.; Atilla, G.E.; Carroll, P.J.; Meggers, E. An Organometallic Inhibitor for Glycogen Synthase Kinase 3. *J. Am. Chem. Soc.* **2004**, *126*, 13594–13595. [CrossRef] [PubMed]
6. Bregman, H.; Williams, D.S.; Meggers, E. Pyrido[2,3-*a*]pyrrolo[3,4-*c*]carbazole-5,7(6*H*)-diones: Synthesis, Cyclometalation, and Protein Kinase Inhibition. *Synthesis* **2005**, 1521–1527. [CrossRef]
7. Chiara, F.; Rasola, A. GSK-3 and Mitochondria in Cancer Cells. *Front. Oncol.* **2013**, *3*, 16. [CrossRef] [PubMed]
8. Cohen, P. Protein Kinases—the Major Drug Targets of the Twenty-First Century? *Nat. Rev. Drug Discov.* **2002**, *1*, 309–315. [CrossRef]
9. Allardyce, C.S.; Dyson, P.J. Ruthenium in Medicine: Current Clinical Uses and Future Prospects. *Platinum Metals Rev.* **2001**, *45*, 62–69.
10. Dyson, P.J. Systematic Design of a Targeted Organometallic Antitumour Drug in Pre-clinical Development. *CHIMIA* **2007**, *61*, 698–703. [CrossRef]
11. Gozzi, M.; Schwarze, B.; Sárosi, M.-B.; Lönnecke, P.; Drača, D.; Maksimović-Ivanić, D.; Mijatović, S.; Hey-Hawkins, E. Antiproliferative Activity of (η^6-Arene)ruthenacarborane Sandwich Complexes Against HCT116 and MCF7 Cell Lines. *Dalton Trans.* **2017**, *46*, 12067–12080. [CrossRef]
12. Süss-Fink, G. Arene Ruthenium Complexes as Anticancer Agents. *Dalton Trans.* **2010**, *39*, 1673–1688. [CrossRef]
13. Gozzi, M.; Murganic, B.; Drača, D.; Popp, J.; Coburger, P.; Maksimović-Ivanić, D.; Mijatović, S.; Hey-Hawkins, E. Quinoline-Conjugated Ruthenacarboranes: Toward Hybrid Drugs with a Dual Mode of Action. *Chem. Med. Chem.* **2019**, *14*, 2061–2074. [CrossRef]
14. Gozzi, M.; Schwarze, B.; Hey-Hawkins, E. Half- and Mixed-sandwich Metallacarboranes for Potential Applications in Medicine. *Pure Appl. Chem.* **2019**, *91*, 563–573. [CrossRef]
15. Bregman, H.; Meggers, E. Ruthenium Half-Sandwich Complexes as Protein Kinase Inhibitors: An *N*-succinimidyl Ester for Rapid Derivatizations of the Cyclopentadienyl Moiety. *Org. Lett.* **2006**, *8*, 5465–5468. [CrossRef]
16. Atilla-Gokcumen, G.E.; Williams, D.S.; Bregman, H.; Pagano, N.; Meggers, E. Organometallic Compounds with Biological Activity: A Very Selective and Highly Potent Cellular Inhibitor for Glycogen Synthase Kinase 3. *ChemBioChem* **2006**, *7*, 1443–1450. [CrossRef]
17. Atilla-Gokcumen, G.E.; Di Costanzo, L.; Meggers, E. Structure of Anticancer Ruthenium Half-Sandwich Complex Bound to Glycogen Synthase Kinase 3β. *J. Biol. Inorg. Chem.* **2011**, *16*, 45–50. [CrossRef]
18. Rüegg, U.T.; Burgess, G.M. Staurosporine, K-252 and UCN-01: Potent but Nonspecific Inhibitors of Protein Kinases. *Trends Pharmacol. Sci.* **1989**, *10*, 218–220. [CrossRef]
19. Scholz, M.; Hey-Hawkins, E. Carbaboranes as Pharmacophores: Properties, Synthesis, and Application Strategies. *Chem. Rev.* **2011**, *111*, 7035–7062. [CrossRef]
20. Issa, F.; Kassiou, M.; Rendina, L.M. Boron in Drug Discovery: Carboranes as Unique Pharmacophores in Biologically Active Compounds. *Chem. Rev.* **2011**, *111*, 5701–5722. [CrossRef]
21. Grimes, R.N. *Carboranes*, 3rd ed.; Elsevier, Academic Press: Amsterdam, The Netherland; Boston, MA, USA; Heidelberg, Germany, 2016; ISBN 9780128018941.
22. Grimes, R.N. Carboranes in the Chemist's Toolbox. *Dalton Trans.* **2015**, *44*, 5939–5956. [CrossRef] [PubMed]
23. Leśnikowski, Z.J. Challenges and Opportunities for the Application of Boron Clusters in Drug Design. *J. Med. Chem.* **2016**, *59*, 7738–7758. [CrossRef] [PubMed]
24. Marta, G.; Benedikt, S.; Evamarie, H.-H. Half- and Mixed-sandwich Metallacarboranes in Catalysis. In *Handbook of Boron Chemistry in Organometallics, Catalysis, Materials and Medicine*, 2nd ed.; Imperial College Press/World Scientific Publishing (UK) Ltd.: London, UK, 2018; Volume 2, pp. 27–80. ISBN 9781786344410.
25. Fujii, S. Expanding the Chemical Space of Hydrophobic Pharmacophores: The Role of Hydrophobic Substructures in the Development of Novel Transcription Modulators. *Med. Chem. Commun.* **2016**, *7*, 1082–1092. [CrossRef]
26. Vincenzi, M.; Bednarska, K.; Leśnikowski, Z.J. Comparative Study of Carborane- and Phenyl-Modified Adenosine Derivatives as Ligands for the A2A and A3 Adenosine Receptors Based on a Rigid in Silico Docking and Radioligand Replacement Assay. *Molecules* **2018**, *23*, 1846. [CrossRef] [PubMed]
27. Stockmann, P.; Gozzi, M.; Kuhnert, R.; Sárosi, M.-B.; Hey-Hawkins, E. New Keys for Old Locks: Carborane-Containing Drugs as Platforms for Mechanism-based Therapies. *Chem. Soc. Rev.* **2019**, *48*, 3497–3512. [CrossRef] [PubMed]

28. Leśnikowski, Z.J. Recent Developments with Boron as a Platform for Novel Drug Design. *Expert Opin. Drug Dis.* **2016**, *11*, 569–578. [CrossRef]
29. Lesnikowski, Z.J. Boron Units as Pharmacophores-New Applications and Opportunities of Boron Cluster Chemistry. *Collect. Czech. Chem. C.* **2007**, *72*, 1646–1658. [CrossRef]
30. Ahrens, V.M.; Frank, R.; Boehnke, S.; Schütz, C.L.; Hampel, G.; Iffland, D.S.; Bings, N.H.; Hey-Hawkins, E.; Beck-Sickinger, A.G. Receptor-mediated Uptake of Boron-rich Neuropeptide Y Analogues for Boron Neutron Capture Therapy. *Chem. Med. Chem.* **2015**, *10*, 164–172. [CrossRef]
31. Armstrong, A.F.; Valliant, J.F. The Bioinorganic and Medicinal Chemistry of Carboranes: From New Drug Discovery to Molecular Imaging and Therapy. *Dalton Trans.* **2007**, 4240–4251. [CrossRef]
32. Scholz, M.; Kaluđerović, G.N.; Kommera, H.; Paschke, R.; Will, J.; Sheldrick, W.S.; Hey-Hawkins, E. Carbaboranes as Pharmacophores: Similarities and Differences between Aspirin and Asborin. *Eur. J. Med. Chem.* **2011**, *46*, 1131–1139. [CrossRef]
33. Scholz, M.; Steinhagen, M.; Heiker, J.T.; Beck-Sickinger, A.G.; Hey-Hawkins, E. Asborin Inhibits Aldo/Keto Reductase 1A1. *Chem. Med. Chem.* **2011**, *6*, 89–93. [CrossRef] [PubMed]
34. Schwarze, B.; Gozzi, M.; Hey-Hawkins, E. Half- and Mixed-sandwich Transition Metal *nido*-Carboranes(-1) and Metallacarboranes for Medicinal Applications. In *Boron-Based Compounds: Potential and Emerging Applications in Medicine*; Hey-Hawkins, E., Viñas Teixidor, C., Eds.; John Wiley & Sons Ltd.: Hoboken, NJ, USA, 2016; pp. 60–108, Online; ISBN 9781119275602.
35. Neumann, W.; Xu, S.; Sárosi, M.B.; Scholz, M.S.; Crews, B.C.; Ghebreselasie, K.; Banerjee, S.; Marnett, L.J.; Hey-Hawkins, E. *nido*-Dicarbaborate Induces Potent and Selective Inhibition of Cyclooxygenase-2. *Chem. Med. Chem.* **2016**, *11*, 175–178. [CrossRef] [PubMed]
36. Gozzi, M.; Schwarze, B.; Coburger, P.; Hey-Hawkins, E. On the Aqueous Solution Behavior of C-Substituted 3,1,2-Ruthenadicarbadodecaboranes. *Inorganics* **2019**, *7*, 91. [CrossRef]
37. Schwarze, B.; Gozzi, M.; Zilberfain, C.; Rüdiger, J.; Birkemeyer, C.; Estrela-Lopis, I.; Hey-Hawkins, E. Nanoparticle-based Formulation of Metallacarboranes with Bovine Serum Albumin for Application in Cell Cultures. *J. Nanopart. Res.* **2020**, *22*, 24. [CrossRef]
38. Brown, D.A.; Fanning, M.O.; Fitzpatrick, N.J. Molecular Orbital Theory of Organometallic Compounds. 15. A Comparative Study of Ferrocene and π-cyclopentadienyl-(3)-1,2-dicarbollyliron. *Inorg. Chem.* **1978**, *17*, 1620–1623. [CrossRef]
39. Hoffmann, R. Building Bridges Between Inorganic and Organic Chemistry (Nobel Lecture). *Angew. Chem. Int. Ed. Engl.* **1982**, *21*, 711–724, *Angew. Chem.* **1982**, *94*, 725–739. [CrossRef]
40. Hosmane, N.S. *Handbook of Boron Science with Applications in Organometallics, Catalysis, Materials and Medicine*; Eagling, R., Ed.; World Scientific Publishing Europe Ltd.: London, UK; Munich, Germany, 2019; ISBN 1786344475.
41. Debreczeni, J.É.; Bullock, A.N.; Atilla, G.E.; Williams, D.S.; Bregman, H.; Knapp, S.; Meggers, E. Ruthenium Half-sandwich Complexes Bound to Protein Kinase Pim-1. *Angew. Chem. Int. Ed.* **2006**, *45*, 1580–1585, *Angew. Chem.* **2006**, *118*, 1610–1615. [CrossRef]
42. Feng, L.; Geisselbrecht, Y.; Blanck, S.; Wilbuer, A.; Atilla-Gokcumen, G.E.; Filippakopoulos, P.; Kräling, K.; Celik, M.A.; Harms, K.; Maksimoska, J.; et al. Structurally Sophisticated Octahedral Metal Complexes as Highly Selective Protein Kinase Inhibitors. *J. Am. Chem. Soc.* **2011**, *133*, 5976–5986. [CrossRef]
43. Hlatky, G.G.; Crowther, D.J. Main Group and Transition Metal Cluster Compounds: 38. 7,8-Dicarbaundecaborane(13). *Inorg. Syn.* **1998**, *32*, 229–231.
44. Smith, G.F.; Cagle, F.W., Jr. The Improved Synthesis of 5-Nitro-1,10-phenanthroline. *J. Org. Chem.* **1947**, *12*, 781–784. [CrossRef]
45. Sergeeva, N.N.; Donnier-Marechal, M.; Vaz, G.M.; Davies, A.M.; Senge, M.O. Stability and Spectral Properties of Europium and Zinc Phenanthroline Complexes as Luminescent Probes in High Content Cell-imaging Analysis. *J. Inorg. Biochem.* **2011**, *105*, 1589–1595. [CrossRef] [PubMed]
46. Wang, C.; Lystrom, L.; Yin, H.; Hetu, M.; Kilina, S.; McFarland, S.A.; Sun, W. Increasing the Triplet Lifetime and Extending the Ground-state Absorption of Biscyclometalated Ir(III) Complexes for Reverse Saturable Absorption and Photodynamic Therapy Applications. *Dalton Trans.* **2016**, *45*, 16366–16378. [CrossRef] [PubMed]
47. Barton, D.H.R.; Kervagoret, J.; Zard, S.Z. An Useful Synthesis of Pyrroles from Nitroolefins. *Tetrahedron* **1990**, *46*, 7587–7598. [CrossRef]

48. Barton, D.H.R.; Zard, S.Z. A New Synthesis of Pyrroles from Nitroalkenes. *J. Chem. Soc. Chem. Commun.* **1985**, 1098–1100. [CrossRef]
49. Villegas, J.M.; Stoyanov, S.R.; Rillema, D.P. Synthesis and Photochemistry of Ru(II) Complexes Containing Phenanthroline-based Ligands with Fused Pyrrole Rings. *Inorg. Chem.* **2002**, *41*, 6688–6694. [CrossRef]
50. Lash, T.D.; Lin, Y.; Novak, B.H.; Parikh, M.D. Porphyrins with Exocyclic Rings. Part 19: Efficient Syntheses of Phenanthrolinoporphyrins. *Tetrahedron* **2005**, *61*, 11601–11614. [CrossRef]
51. Lash, T.D.; Novak, B.H.; Lin, Y. Synthesis of Phenanthropyrroles and Phenanthrolinopyrroles from Isocyanoacetates: An Extension of the Barton-Zard Pyrrole Condensation. *Tetrahedron Lett.* **1994**, *35*, 2493–2494. [CrossRef]
52. Lin, Y.; Lash, T.D. Porphyrin Synthesis by the "3+1" Methodology: A Superior Approach for the Preparation of Porphyrins with Fused 9,10-Phenanthroline Subunits. *Tetrahedron Lett.* **1995**, *36*, 9441–9444. [CrossRef]
53. Ono, N.; Hironaga, H.; Ono, K.; Kaneko, S.; Murashima, T.; Ueda, T.; Tsukamura, C.; Ogawa, T. A New Synthesis of Pyrroles and Porphyrins Fused with Aromatic Rings. *J. Chem. Soc. Perkin Trans. 1* **1996**, 417–423. [CrossRef]
54. Ono, N.; Hironaga, H.; Simizu, K.; Ono, K.; Kuwano, K.; Ogawa, T. Synthesis of Pyrroles Annulated with Polycyclic Aromatic Compounds; Precursor Molecules for Low Band Gap Polymers. *J. Chem. Soc. Chem. Commun.* **1994**, 1019–1020. [CrossRef]
55. Anderson, S.; Mullica, D.F.; Sappenfield, E.L.; Stone, F.G.A. Carborane Complexes of Ruthenium: A Convenient Synthesis of [Ru(CO)$_3$(η^5-7,8-C$_2$B$_9$H$_{11}$)] and a Study of Reactions of This Complex. *Organometallics* **1995**, *14*, 3516–3526. [CrossRef]
56. Behnken, P.E.; Hawthorne, M.F. Reactions at the Metal Vertex of a Ruthenacarborane Cluster. Activation of Carbon Monoxide by *closo*-3,3,3-(CO)$_3$-3,1,2-RuC$_2$B$_9$H$_{11}$. *Inorg. Chem.* **1984**, 3420–3423. [CrossRef]
57. Siedle, A.R. Dicarbollide Complexes of Rhodium and Ruthenium. *J. Organomet. Chem.* **1975**, *90*, 249–256. [CrossRef]
58. Jelliss, P.A.; Mason, J.; Nazzoli, J.M.; Orlando, J.H.; Vinson, A.; Rath, N.P.; Shaw, M.J. Synthesis and Characterization of Ruthenacarborane Complexes Incorporating Chelating N-donor Ligands: Unexpected Luminescence from the Complex 3-CO-3,3-κ^2-Me$_2$N(CH$_2$)$_2$NMe$_2$-*closo*-3,1,2-RuC$_2$B$_9$H$_{11}$. *Inorg. Chem.* **2006**, *45*, 370–385. [CrossRef] [PubMed]
59. Dyson, P.J.; McIndoe, J.S. *Transition Metal Carbonyl Cluster Chemistry*; Gordon and Breach Science Publishers: Amsterdam, The Netherlands, 2000; ISBN 90-5699-289-9.
60. Paw, W.; Eisenberg, R. Synthesis, Characterization, and Spectroscopy of Dipyridocatecholate Complexes of Platinum. *Inorg. Chem.* **1997**, *36*, 2287–2293. [CrossRef]
61. Pearsons, R.G. Hard and Soft Acids and Bases. *J. Am. Chem. Soc.* **1963**, *85*, 3533–3539. [CrossRef]
62. Alamgir, M.; Mitchell, P.S.R.; Bowyer, P.K.; Kumar, N.; Black, D.S. Synthesis of 4,7-Indoloquinones from Indole-7-carbaldehydes by Dakin Oxidation. *Tetrahedron* **2008**, *64*, 7136–7142. [CrossRef]
63. Howard, J.K.; Hyland, C.J.T.; Just, J.; Smith, J.A. Controlled Oxidation of Pyrroles: Synthesis of Highly Functionalized γ-Lactams. *Org. Lett.* **2013**, *15*, 1714–1717. [CrossRef]
64. *MestReNova*; v12.0.0-20080; Mestrelab Research S.L.: Santiago de Compostela, Spain, 2017.
65. *Diamond*; v4.6.2; Crystal Impact GbR: Bonn, Germany, 1997–2020.

Sample Availability: Samples of the compounds are not available from the authors.

© 2020 by the authors. Licensee MDPI, Basel, Switzerland. This article is an open access article distributed under the terms and conditions of the Creative Commons Attribution (CC BY) license (http://creativecommons.org/licenses/by/4.0/).

Article

Deboronation-Induced Ratiometric Emission Variations of Terphenyl-Based *Closo-o*-Carboranyl Compounds: Applications to Fluoride-Sensing

Hyunhee So [†], Min Sik Mun [†], Mingi Kim [†], Jea Ho Kim, Ji Hye Lee, Hyonseok Hwang, Duk Keun An * and Kang Mun Lee *

Department of Chemistry, Institute for Molecular Science and Fusion Technology, Kangwon National University, Chuncheon 24341, Korea; shh6353@naver.com (H.S.); bbcisgj2002@kangwon.ac.kr (M.S.M.); kmg6523@kangwon.ac.kr (M.K.); syoil12@daum.net (J.H.K.); jhlee81@kangwon.ac.kr (J.H.L.); hhwang@kangwon.ac.kr (H.H.)
* Correspondence: dkan@kangwon.ac.kr (D.K.A.); kangmunlee@kangwon.ac.kr (K.M.L.);
  Tel.: +82-33-250-8494 (D.K.A.); +82-33-250-8499 (K.M.L.)
† These authors contributed equally to the work.

Academic Editor: Ashok Kakkar
Received: 25 April 2020; Accepted: 21 May 2020; Published: 21 May 2020

Abstract: *Closo-o*-carboranyl compounds bearing the *ortho*-type perfectly distorted or planar terphenyl rings (*closo*-**DT** and *closo*-**PT**, respectively) and their *nido*-derivatives (*nido*-**DT** and *nido*-**PT**, respectively) were synthesized and fully characterized using multinuclear NMR spectroscopy and elemental analysis. Although the emission spectra of both *closo*-compounds exhibited intriguing emission patterns in solution at 298 and 77 K, in the film state, *closo*-**DT** mainly exhibited a π-π* local excitation (LE)-based emission in the high-energy region, whereas *closo*-**PT** produced an intense emission in the low-energy region corresponding to an intramolecular charge transfer (ICT) transition. In particular, the positive solvatochromic effect of *closo*-**PT** and theoretical calculation results at the first excited (S_1) optimized structure of both *closo*-compounds strongly suggest that these dual-emissive bands at the high- and low-energy can be assigned to each π-π* LE and ICT transition. Interestingly, both the *nido*-compounds, *nido*-**DT** and *nido*-**PT**, exhibited the only LE-based emission in solution at 298 K due to the anionic character of the *nido-o*-carborane cages, which cannot cause the ICT transitions. The specific emissive features of *nido*-compounds indicate that the emissive color of *closo*-**PT** in solution at 298 K is completely different from that of *nido*-**PT**. As a result, the deboronation of *closo*-**PT** upon exposure to increasing concentrations of fluoride anion exhibits a dramatic ratiometric color change from orange to deep blue via turn-off of the ICT-based emission. Consequently, the color change response of the luminescence by the alternation of the intrinsic electronic transitions via deboronation as well as the structural feature of terphenyl rings indicates the potential of the developed *closo-o*-carboranyl compounds that exhibit the intense ICT-based emission, as naked-eye-detectable chemodosimeters for fluoride ion sensing.

Keywords: *closo-o*-carborane; *nido-o*-carborane; intramolecular charge transfer; deboronation; color change

1. Introduction

Closo-ortho-carboranes (1,2-dicarba-*closo-o*-dodecaboranes, o-1,2-$C_2B_{10}H_{12}$) are well-known boron-cluster components of three-dimensional (3D) icosahedral analogs. Recently, *closo-ortho*-carboranes have attracted significant attention as new molecular scaffolds of steric and electronic substituents for luminescent organic and organometallic compounds due to their unique photophysical properties and reasonable thermal and electrochemical stabilities originating from the *o*-carborane unit [1–28]. These electronic features are imparted by the electron-withdrawing properties of the carbon atoms, and the high polarizability of the σ-aromaticity of the organic and organometallic luminophores that comprise the *o*-carborane moiety. These characteristics lead to the formation of electronic donor-acceptor dyad systems that induce intrinsic intramolecular charge transfer (ICT) transitions between the π-conjugated aromatic groups and the *o*-carborane cage [29–53]. Such ICT characteristics can induce unique luminescence behavior in various *o*-carborane-based organic luminophores [29–70]. Interestingly, such an intramolecular radiative mechanism activated by the ICT transitions in the *o*-carboranyl luminophores has been found amenable to modifications via variations to the structure of the *o*-carborane cages or appended aryl groups [32,43–53,70] and their molecular geometries [66–69]. Furthermore, the direct control of the ICT-based emission in the *closo-o*-carboranyl compounds involves the conversion of *closo-o*-carboranes to *nido-o*-species (*o*-1,2-$C_2B_9H_{12}^-$, one boron atom removed analog of the *closo-o*-carborane cage) by reaction with nucleophilic anions. This powerful process can cause dramatic changes in the inherent electronic environment because of the strong electron-donating property of *nido-o*-carboranes [71], leading to the alteration of their luminescent features [72–78]. For example, Carter et al. reported a fluorene-based dimer bearing an *o*-carborane, which exhibited a visible fluorescence change (orange to bright blue) by degradation to the *nido*-species [73]. Furthermore, Núñez et al. reported photoluminescent *closo*- and *nido*-di-carboranyl and tetra-carboranyl derivatives, which possessed intrinsic ICT electronic transitions and demonstrated differences in the emission band maxima of the two species [75]. We recently reported polyolefins bearing pendant *o*-carborane moieties that exhibit strong blue emissions in the solid state. Notably, the observed emissions disappeared after degradation of the carborane cage upon reaction with hydroxyl ions [76]. Additionally, 1,3,5-tris-(*closo-o*-carboranyl-methyl)benzene displayed ratiometric emissive color change via deboronation to the corresponding *nido-o*-species [77]. The degradation of the *closo-o*-carborane–triarylborane dyad to the *nido-o*-carboranyl compound exhibited a turn-on fluorescence response toward fluorides [78]. Thus, these *closo-o*-carboranyl derivatives exhibited great potential as polymeric or single-molecular chemodosimeters for sensing nucleophilic anions.

On the basis of inducing significant changes in the electronic properties through the conversion of *closo-o*-carborane to its anionic *nido-o*-species, we sought to investigate in detail the impact of the deboronation of the *closo-o*-carboranyl compounds on their ICT-based emission. For this study, we designed two simple terphenyl-based *o*-carboranyl compounds based on our previous results [69] (Figure 1). The first is 2′,5′-dimethyl-1,1′:4′,1″-terphenyl-based *closo-o*-carborane (**DT**), with distorted *o*-terphenyl rings, which showed a weak ICT-based emission transition, and the second is 6,6,12,12-tetramethyl-6,12-dihydroindeno[1,2-*b*]fluorene-based *closo-o*-carboranyl compound (**PT**), with planar *o*-terphenyl rings, which possessed an intense emission from the ICT transition. Subsequently, the designed *closo-o*-compounds, as well as the *nido-o*-carboranyl compounds, were prepared and fully characterized. The comparison of the photophysical properties of these *closo*- and *nido*-compounds indicated that the deboronation of the *o*-carborane moiety and the structural feature of the appended terphenyl rings may deactivate the ICT transition and also quenching of the emission, thereby providing a novel method for fluoride sensing.

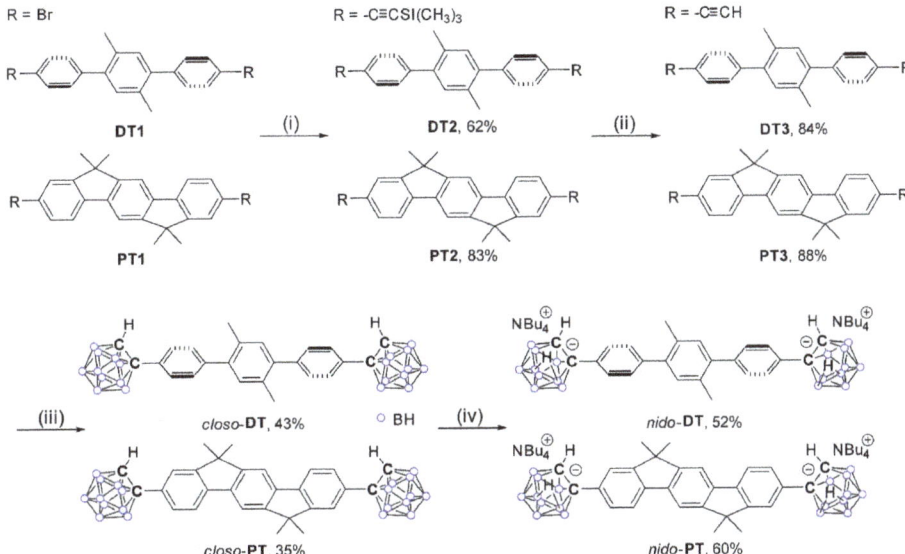

Figure 1. Synthetic routes to the terphenyl-based *closo*- and *nido*-*o*-carboranyl complexes, *closo*-**DT**, *closo*-**PT**, *nido*-**DT**, and *nido*-**PT**. Reaction conditions: (i) Ethynyltrimethylsilane, CuI, Pd(PPh$_3$)$_2$Cl$_2$, NEt$_3$/toluene, r.t., 24 h. (ii) K$_2$CO$_3$, methanol, r.t., 2 h. (iii) B$_{10}$H$_{14}$, Et$_2$S, toluene, 110 °C, 72 h. (iv) *n*-tetrabutylammonium fluoride (TBAF), THF, 60 °C, 2 h.

2. Results and Discussion

2.1. Synthesis and Characterization

The synthetic routes for the terphenyl-based *closo*-*o*- (*closo*-**DT** and *closo*-**PT**) and *nido*-*o*- (*nido*-**DT** and *nido*-**PT**) carboranyl compounds, where the *o*-carborane cages are substituted at both the ends by terphenyl moieties, are outlined in Figure 1. The Sonogashira coupling reaction between ethynyltrimethylsilane and the bromo-precursors **DT1** and **PT1** produced the ethynyltrimethylsilane-substituted terphenyl compounds **DT2** and **PT2**, respectively, in high yields (62% for **DT2** and 83% for **PT2**). The mild base (K$_2$CO$_3$)-mediated deprotection of the trimethylsilyl protecting groups of **DT2** and **PT2** furnished **DT3** and **PT3**, respectively, which were then subjected to decaborane (B$_{10}$H$_{14}$)-promoted cage-forming reactions in the presence of Et$_2$S (Figure 1) [79–81] to prepare the *closo*-*o*-carborane-substituted terphenyl compounds *closo*-**DT** and *closo*-**PT**, respectively. The dimethyl groups of *closo*-**PT** were introduced to achieve good solubility in a range of organic solvents. Subsequent treatment of *closo*-**DT** and *closo*-**PT** with excess *n*-tetrabutylammonium fluoride (NBu$_4$F, TBAF) in THF at 60 °C led to the conversion of the *closo*-carboranes to the *nido*-species; *nido*-**DT** is the (NBu$_4$)$_2$-salt of the *nido*-form of *closo*-**DT**, and *nido*-**PT** is the (NBu$_4$)$_2$-salt of the *nido*-form of *closo*-**PT** (Figure 1).

All of the prepared *closo*- and *nido*-*o*-carboranyl compounds were fully characterized using multinuclear (^{1}H{^{11}B}, ^{13}C, and ^{11}B{^{1}H}) NMR spectroscopy (Figures S1–S12 in the Supplementary Material) and elemental analysis. The ^{1}H and ^{13}C NMR spectra of *closo*-**DT** and *closo*-**PT** exhibited resonances corresponding to the terphenyl moieties. In addition, five broad singlet peaks were observed between −2 and −15 ppm in the ^{11}B{^{1}H} NMR spectra of both *closo*-**DT** and *closo*-**PT**, which confirmed the presence of the *closo*-*o*-carborane cage. Furthermore, signals were observed at ~78 and ~61 ppm in the ^{13}C-NMR spectra, which were attributed to the two carbon atoms of the *closo*-*o*-carboranyl groups. Unlike the neutral *closo*-**DT** and *closo*-**PT**, the broad singlets (δ = −2.3 and −2.4 ppm) in the

^{1}H{^{11}B} NMR spectra of both *nido*-**DT** and *nido*-**PT** are characteristic of the B–H–B bridge protons of *nido-o*-carborane moieties. The ^{11}B{^{1}H} NMR signals of *nido*-**DT** and *nido*-**PT** at δ ca. −8 to −37 ppm, which are shifted significantly upfield due to the anionic character of the *nido-o*-carboranes, clearly confirmed the presence of the *nido-o*-carboranyl boron atoms.

2.2. Photophysical Properties of the Closo- and Nido-o-Carboranyl Compounds

The photophysical properties of all terphenyl-based *closo-* and *nido-o*-carboranyl were investigated using UV/Vis absorption and photoluminescence (PL) spectroscopies (Figure 2 and Table 1). The *closo-o*-carboranyl compounds, *closo*-**DT** and *closo*-**PT**, displayed major absorption bands at λ_{abs} = ~268 and 336 nm, respectively, with structureless vibronic features. These bands were attributed to spin-allowed π–π* LE transitions of the central terphenylene groups [82] and typical ICT transitions between the *o*-carborane units and the central phenyl rings (see the time-dependent density functional theory (TD-DFT) results *vide infra*). Indeed, these ICT-based low-energy absorption bands were not present in the spectra of *nido*-**DT** and *nido*-**PT**, due to which those absorption spectra were slightly blue-shifted (λ_{abs} = 254 and 323 nm, respectively, Table 1) compared with those of the *closo*-compounds. These findings imply that the deboronation of the *o*-carborane cages in the *nido*-species quenches the ICT transitions involving the *o*-carborane unit.

To gain insight into the intrinsic photophysical properties of all *o*-carboranyl compounds, the emissive properties of *closo-o*-compounds were examined by PL under a variety of conditions, and further, the emissions of *nido-o*-compounds in THF at 298 K were investigated (Figure 2 and Table 1). Although the PL spectra of both *closo*-**DT** and *closo*-**PT** in THF exhibited intriguing emission patterns in all states upon excitation at 292 and 345 nm, respectively, the *closo*-**DT** emission was focused in the high-energy region centered at λ_{em} = ~350 nm, whereas *closo*-**PT** exhibited an intense low-energy emission in the 500 to 600 nm range, which tailed off at 650 nm. With reference to the results of the TD-DFT computational study (vide infra), this high-energy emission appears to originate from the π–π* LE transitions of the central terphenyl moieties. In contrast, the low-energy emission is closely associated with ICT transitions between the *o*-carborane cages and the terphenyl rings. Furthermore, the emission spectrum of *closo*-**DT** in THF at 298 K exhibited an intense emission in the high-energy region at λ_{em} = 350 nm due to π–π* LE transitions based on the central phenyl rings. The fact that the high-energy emission band of *closo*-**DT** was consistently maintained in a variety of solvents of different polarities (λ_{em} = 349–350 nm, Table 1 and Figure S14a in the Supplementary Material) and that the low-energy emission of *closo*-**PT** was dramatically altered (Table 1 and Figure S14b), strongly indicates that *closo*-**DT** and *closo*-**PT** exhibit LE- and ICT-based emissive characteristics, respectively. These intriguing features are clear evidence that the planarity of the terphenyl rings plays an important role in the alternation of the intramolecular electronic transitions as well as the corresponding radiative decay mechanism [69]. Moreover, *closo*-**DT** exhibited only a trace ICT-based emission in solution (THF solution at 298 K), and the PL spectra in the rigid state (THF at 77 K and in the film state, i.e., 5 wt% doped on poly(methyl methacrylate) (PMMA)) showed an enhanced low energy emission (λ_{em} = 482 nm in THF at 77 K and λ_{em} = 492 nm in the film) that tailed to 550 nm. The emission band for *closo*-**PT** around 500 nm was also significantly increased in the rigid state (THF at 77 K and in film), indicating that the electronic transition for both *closo*-compounds are governed by non-radiative process in solution state at 298 K. This behavior originates from the increased efficiency of the radiative decay associated with the ICT transition in the rigid molecular state, which restricts structural fluctuations such as C–C bond variations in the *o*-carborane cage [9,38,66–69]. In addition, the PL spectra of the two *nido-o*-compounds in THF at 298 K exhibited identical emission patterns in the high-energy region (λ_{em} = 343 nm for *nido*-**DT** and 390 nm for *nido*-**PT**, respectively) alone, and each spectrum of the *closo*-compounds corresponded to the terphenyl-centered π–π* LE transition. Accordingly, these phenomena demonstrate that the CT-based emission can be quenched by the anionic character of *nido-o*-carborane as well as the distortion of the terphenyl rings, which inhibits the ICT transitions. Such features suggest that *closo-o*-carboranyl compounds that exhibit the intense ICT-based emission, such as *closo*-**PT**, can cause dramatic emission color changes via deboronation of the *o*-carborane cage, owing to the interruption of the ICT transition corresponding to the *o*-carborane and conservation of the LE transition. This phenomenon was verified by spectral changes in the emission of *closo*-**PT** in the presence of TBAF (vide infra).

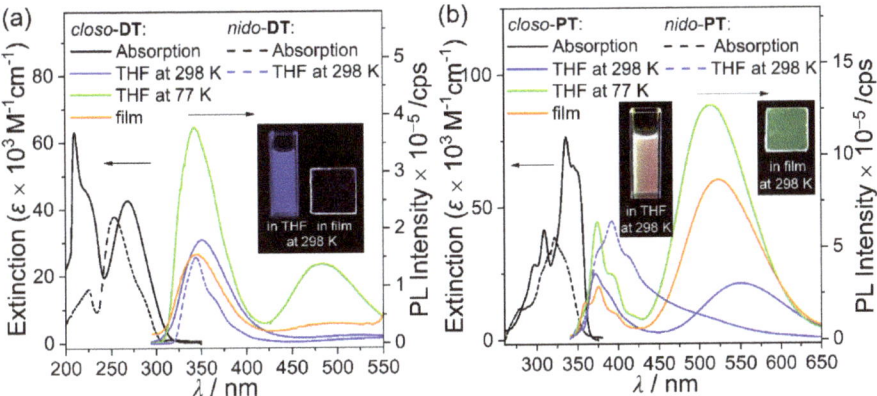

Figure 2. UV–Vis absorption and photoluminescence (PL) spectra for (**a**) *closo*- and *nido*-**DT** (λ_{ex} = 292 nm) and (**b**) *closo*- and *nido*-**PT** (λ_{ex} = 345 nm). Black-solid: absorption spectra in THF (30 µM) for *closo*-species. Black-dash: absorption spectra in THF (30 µM) for *nido*-species. Blue-solid: PL spectra in THF (30 µM) at 298 K for *closo*-species. Blue-dash: PL spectra in THF (30 µM) at 298 K for *nido*-species. Green-solid: PL spectra in THF (30 µM) at 77 K for *closo*-species. Orange-solid: PL spectra of the films (5 wt% doped on PMMA) at 298 K for *closo*-species. Inset figures show the emission color in each state of *closo*-species under irradiation by a hand-held UV lamp (λ_{ex} = 295 nm for *closo*-**DT** and 365 nm for *closo*-**PT**).

Table 1. Absorption and emission data for terphenyl-based *o*-carboranyl compounds.

Compound	λ_{abs} [1]/nm ($\varepsilon \times 10^{-3}$ M^{-1} cm^{-1})	λ_{ex}/nm	λ_{em}/nm				
			Tol [2]	THF [2]	DCM [2]	77 K [1]	Film [3]
closo-**DT**	268 (42.6)	292	349	350	349	343, 482	345, 492(sh)
nido-**DT**	254 (37.8)	292	-	343	-	-	-
closo-**PT**	336 (84.8)	345	376, 521	374, 549	375, 560	374, 514	375, 524
nido-**PT**	323 (38.2)	345	-	390	-	-	-

[1] c = 30 µM in THF. [2] c = 30 µM, observed at 298 K. [3] Measured in the film state (5 wt% doped on PMMA) at 298 K.

2.3. Computational Chemistry and Orbital Analyses for Closo-o-Carboranyl Compounds

To elucidate the nature of the electronic transitions and to analyze the orbitals of *closo*-**DT** and *closo*-**PT**, their S_0- and S_1-optimized structures were subjected to TD-DFT calculations using the B3LYP functional (Figure 3 and Table 2). To include the effects of the THF solvent [83,84], a conductor-like polarizable continuum model was chosen. The computational data for the S_0 state showed that HOMO → LUMO transitions are the major lowest-energy electronic transitions in both *closo*-*o*-carboranyl compounds. The HOMO of each compound is entirely localized on the central terphenyl group (>96%; Tables S2 and S4 in the Supplementary Material), whereas the orbital contribution of the *o*-carborane unit to each LUMO is slightly higher, at >16%. These results indicate that the lowest-energy absorptions of both *closo*-compounds are attributable to the π–π* transitions on the central terphenyl moieties, with minor contributions from the ICT transitions between the *o*-carborane and terphenyl groups as well. All calculated results based on the optimized S_0 structures are in good agreement with the experimentally observed UV/Vis absorption spectra.

In contrast, the calculated results for the S$_1$ states of *closo*-**DT** and *closo*-**PT** indicate that the major transitions associated with the low-energy emissions involve both HOMO → LUMO and HOMO → LUMO+1 transitions (Figure 3 and Table 2). Although the LUMO of each compound is significantly localized on the *o*-carborane moiety (~80%; Tables S2 and S4), each HOMO is predominantly located on the central terphenyl group (>92%). These results strongly suggest that the experimentally observed emissions in the low-energy regions mainly originate from ICT transitions between the *o*-carborane and terphenyl moieties. In addition, each LUMO+1 is mainly located on the central terphenyl group (>86%; Tables S2 and S4), strongly indicating that the intense emissions observed in the high-energy region, centered at ~350 nm for *closo*-**DT** and ~370 nm for *closo*-**PT**, originate from π–π* transitions in the terphenyl moieties, i.e., LE-based emissions. Consequently, the electronic transitions that occur in each *o*-carboranyl compound were precisely predicted using computational methods.

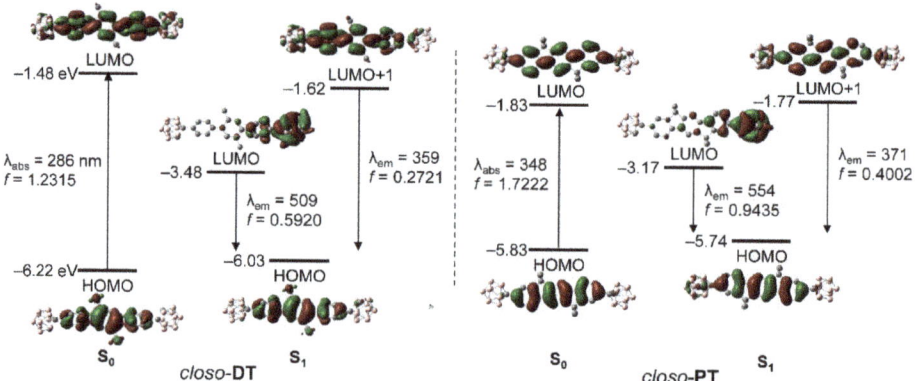

Figure 3. Frontier molecular orbitals of *closo*-**DT** and *closo*-**PT** in their ground states (S$_0$) and first excited singlet states (S$_1$), and their relative energies calculated by DFT (isovalue = 0.04). The transition energy (in nm) was calculated using the TD-B3LYP/6-31G(d) level of theory.

Table 2. Major low-energy electronic transitions in *closo*-**DT** and *closo*-**PT** involving their ground states (S$_0$) and first excited singlet states (S$_1$) calculated using the TD-B3LYP/6-31G(d) level of theory [1].

	State	λ_{calc}/nm	f_{calc}	Assignment
closo-**DT**	S$_0$	285.7	1.2315	HOMO → LUMO (98.0%)
	S$_1$	509.37	0.592	HOMO → LUMO (99.6%)
		359.14	0.2721	HOMO → LUMO+1 (87.7%)
closo-**PT**	S$_0$	348.17	1.7222	HOMO → LUMO (98.8%)
	S$_1$	554.44	0.9435	HOMO → LUMO (99.7%)
		371.34	0.4002	HOMO → LUMO+1 (78.9%)

[1] Singlet energies for vertical transitions were calculated using optimized S$_1$ geometries.

2.4. Emission-Color Changes of Closo-o-Carboranyl Compounds Via Treatment of Fluoride Anion

Finally, to clarify the changes in the photoluminescence properties exhibited during the conversion of both *closo*-**DT** and *closo*-**PT** to the *nido*-species, we investigated the changes in the emissive patterns of both *closo*-compounds as a function of increasing amounts of TBAF in THF. These conversion processes of both *closo*-compounds to the respective *nido*-species by reaction with the fluoride anion occur consecutively, as clearly evidenced from the changes in the specific peaks of the ^{1}H-NMR spectra in THF-d_8 (Figure 4). The aryl protons of both *closo*-compounds in the region from 8.0 to 7.0 ppm shifted steadily to the upfield region upon increasing the concentration of TBAF, and finally, these peaks merged with the corresponding peaks in the spectra of each *nido*-compound in THF-d_8, respectively. In particular, the broad singlet peaks around δ = −2.0 and −2.5 ppm, which were assigned to the B–H–B bridge protons of the *nido-o*-carborane, could be gradually monitored by increasing the concentration of TBAF. The results of ^{1}H-NMR spectral changes indicate that the conversion of the *closo*-compounds to the *nido*-species almost reached full conversion to that of corresponding pure *nido*-compounds when 5 equivalents of TBAF was used for the deboronation process.

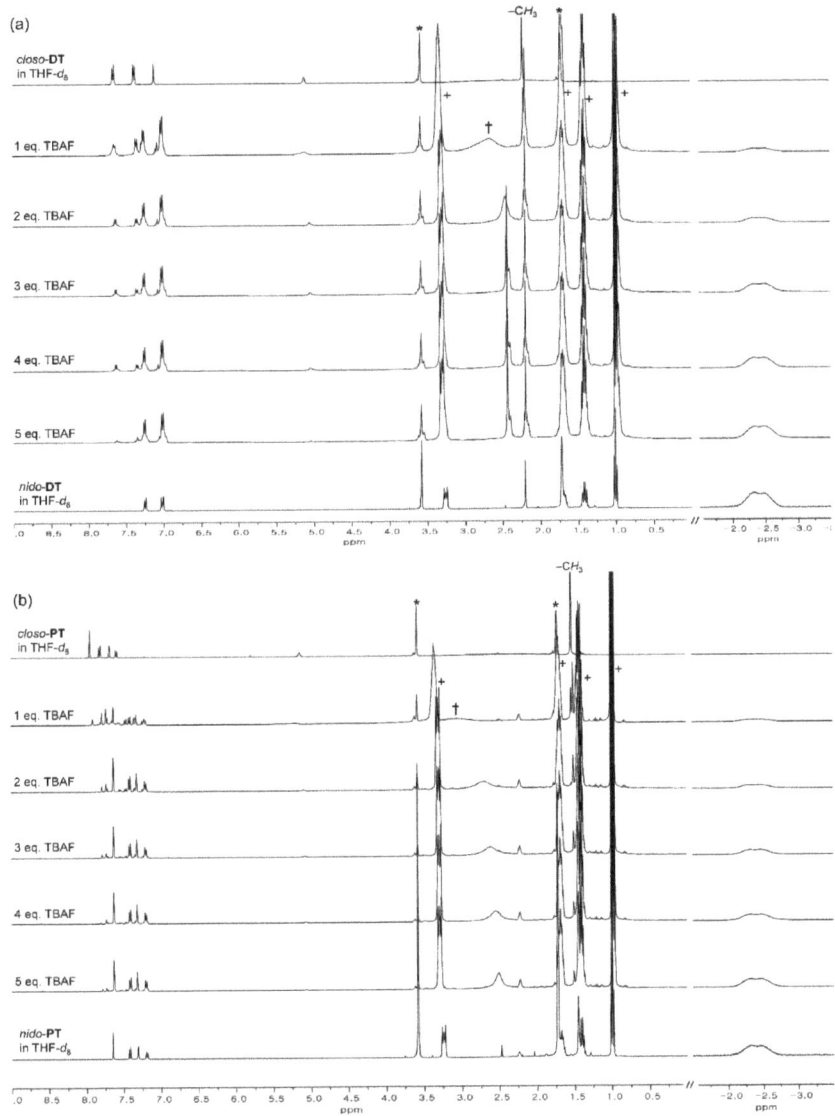

Figure 4. ^{1}H-NMR spectral changes of (**a**) *closo*-**DT** and (**b**) *closo*-**PT** upon increasing the amount of added fluoride anions and comparison with those of *nido*-**DT** and *nido*-**PT** (∗ from residual THF in THF-d^8, † from *n*-butyl group of excess TBAF, and + from *n*-butyl group for each *nido*-compound).

As illustrated in Figure 5, upon addition of incremental amounts of TBAF (0–5 equivalents) into the respective solutions of *closo*-**DT** and *closo*-**PT**, followed by heating at 60 °C for 2 h, the LE-based emission for *closo*-**DT** (λ_{em} = ~350 nm) did not change significantly, whereas the ICT-based emission for *closo*-**PT** (λ_{em} = ~550 nm) underwent gradual quenching, and eventually, a slightly enhanced LE-based emission ($\lambda_{em} \approx$ 380–410 nm) remained. In particular, the emission intensities and band shapes of each *closo*-compound after treatment with 5 equivalents of TBAF were mostly similar to those (Figure 5, red-solid lines) of the *nido*-compounds. Consequently, the conversion of *closo*-**PT** to *nido*-**PT** exhibited a vivid emission color change from orange to deep-blue (insets in Figure 5b), whereas *closo*-**DT** did not display any color changes from the emission in spite of the deboronation (insets in Figure 5a). These results demonstrate that degradation to the *nido*-form can not only prevent the ICT transition in the o-carboranyl compounds, but also reinforce the π-π*–LE transition, which induces the emission color changes. Consequently, the luminescence-based color change response due to the alternation of the intrinsic electronic transitions caused by the reaction with fluoride anion and the structural feature of central terphenyl groups, indicates the potential of *closo*-**PT** as a naked-eye-detectable chemodosimeter for fluoride ion sensing.

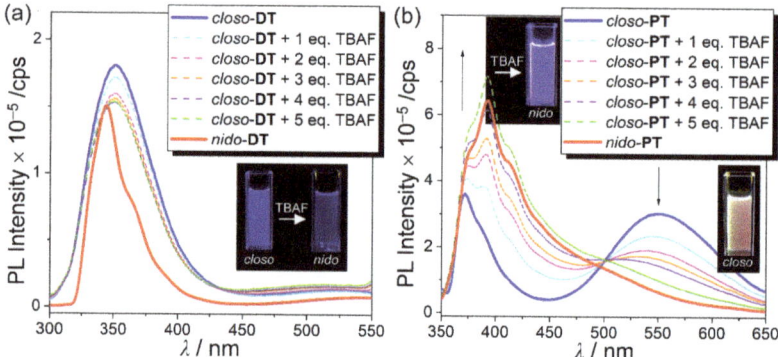

Figure 5. Spectral changes in the emission of (**a**) *closo*-**DT** (3.0×10^{-5} M, λ_{ex} = 292 nm) and (**b**) *closo*-**PT** (3.0×10^{-5} M, λ_{ex} = 345 nm) in THF in the presence of different amounts of TBAF, upon heating at 60 °C for 2 h. Insets are photographs of each *closo*- and *nido*-type (3.0×10^{-5} M in THF) under a UV lamp (λ_{ex} = 295 nm for **DT** derivatives and 365 nm for **PT** derivatives).

3. Materials and Methods

3.1. General Considerations

All operations were performed under an inert nitrogen atmosphere using standard Schlenk and glove-box techniques. Anhydrous solvents (toluene, trimethylamine (NEt$_3$), and methanol; Aldrich) were dried by passing through an activated alumina column and stored over activated molecular sieves (5 Å). Spectrophotometric-grade solvents (tetrahydrofuran (THF), toluene, dichloromethane (DCM), methanol, and *n*-hexane) were used as received from Alfa Aesar (Ward Hill, MA, USA). Commercial reagents were used without any further purification after purchase from Sigma-Aldrich (potassium carbonate (K$_2$CO$_3$), magnesium sulfate (MgSO$_4$) St. Louis, MO, USA), bis(triphenylphosphine)palladium(II) dichloride (Pd(PPh$_3$)$_2$Cl$_2$), copper(I) iodide (CuI), diethyl sulfide (Et$_2$S), ethynyltrimethylsilane, and poly(methyl methacrylate) (PMMA)). Decaborane (B$_{10}$H$_{14}$) was purchased from Alfa Aesar. The dibromo precursors, 4,4''-dibromo-2',5'-dimethyl-1,1':4',1''-terphenyl (**DT1**) and 2,8-dibromo-6,6,12,12-tetramethyl-6,12-dihydroindeno[1,2-*b*]fluorene (**PT1**), were prepared as reported in the literature [69]. CD$_2$Cl$_2$ and THF-d_8, purchased from Cambridge Isotope Laboratories, were dried over activated molecular sieves (5 Å). All nuclear magnetic resonance (NMR) spectra were recorded on a Bruker Avance 400 spectrometer (400.13 MHz for ^{1}H, 100.62 MHz for ^{13}C, and 128.38 MHz for ^{11}B, Bruker, Billerica, MA, USA) at ambient temperature. Chemical shifts are given in ppm and are referenced against external Me$_4$Si (^{1}H and ^{13}C) or BF$_3$·Et$_2$O (^{11}B). Elemental analysis was performed on an EA3000 instrument (Eurovector) at the Central Laboratory of Kangwon National University. UV–Vis absorption and photoluminescence (PL) spectra were recorded on Jasco V-530 (Jasco, Easton, MD, USA) and Horiba FluoroMax-4P spectrophotometers (HORIBA, Edison, NJ, USA), respectively. Fluorescence decay lifetimes (τ_{obs}) were measured using a time-correlated single-photon counting spectrometer (FLS920, at the Central Laboratory of Kangwon National University, Edinburgh Instruments Ltd., Livingston, UK) equipped with an EPL 375 ps pulsed semiconductor diode laser as the excitation source and a microchannel plate photomultiplier tube (200–850 nm) as the detector, at 298 K. The absolute PL quantum yields (ϕ_{em}) were obtained with an absolute PL quantum yield spectrophotometer (HORIBA FluoroMax-4P equipped with an FM-SPHERE 3.2-inch internal integrating sphere, HORIBA, Edison, NJ, USA) at 298 K.

3.2. Synthesis of DT2

Triethylamine (16 mL) was added via cannulation to a mixture of **DT1** (0.42 g, 1.0 mmol), copper iodide (15 mg), and Pd(PPh$_3$)$_2$Cl$_2$ (62 mg) at 25 °C. After stirring for 15 min, ethynyltrimethylsilane (0.55 mL, 4.0 mmol) was added, and the reaction mixture was heated at 90 °C with stirring for 24 h. After cooling to 25 °C, the volatiles were removed by rotary evaporation to afford a dark brown residue. The crude product was purified by column chromatography on silica gel (eluent: DCM/n-hexane = 1/10, v/v) to yield **DT2** as a yellow solid, 0.28 g (yield = 62%). ^{1}H-NMR (CD$_2$Cl$_2$): δ 7.51 (d, J = 8.3 Hz, 4H), 7.32 (d, J = 8.4 Hz, 4H), 7.13 (s, 2H), 2.25 (s, 6H, –CH$_3$), 0.27 (s, 18H, –Si(CH$_3$)$_3$). ^{13}C-NMR (CD$_2$Cl$_2$): δ 142.35, 140.71, 133.04, 132.05, 131.97, 129.60, 121.97, 105.26 (acetylene-C), 94.83 (acetylene-C), 20.01 (–CH$_3$), 0.03 (–Si(CH$_3$)$_3$). Anal. Calcd. for C$_{30}$H$_{34}$Si$_2$: C, 79.94; H, 7.60. Found: C, 79.87; H, 7.49.

3.3. Synthesis of PT2

PT2 was prepared according to a procedure analogous to that used for **DT2**, with **PT1** (0.47 g, 1.0 mmol), copper iodide (15 mg), Pd(PPh$_3$)$_2$Cl$_2$ (62 mg), and ethynyltrimethylsilane (0.55 mL, 4.0 mmol), and was isolated as a yellow solid (0.42 g; yield = 83%). ^{1}H-NMR (CD$_2$Cl$_2$): δ 7.78 (s, 2H), 7.71 (d, J = 7.9 Hz, 2H), 7.55 (s, 2H), 7.45 (d, J = 7.8 Hz, 2H), 1.53 (s, 12H, –CH$_3$), 0.27 (s, 18H, –Si(CH$_3$)$_3$). ^{13}C-NMR (CD$_2$Cl$_2$): δ 154.55, 154.16, 140.07, 138.92, 131.40, 126.63, 121.79, 120.08, 114.99, 106.24 (acetylene-C), 94.53 (acetylene-C), 46.96 (–C(CH$_3$)$_2$), 27.37 (–CH$_3$), 0.07 (–Si(CH$_3$)$_3$). Anal. Calcd. for C$_{34}$H$_{38}$Si$_2$: C, 81.21; H, 7.62. Found: C, 80.99; H, 7.55.

3.4. Synthesis of DT3

K$_2$CO$_3$ (0.28 g, 2.0 mmol) was dissolved in methanol (10 mL) and added to a solution of **DT2** (0.23 g, 0.5 mmol) in DCM (5 mL). After stirring for 2 h at 25 °C, the resulting mixture was treated with DCM (50 mL) and the organic layer was separated. The aqueous layer was further extracted with DCM (20 × 2 mL). The combined organic extracts were dried over MgSO$_4$, filtered, and evaporated to dryness to afford a white residue. The crude product was purified by washing with n-hexane (10 mL) to yield **DT3** as a white solid, 0.13 g (yield = 84%). ^{1}H-NMR (CD$_2$Cl$_2$): δ 7.56 (d, J = 8.0 Hz, 4H), 7.34 (d, J = 8.0 Hz, 4H), 7.13 (s, 2H), 3.18 (s, 2H, –CCH), 2.26 (s, 6H, –CH$_3$). ^{13}C-NMR (CD$_2$Cl$_2$): δ 142.68, 140.68, 133.06, 132.24, 132.07, 129.66, 120.90, 83.81 (acetylene-C), 77.69 (acetylene-C), 20.01 (–CH$_3$). Anal. Calcd. for C$_{24}$H$_{18}$: C, 94.08; H, 5.92. Found: C, 93.77; H, 5.62.

3.5. Synthesis of PT3

PT3 was prepared according to a procedure analogous to that used for **DT3** with **PT2** (0.40 g, 0.8 mmol) and K$_2$CO$_3$ (0.44 g, 3.2 mmol), and was isolated as a white solid (0.25 g; yield = 88%). ^{1}H-NMR (CD$_2$Cl$_2$): δ 7.80 (s, 2H), 7.74 (d, J = 7.9 Hz, 2H), 7.59 (s, 2H), 7.50 (d, J = 7.8 Hz, 2H), 3.20 (s, 2H, –CCH), 1.54 (s, 12H, –CH$_3$). ^{13}C-NMR (CD$_2$Cl$_2$): δ 154.59, 154.15, 140.34, 138.91, 131.64, 126.88, 120.72, 120.15, 115.06, 84.70 (acetylene-C), 77.44 (acetylene-C), 46.98 (–C(CH$_3$)$_2$), 27.36 (–CH$_3$). Anal. Calcd. for C$_{28}$H$_{22}$: C, 93.81; H, 6.19. Found: C, 93.77; H, 6.04.

3.6. Synthesis of closo-DT

Excess Et$_2$S (2.5 equiv., 1.2 mmol) was added at 25 °C to a solution of decaborane (B$_{10}$H$_{14}$, 0.52 mmol) and **DT3** (61 mg, 0.20 mmol) in toluene (20 mL). After heating to reflux, the reaction mixture was further stirred for 72 h. The solvent and volatiles were removed under vacuum and methanol (10 mL) was added. The resulting solid was filtered and redissolved in toluene. The crude product upon washing with *n*-hexane (15 mL), afforded *closo*-**DT** as a white solid (47 mg. Yield = 43%). ^{1}H{^{11}B} NMR (THF-d_8): δ 7.66 (d, *J* = 8.2 Hz, 4H), 7.38 (d, *J* = 8.1 Hz, 4H), 7.12 (s, 2H), 5.13 (s, 2H, CB-C*H*), 2.54 (br s, 8H, CB-B*H*), 2.39 (br s, 3H, CB-B*H*), 2.30 (br s, 9H, CB-B*H*), 2.23 (s, 6H, –CH$_3$). ^{13}C-NMR (THF-d_8): δ 144.05, 140.59, 133.27, 133.24, 132.34, 130.14, 128.04, 77.58 (CB-C), 61.49 (CB-C), 19.74 (–CH$_3$). ^{11}B{^{1}H} NMR (THF-d_8): δ −4.44 (3B), −6.57 (1B), −10.84 (5B), −12.77 (7B), −14.67 (4B). Anal. Calcd. for C$_{28}$H$_{38}$B$_{20}$: C, 56.92; H, 6.48. Found: C, 56.79; H, 6.33.

3.7. Synthesis of closo-PT

Closo-**PT** was prepared according to a procedure analogous to that used for *closo*-**DT**, with decaborane (B$_{10}$H$_{14}$, 0.52 mmol), **PT3** (78 mg, 0.20 mmol), and Et$_2$S (2.5 equiv.). The crude product upon washing with *n*-hexane (15 mL), afforded *closo*-**PT** as a white solid (42 mg, Yield = 35%). ^{1}H{^{11}B} NMR (THF-d_8): δ 7.94 (s, 2H), 7.81 (d, *J* = 8.1 Hz, 2H), 7.67 (s, 2H), 7.58 (d, *J* = 7.9 Hz, 2H), 5.14 (s, 2H, CB-C*H*), 2.56 (br s, 7H, CB-B*H*), 2.50 (br s, 1H, CB-B*H*), 2.39 (br s, 2H, CB-B*H*), 2.30 (br s, 10H, CB-B*H*), 1.54 (s, 12H, –CH$_3$). ^{13}C-NMR (THF-d_8): δ 155.41, 154.80, 141.67, 138.98, 133.36, 127.39, 122.52, 120.66, 115.65, 78.45 (CB-C), 61.55 (CB-C), 47.50 (–C(CH$_3$)$_2$), 27.08 (–CH$_3$). ^{11}B{^{1}H} NMR (THF-d_8): δ −2.73 (3B), −4.64 (1B), −9.09 (5B), −10.77 (7B), −12.86 (4B). Anal. Calcd. for C$_{32}$H$_{42}$B$_{20}$: C, 59.79; H, 6.59. Found: C, 59.87; H, 6.45.

3.8. Synthesis of nido-DT

Closo-**DT** (0.027 g, 0.05 mmol) was dissolved in 0.3 mL of a 0.2 M solution of *n*-tetrabutylammonium fluoride (TBAF) in THF at 25 °C. The reaction mixture was heated to reflux (60 °C) and stirred for 2 h. After cooling to 25 °C, the resulting mixture was treated with 50mL of distilled water and 50 mL of DCM, and the organic portion was separated. The aqueous layer was further extracted with DCM (20 mL). The combined organic portions were dried over MgSO$_4$, filtered, and concentrated to dryness, affording a pale yellow residue. The crude product upon washing with methanol (15 mL), afforded *nido*-**DT** as a white solid (26 mg, Yield = 52%). ^{1}H{^{11}B} NMR: δ 7.28 (d, *J* = 8.0 Hz, 4H), 7.13 (d, *J* = 7.9 Hz, 4H), 7.07 (s, 2H), 3.11 (m, 16H, *n*-butyl-CH$_2$), 2.36 (s, 2H, CB-C*H*), 2.25 (s, 6H, –CH$_3$), 2.12 (br s, 4H, CB-B*H*), 1.88 (br s, 4H, CB-B*H*), 1.82 (br s, 4H, CB-B*H*), 1.62 (m, 16H, *n*-butyl-CH$_2$), 1.43 (m, 16H, *n*-butyl-CH$_2$), 1.26 (br s, 6H, CB-B*H*), 1.02 (t, *J* = 7.2 Hz, 24H, *n*-butyl-CH$_3$), −2.36 (br s, 2H, B-*H*-B). ^{13}C NMR (CD$_2$Cl$_2$): δ 144.65, 140.78, 138.40, 132.83, 132.21, 128.64, 126.71, 59.43 (*n*-butyl-CH$_2$), 24.29 (*n*-butyl-CH$_2$), 20.18 (–CH$_3$), 20.10 (*n*-butyl-CH$_2$), 13.76 (*n*-butyl-CH$_3$). ^{11}B{^{1}H} NMR (CD$_2$Cl$_2$): δ −8.97 (3B), −10.43 (2B), −13.79 (1B), −18.28 (3B), −19.50 (1B), −23.00 (1B), −32.95 (3B), −36.10 (4B). Anal. Calcd. for C$_{60}$H$_{110}$B$_{18}$N$_2$: C, 68.37; H, 10.52; N, 2.66. Found: C, 68.11; H, 10.42; N, 2.54.

3.9. Synthesis of nido-PT

A procedure analogous to that for *nido*-**DT** was employed using *closo*-**PT** (0.027 g, 0.04 mmol) and 0.23 mL of a 0.2 M solution of TBAF in THF. The crude product, upon washing with methanol (15 mL), afforded *nido*-**PT** as a white solid (26 mg, Yield = 60%). ^{1}H{^{11}B} NMR (CD$_2$Cl$_2$): δ 7.65 (s, 2H), 7.49 (d, *J* = 7.9 Hz, 2H), 7.32 (s, 2H), 7.23 (d, *J* = 7.9 Hz, 2H), 3.08 (m, 16H, *n*-butyl-CH$_2$), 2.39 (s, 2H, CB-C*H*), 2.12 (br s, 4H, CB-B*H*), 2.00 (br s, 1H, CB-B*H*), 1.89 (br s, 5H, CB-B*H*), 1.60 (m, 16H, *n*-butyl-CH$_2$), 1.48 (s, 12H, –CH$_3$), 1.41 (m, 16H, *n*-butyl-CH$_2$), 1.31 (br s, 4H, CB-B*H*), 1.26 (br s, 4H, CB-B*H*), 1.00 (t, *J* = 7.2 Hz, 24H, *n*-butyl-CH$_3$), −2.34 (br s, 2H, B-*H*-B). ^{13}C-NMR (CD$_2$Cl$_2$): δ 153.80, 153.54, 145.25, 138.62, 136.55, 126.10, 121.40, 118.72, 114.00, 59.40 (*n*-butyl-CH$_2$), 46.71 (–C(CH$_3$)$_2$), 27.74 (–CH$_3$), 24.27 (*n*-butyl-CH$_2$), 20.09 (*n*-butyl-CH$_2$), 13.75 (*n*-butyl-CH$_3$). ^{11}B{^{1}H} NMR (CD$_2$Cl$_2$): δ −8.91

(3B), −10.45 (2B), −13.68 (1B), −18.46 (3B), −19.47 (1B), −23.09 (1B), −32.95 (3B), −36.05 (4B). Anal. Calcd. for $C_{64}H_{114}B_{18}N_2$: C, 69.49; H, 10.39; N, 2.53. Found: C, 69.30; H, 10.16; N, 2.39.

3.10. UV/Vis Absorption and Photoluminescence (PL) Experiments

The solution-phase UV–Vis absorption and PL measurements of the *closo*- and *nido-o*-carbornyl compounds were performed in degassed organic solvents with a 1 cm quartz cuvette (3.0×10^{-5} M) at 298 K. PL measurements for the *closo*-compounds were also performed in THF at 77 K and in the film state (5 wt% doped in PMMA) on 1.5×1.5 cm quartz plates (thickness = 1 mm) at 298 K.

3.11. Computational Studies

The optimized geometries for the ground (S_0) and first excited (S_1) states of both *closo-o*-carboranyl compounds (*closo*-**DT** and *closo*-**PT**) in THF were obtained using the B3LYP/6-31G(d,p) [85] level of theory. The vertical excitation energies at the optimized S_0 geometries as well as the optimized geometries of the S_1 states were calculated using time-dependent density functional theory (TD-DFT) [86] at the same level of theory. Solvent effects were included using the conductor-like polarizable continuum model (CPCM) [83,84]. All geometry optimizations were performed using the Gaussian 16 program [87]. The percent contribution of a group in a molecule to each molecular orbital was calculated with the GaussSum 3.0 program [88]. Visualizations were prepared using GaussView 6 [89].

4. Conclusions

We herein reported the preparation and characterization of distorted and planar terphenyl-based *closo*- (*closo*-**DT** and *closo*-**PT**) and *nido*- (*nido*-**DT** and *nido*-**PT**) *o*-carboranyl compounds. Although *closo*-**DT** exhibited strong π–π* LE-based emission in THF at 298 K in the high-energy region, *closo*-**PT** demonstrated intense emission in the low-energy region that was attributable to the ICT transitions involving the *o*-carborane cage. Interestingly, both *nido*-compounds exhibited LE-based emission alone in the same condition due to the anionic character of the *nido-o*-carborane cages, which cannot cause the ICT transitions. Consequently, the successful deboronation of *closo*-**PT** to *nido*-**PT** upon exposure to increasing concentration of fluoride anion leads to ratiometric emission color change from orange to deep-blue in solution. Such results strongly imply that the fine-tuning of electronic and structural features, which can control the ICT-based emission, shows the potential of *closo-o*-carboranyl compounds as candidates for naked-eye-detectable chemodosimeters for fluoride ion-sensing.

Supplementary Materials: The following are available online. Multinuclear NMR spectra (^{1}H, ^{13}C, and ^{11}B) of the *closo*- and *nido-o*-carboranyl compounds and their precursors (Figures S1–S12), and photophysical (Figures S13–S14) and computational data (Figures S15–S16 and Tables S1–S8).

Author Contributions: H.S., M.S.M., M.K., and J.H.K. synthesized the compounds and analyzed the data. D.K.A. and K.M.L. analyzed the data and wrote the paper. J.H.L. and H.H. conducted the computational study, analyzed the data, and wrote the paper. All authors have read and agreed to the published version of the manuscript.

Funding: This research was funded by the National Research Foundation of Korea (NRF) grant (NRF-2016M3A7B4909246, NRF-2017R1D1A1B03035412, NRF-2018R1D1A1B07040387, and NRF-2020R1A2C1006400) funded by the Ministry of Science and ICT and the Ministry of Education.

Acknowledgments: We thank Seonah Kim and Chan Hee Ryu (Department of Chemistry, Institute for Molecular Science and Fusion Technology, Kangwon National University) for assistance with design of synthetic routes for *o*-carboranyl compounds.

Conflicts of Interest: The authors declare no conflicts of interest.

References

1. Bregadze, V.I. Dicarba-closo-dodecaboranes $C_2B_{10}H_{12}$ and their derivatives. *Chem. Rev.* **1992**, *92*, 209–223. [CrossRef]
2. González-Campo, A.; Juárez-Pérez, E.J.; Viñas, C.; Boury, B.; Sillanpää, R.; Kivekäs, R.; Núñez, R. Carboranyl Substituted Siloxanes and Octasilsesquioxanes: Synthesis, Characterization, and Reactivity. *Macromolecules* **2008**, *41*, 8458–8466. [CrossRef]
3. Issa, F.; Kassiou, M.; Rendina, L.M. Boron in Drug Discovery: Carboranes as Unique Pharmacophores in Biologically Active Compounds. *Chem. Rev.* **2011**, *111*, 5701–5722. [CrossRef]
4. Wee, K.-R.; Cho, Y.-J.; Jeong, S.; Kwon, S.; Lee, J.-D.; Suh, I.-H.; Kang, S.O. Carborane-Based Optoelectronically Active Organic Molecules: Wide Band Gap Host Materials for Blue Phosphorescence. *J. Am. Chem. Soc.* **2012**, *134*, 17982–17990. [CrossRef] [PubMed]
5. Ferrer-Ugalde, A.; Juárez-Pérez, E.J.; Teixidor, F.; Viñas, C.; Núñez, R. Synthesis, Characterization, and Thermal Behavior of Carboranyl–Styrene Decorated Octasilsesquioxanes: Influence of the Carborane Clusters on Photoluminescence. *Chem. Eur. J.* **2013**, *19*, 17021–17030. [CrossRef] [PubMed]
6. Kim, T.; Kim, H.; Lee, K.M.; Lee, Y.S.; Lee, M.H. Phosphorescence Color Tuning of Cyclometalated Iridium Complexes by *o*-Carborane Substitution. *Inorg. Chem.* **2013**, *52*, 160–168. [CrossRef]
7. Bae, H.J.; Chung, J.; Kim, H.; Park, J.; Lee, K.M.; Koh, T.-W.; Lee, Y.S.; Yoo, S.; Do, Y.; Lee, M.H. Deep Red Phosphorescence of Cyclometalated Iridium Complexes by *o*-Carborane Substitution. *Inorg. Chem.* **2014**, *53*, 128–138. [CrossRef]
8. Asay, M.J.; Fisher, S.P.; Lee, S.E.; Tham, F.S.; Borchardt, D.; Lavallo, V. Synthesis of unsymmetrical N-carboranyl NHCs: Directing effect of the carborane anion. *Chem. Commun.* **2015**, *51*, 5359–5362. [CrossRef]
9. Lee, Y.H.; Park, J.; Jo, S.-J.; Kim, M.; Lee, J.; Lee, S.U.; Lee, M.H. Manipulation of Phosphorescence Efficiency of Cyclometalated Iridium Complexes by Substituted *o*-Carboranes. *Chem. Eur. J.* **2015**, *21*, 2052–2061. [CrossRef]
10. Núñez, R.; Tarrés, M.; Ferrer-Ugalde, A.; Fabrizi de Biani, F.; Teixidor, F. Electrochemistry and Photoluminescence of Icosahedral Carboranes, Boranes, Metallacarboranes, and Their Derivatives. *Chem. Rev.* **2016**, *116*, 14307–14378. [CrossRef]
11. Mukherjee, S.; Thilagar, P. Boron clusters in luminescent materials. *Chem. Commun.* **2016**, *52*, 1070–1093. [CrossRef]
12. Dziedzic, R.M.; Saleh, L.M.A.; Axtell, J.C.; Martin, J.L.; Stevens, S.L.; Royappa, A.T.; Rheingold, A.L.; Spokoyny, A.M. B–N, B–O, and B–CN Bond Formation via Palladium-Catalyzed Cross-Coupling of B-Bromo-Carboranes. *J. Am. Chem. Soc.* **2016**, *138*, 9081–9084. [CrossRef]
13. Kirlikovali, K.O.; Axtell, J.C.; Gonzalez, A.; Phung, A.C.; Khan, S.I.; Spokoyny, A.M. Luminescent metal complexes featuring photophysically innocent boron cluster ligands. *Chem. Sci.* **2016**, *7*, 5132–5138. [CrossRef]
14. Saleh, L.M.A.; Dziedzic, R.M.; Khan, S.I.; Spokoyny, A.M. Forging Unsupported Metal–Boryl Bonds with Icosahedral Carboranes. *Chem. Eur. J.* **2016**, *22*, 8466–8470. [CrossRef]
15. Eleazer, B.J.; Smith, M.D.; Popov, A.A.; Peryshkov, D.V. (BB)-Carboryne Complex of Ruthenium: Synthesis by Double B–H Activation at a Single Metal Center. *J. Am. Chem. Soc.* **2016**, *138*, 10531–10538. [CrossRef]
16. Wong, Y.O.; Smith, M.D.; Peryshkov, D.V. Synthesis of the First Example of the 12-Vertex-*closo*/12-Vertex-*nido* Biscarborane Cluster by a Metal-Free B–H Activation at a Phosphorus(III) Center. *Chem. Eur. J.* **2016**, *22*, 6764–6767. [CrossRef] [PubMed]
17. Chan, A.L.; Estrada, J.; Kefalidis, C.E.; Lavallo, V. Changing the Charge: Electrostatic Effects in Pd-Catalyzed Cross-Coupling. *Organometallics* **2016**, *35*, 3257–3260. [CrossRef]
18. Fisher, S.P.; El-Hellani, A.; Tham, F.S.; Lavallo, V. Anionic and zwitterionic carboranyl N-heterocyclic carbene Au(I) complexes. *Dalton Trans.* **2016**, *45*, 9762–9765. [CrossRef] [PubMed]
19. Kim, Y.; Park, S.; Lee, Y.H.; Jung, J.; Yoo, S.; Lee, M.H. Homoleptic Tris-Cyclometalated Iridium Complexes with Substituted *o*-Carboranes: Green Phosphorescent Emitters for Highly Efficient Solution-Processed Organic Light-Emitting Diodes. *Inorg. Chem.* **2016**, *55*, 909–917. [CrossRef]
20. Tu, D.; Leong, P.; Guo, S.; Yan, H.; Lu, C.; Zhao, Q. Highly Emissive Organic Single-Molecule White Emitters by Engineering o-Carborane-Based Luminophores. *Angew. Chem. Int. Ed.* **2017**, *56*, 11370–11374. [CrossRef]

21. Kirlikovali, K.O.; Axtell, J.C.; Anderson, K.; Djurovich, P.I.; Rheingold, A.L.; Spokoyny, A.M. Fine-Tuning Electronic Properties of Luminescent Pt(II) Complexes via Vertex-Differentiated Coordination of Sterically Invariant Carborane-Based Ligands. *Organometallics* **2018**, *37*, 3122–3131. [CrossRef]
22. Nar, I.; Atsay, A.; Altındal, A.; Hamuryudan, E. *o*-Carborane, Ferrocene, and Phthalocyanine Triad for High-Mobility Organic Field-Effect Transistors. *Inorg. Chem.* **2018**, *57*, 2199–2208. [CrossRef] [PubMed]
23. Grimes, R.N. *Carboranes*, 2nd ed.; Academic Press: London, UK, 2011.
24. Spokoyny, A.M. New ligand platforms featuring boron-rich clusters as organomimetic substituents. *Pure Appl. Chem.* **2013**, *85*, 903–919. [CrossRef]
25. Poater, J.; Solà, M.; Viñas, C.; Teixidor, F. π Aromaticity and Three-Dimensional Aromaticity: Two sides of the Same Coin? *Angew. Chem. Int. Ed.* **2014**, *53*, 12191–12195. [CrossRef]
26. Poater, J.; Solà, M.; Viñas, C.; Teixidor, F. Hückel's Rule of Aromaticity Categorizes Aromatic closo Boron Hydride Clusters. *Chem. Eur. J.* **2016**, *22*, 7437–7443. [CrossRef] [PubMed]
27. Núñez, R.; Romero, I.; Teixidor, F.; Viñas, C. Icosahedral boron clusters: A perfect tool for the enhancement of polymer features. *Chem. Soc. Rev.* **2016**, *45*, 5147–5173. [CrossRef]
28. Cabrera-González, J.; Sánchez-Arderiu, V.; Viñas, C.; Parella, T.; Teixidor, F.; Náñez, R. Redox-Active Metallacarborane-Decorated Octasilsesquioxanes. Electrochemical and Thermal Properties. *Inorg. Chem.* **2016**, *55*, 11630–11634. [CrossRef]
29. Kokado, K.; Chujo, Y. Multicolor Tuning of Aggregation-Induced Emission through Substituent Variation of Diphenyl-*o*-carborane. *J. Org. Chem.* **2011**, *76*, 316–319. [CrossRef]
30. Dash, B.P.; Satapathy, R.; Gaillard, E.R.; Norton, K.M.; Maguire, J.A.; Chug, N.; Hosmane, N.S. Enhanced π-Conjugation and Emission via Icosahedral Carboranes: Synthetic and Spectroscopic Investigation. *Inorg. Chem.* **2011**, *50*, 5485–5493. [CrossRef]
31. Wee, K.-R.; Han, W.-S.; Cho, D.W.; Kwon, S.; Pac, C.; Kang, S.O. Carborane photochemistry triggered by aryl substitution: Carborane-based dyads with phenyl carbazoles. *Angew. Chem. Int. Ed.* **2012**, *51*, 2677–2680. [CrossRef]
32. Weber, L.; Kahlert, J.; Brockhinke, R.; Böhling, L.; Brockhinke, A.; Stammler, H.-G.; Neumann, B.; Harder, R.A.; Fox, M.A. Luminescence Properties of C-Diazaborolyl-*ortho*-Carboranes as Donor–Acceptor Systems. *Chem. Eur. J.* **2012**, *18*, 8347–8357. [CrossRef] [PubMed]
33. Bae, H.J.; Kim, H.; Lee, K.M.; Kim, T.; Eo, M.; Lee, Y.S.; Do, Y.; Lee, M.H. Heteroleptic tris-cyclometalated iridium(III) complexes supported by an *o*-carboranyl-pyridine ligand. *Dalton Trans.* **2013**, *42*, 8549–8552. [CrossRef] [PubMed]
34. Weber, L.; Kahlert, J.; Brockhinke, R.; Böhling, L.; Halama, J.; Brockhinke, A.; Stammler, H.-G.; Neumann, B.; Nervi, C.; Harder, R.A.; et al. C,C'-Bis(benzodiazaborolyl)dicarba-*closo*-dodecaboranes: Synthesis, structures, photophysics and electrochemistry. *Dalton Trans.* **2013**, *42*, 10982–10996. [CrossRef] [PubMed]
35. Weber, L.; Kahlert, J.; Böhling, L.; Brockhinke, A.; Stammler, H.-G.; Neumann, B.; Harder, R.A.; Low, P.J.; Fox, M.A. Electrochemical and spectroelectrochemical studies of C-benzodiazaborolyl-*ortho*-carboranes. *Dalton Trans.* **2013**, *42*, 2266–2281. [CrossRef] [PubMed]
36. Kwon, S.; Wee, K.-R.; Cho, Y.-J.; Kang, S.O. Carborane Dyads for Photoinduced Electron Transfer: Photophysical Studies on Carbazole and Phenyl-*o*-carborane Molecular Assemblies. *Chem. Eur. J.* **2014**, *20*, 5953–5960. [CrossRef]
37. Ferrer-Ugalde, A.; González-Campo, A.; Viñas, C.; Rodríguez-Romero, J.; Santillan, R.; Farfán, N.; Sillanpää, R.; Sousa-Pedrares, A.; Núñez, R.; Teixidor, F. Fluorescence of New *o*-Carborane Compounds with Different Fluorophores: Can it be Tuned? *Chem. Eur. J.* **2014**, *20*, 9940–9951. [CrossRef]
38. Bae, H.J.; Kim, H.; Lee, K.M.; Kim, T.; Lee, Y.S.; Do, Y.; Lee, M.H. Through-space charge transfer and emission color tuning of di-*o*-carborane substituted benzene. *Dalton Trans.* **2014**, *43*, 4978–4985. [CrossRef]
39. Lee, Y.H.; Park, J.; Lee, J.; Lee, S.U.; Lee, M.H. Iridium Cyclometalates with Tethered *o*-Carboranes: Impact of Restricted Rotation of *o*-Carborane on Phosphorescence Efficiency. *J. Am. Chem. Soc.* **2015**, *137*, 8018–8021. [CrossRef]
40. Naito, H.; Morisaki, Y.; Chujo, Y. *o*-Carborane-Based Anthracene: A Variety of Emission Behaviors. *Angew. Chem. Int. Ed.* **2015**, *54*, 5084–5087. [CrossRef]
41. Kim, T.; Lee, J.; Lee, S.U.; Lee, M.H. *o*-Carboranyl–Phosphine as a New Class of Strong-Field Ancillary Ligand in Cyclometalated Iridium(III) Complexes: Toward Blue Phosphorescence. *Organometallics* **2015**, *34*, 3455–3458. [CrossRef]

42. Choi, B.H.; Lee, J.H.; Hwang, H.; Lee, K.M.; Park, M.H. Novel Dimeric o-Carboranyl Triarylborane: Intriguing Ratiometric Color-Tunable Sensor via Aggregation-Induced Emission by Fluoride Anions. *Organometallics* **2016**, *35*, 1771–1777. [CrossRef]
43. Wee, K.-R.; Cho, Y.-J.; Song, J.K.; Kang, S.O. Multiple photoluminescence from 1,2-dinaphthyl-ortho-carborane. *Angew. Chem. Int. Ed.* **2013**, *52*, 9682–9685. [CrossRef] [PubMed]
44. Naito, H.; Nishino, K.; Morisaki, Y.; Tanaka, K.; Chujo, Y. Solid-State Emission of the Anthracene-o-Carborane Dyad from the Twisted-Intramolecular Charge Transfer in the Crystalline State. *Angew. Chem. Int. Ed.* **2017**, *56*, 254–259. [CrossRef] [PubMed]
45. Wu, X.; Guo, J.; Cao, Y.; Zhao, J.; Jia, W.; Chen, Y.; Jia, D. Mechanically triggered reversible stepwise tricolor switching and thermochromism of anthracene-o-carborane dyad. *Chem. Sci.* **2018**, *9*, 5270–5277. [CrossRef] [PubMed]
46. Li, J.; Yang, C.; Peng, X.; Chen, Y.; Qi, Q.; Luo, X.; Lai, W.-Y.; Huang, W. Stimuli-responsive solid-state emission from o-carborane–tetraphenylethene dyads induced by twisted intramolecular charge transfer in the crystalline state. *J. Mater. Chem. C* **2018**, *6*, 19–28. [CrossRef]
47. Nishino, K.; Yamamoto, H.; Tanaka, K.; Chujo, Y. Development of Solid-State Emissive Materials Based on Multifunctional o-Carborane–Pyrene Dyads. *Org. Lett.* **2016**, *18*, 4064–4067. [CrossRef] [PubMed]
48. Wu, X.; Guo, J.; Zhao, J.; Che, Y.; Jia, D.; Chen, Y. Multifunctional luminescent molecules of o-carborane-pyrene dyad/triad: Flexible synthesis and study of the photophysical properties. *Dyes and Pigm.* **2018**, *154*, 44–51. [CrossRef]
49. Marsh, A.V.; Cheetham, N.J.; Little, M.; Dyson, M.; White, A.J.P.; Beavis, P.; Warriner, C.N.; Swain, A.C.; Stavrinou, P.N.; Heeney, M. Carborane-Induced Excimer Emission of Severely Twisted Bis-o-Carboranyl Chrysene. *Angew. Chem. Int. Ed.* **2018**, *57*, 10640–10645. [CrossRef]
50. Kim, S.-Y.; Cho, Y.-J.; Jin, G.F.; Han, W.-S.; Son, H.-J.; Cho, D.W.; Kang, S.O. Intriguing emission properties of triphenylamine–carborane systems. *Phys. Chem. Chem. Phys.* **2015**, *17*, 15679–15682. [CrossRef]
51. Wan, Y.; Li, J.; Peng, X.; Huang, C.; Qi, Q.; Lai, W.-Y.; Huang, W. Intramolecular charge transfer induced emission from triphenylamine-o-carborane dyads. *RSC Adv.* **2017**, *7*, 35543–35548. [CrossRef]
52. Nishino, K.; Uemura, K.; Gon, M.; Tanaka, K.; Chujo, Y. Enhancement of Aggregation-Induced Emission by Introducing Multiple o-Carborane Substitutions into Triphenylamine. *Molecules* **2017**, *22*, 2009. [CrossRef] [PubMed]
53. Nishino, K.; Uemura, K.; Tanaka, K.; Morisaki, Y.; Chujo, Y. Modulation of the cis- and trans-Conformations in Bis-ocarborane Substituted Benzodithiophenes and Emission Enhancement Effect on Luminescent Efficiency by Solidification. *Eur. J. Org. Chem.* **2018**, *12*, 1507–1512. [CrossRef]
54. Ferrer-Ugalde, A.; Juárez-Pérez, E.J.; Teixidor, F.; Viñas, C.; Sillanpää, R.; Pérez-Inestrosa, E.; Núñez, R. Synthesis and Characterization of New Fluorescent Styrene-Containing Carborane Derivatives: The Singular Quenching Role of a Phenyl Substituent. *Chem. Eur. J.* **2012**, *18*, 544–553. [CrossRef] [PubMed]
55. Tu, D.; Leong, P.; Li, Z.; Hu, R.; Shi, C.; Zhang, K.Y.; Yan, H.; Zhao, Q. A carborane-triggered metastable charge transfer state leading to spontaneous recovery of mechanochromic luminescence. *Chem. Commun.* **2016**, *52*, 12494–12497. [CrossRef] [PubMed]
56. Ferrer-Ugalde, A.; Cabrera-González, J.; Juárez-Pérez, E.J.; Teixidor, F.; Pérez-Inestrosa, E.; Montenegro, J.M.; Sillanpää, R.; Haukka, M.; Núñez, R. Carborane–stilbene dyads: The influence of substituents and cluster isomers on photoluminescence properties. *Dalton Trans.* **2017**, *46*, 2091–2104. [CrossRef]
57. Li, X.; Yin, Y.; Yan, H.; Lu, C. Aggregation-Induced Emission Characteristics of o-Carborane-Functionalized Tetraphenylethylene Luminogens: The Influence of Carborane Cages on Photoluminescence. *Chem. Asian J.* **2017**, *12*, 2207–2210. [CrossRef]
58. Kaiser, R.P.; Mosinger, J.; Císařová, I.; Kotora, M. Synthesis of selectively 4-substituted 9,9'-spirobifluorenes and modulation of their photophysical properties. *Org. Biomol. Chem.* **2017**, *15*, 6913–6920. [CrossRef]
59. Santos, W.G.; Budkina, D.S.; Deflon, V.M.; Tarnovsky, A.N.; Cardoso, D.R.; Forbes, M.D.E. Photoinduced Charge Shifts and Electron Transfer in Viologen–Tetraphenylborate Complexes: Push–Pull Character of the Exciplex. *J. Am. Chem. Soc.* **2017**, *139*, 7681–7684. [CrossRef]
60. Naito, H.; Nishino, K.; Morisaki, Y.; Tanaka, K.; Chujo, Y. Luminescence Color Tuning from Blue to Near Infrared of Stable Luminescent Solid Materials Based on Bis-o-Carborane-Substituted Oligoacenes. *Chem. Asian J.* **2017**, *12*, 2134–2138. [CrossRef]

61. Naito, H.; Nishino, K.; Morisaki, Y.; Tanaka, K.; Chujo, Y. Highly-efficient solid-state emissions of anthracene–o-carborane dyads with various substituents and their thermochromic luminescence properties. *J. Mater. Chem. C.* **2017**, *5*, 10047–10054. [CrossRef]
62. Wu, X.; Guo, J.; Quan, Y.; Jia, W.; Jia, D.; Chen, Y.; Xie, Z. Cage carbon-substitute does matter for aggregation-induced emission features of o-carborane-functionalized anthracene triads. *J. Mater. Chem. C.* **2018**, *6*, 4140–4149. [CrossRef]
63. Mori, H.; Nishino, K.; Wada, K.; Morisaki, Y.; Tanaka, K.; Chujo, Y. Modulation of luminescence chromic behaviors and environment-responsive intensity changes by substituents in bis-o-carborane-substituted conjugated molecules. *Mater. Chem. Front.* **2018**, *2*, 573–579. [CrossRef]
64. Chen, Y.; Guo, J.; Wu, X.; Jia, D.; Tong, F. Color-tuning aggregation-induced emission of o-Carborane-bis(1,3,5-triaryl-2-pyrazoline) triads: Preparation and investigation of the photophysics. *Dyes Pigm.* **2018**, *148*, 180–188. [CrossRef]
65. Kim, S.-Y.; Lee, J.-D.; Cho, Y.-J.; Son, M.R.; Son, H.-J.; Cho, D.W.; Kang, S.O. Excitation spectroscopic and synchronous fluorescence spectroscopic analysis of the origin of aggregation-induced emission in N,N-diphenyl-1-naphthylamine-o-carborane derivatives. *Phys. Chem. Chem. Phys.* **2018**, *20*, 17458–17463. [CrossRef] [PubMed]
66. Shin, N.; Yu, S.; Lee, J.H.; Hwang, H.; Lee, K.M. Biphenyl- and Fluorene-Based o-Carboranyl Compounds: Alteration of Photophysical Properties by Distortion of Biphenyl Rings. *Organometallics* **2017**, *36*, 1522–1529. [CrossRef]
67. Jin, H.; Bae, H.J.; Kim, S.; Lee, J.H.; Hwang, H.; Park, M.H.; Lee, K.M. 2-Phenylpyridine- and 2-(benzo[b]thiophen-2-yl)pyridine-based o-carboranyl compounds: Impact of the structural formation of aromatic rings on photophysical properties. *Dalton Trans.* **2019**, *48*, 1467–1476. [CrossRef]
68. Jin, H.; Kim, S.; Bae, H.J.; Lee, J.H.; Hwang, H.; Park, M.H.; Lee, K.M. Effect of Planarity of Aromatic Rings Appended to o-Carborane on Photophysical Properties: A Series of o-Carboranyl Compounds Based on 2-Phenylpyridine- and 2-(Benzo[b]thiophen-2-yl)pyridine. *Molecules* **2019**, *24*, 201. [CrossRef]
69. So, H.; Kim, J.H.; Lee, J.H.; Hwang, H.; An, D.K.; Lee, K.M. Planarity of terphenyl rings possessing o-carborane cages: Turning on intramolecular-charge-transfer-based emission. *Chem. Commun.* **2019**, *55*, 14518. [CrossRef]
70. Kim, S.; Lee, J.H.; So, H.; Ryu, J.; Lee, J.; Hwang, H.; Kim, Y.; Park, M.H.; Lee, K.M. Spirobifluorene-Based o-Carboranyl Compounds: Insights into the Rotational Effect of Carborane Cages on Photoluminescence. *Chem. Eur. J.* **2020**, *26*, 548. [CrossRef]
71. Nishino, K.; Morisaki, Y.; Tanaka, K.; Chujo, Y. Electron-donating abilities and luminescence properties of tolane-substituted *nido*-carboranes. *New J. Chem.* **2017**, *41*, 10550. [CrossRef]
72. Fox, M.A.; Gill, W.R.; Herbertson, P.L.; MacBride, J.A.H.; Wade, K.; Colquhoun, H.M. Deboronation of C-substituted *ortho*- and *meta-closo*-carboranes using "wet" fluoride ion solutions. *Polyhedron* **1996**, *15*, 565. [CrossRef]
73. Yoo, J.; Hwang, J.-W.; Do, Y. Facile and Mild Deboronation of o-Carboranes Using Cesium Fluoride. *Inorg. Chem.* **2001**, *40*, 568. [CrossRef]
74. Peterson, J.J.; Werre, M.; Simon, Y.C.; Coughlin, E.B.; Carter, K.R. Carborane-Containing Polyfluorene: O-Carborane in the Main Chain. *Macromolecules* **2009**, *42*, 8594. [CrossRef]
75. Lerouge, F.; Ferrer-Ugalde, A.; Viñas, C.; Teixidor, F.; Sillanpää, R.; Abreu, A.; Xochitiotzi, E.; Farfán, N.; Santillan, R.; Núñez, R. Synthesis and fluorescence emission of neutral and anionic di- and tetra-carboranyl compounds. *Dalton Trans.* **2011**, *40*, 7541. [CrossRef]
76. Park, M.H.; Lee, K.M.; Kim, T.; Do, Y.; Lee, M.H. *Ortho*-Carborane-Functionalized Luminescent Polyethylene: Potential Chemodosimeter for the Sensing of Nucleophilic Anions. *Chem. Asian J.* **2011**, *6*, 1362. [CrossRef]
77. You, D.K.; Lee, J.H.; Hwang, H.; Kwon, H.; Park, M.H.; Lee, K.M. Deboronation-induced ratiometric emission sensing of fluoride by 1,3,5-tris-(o-carboranyl-methyl)benzene. *Tetrahedron Lett.* **2017**, *58*, 3246. [CrossRef]
78. Song, K.C.; Kim, H.; Lee, K.M.; Lee, Y.S.; Do, Y.; Lee, M.H. Dual sensing of fluoride ions by the o-carborane–triarylborane dyad. *Dalton Trans.* **2013**, *42*, 2351. [CrossRef]
79. Hawthorne, M.F.; Berry, T.E.; Wegner, P.A. The Electronic Properties of the 1,2- and 1,7-Dicarbaclovododecaborane(12) Groups Bonded at Carbon. *J. Am. Chem. Soc.* **1965**, *87*, 4746–4750. [CrossRef]

80. Paxson, T.E.; Callahan, K.P.; Hawthorne, M.F. Improved synthesis of biscarborane and its precursor ethynylcarborane. *Inorg. Chem.* **1973**, *12*, 708–709. [CrossRef]
81. Jiang, W.; Knobler, C.B.; Hawthorne, M.F. Synthesis and Structural Characterization of Bis- and Tris(*closo*-1,2-$C_2B_{10}H_{11}$-1-yl)-Substituted Biphenyl and Benzene. *Inorg. Chem.* **1996**, *35*, 3056–3058. [CrossRef]
82. Lee, J.H.; Hwang, H.; Lee, K.M. *p*-Terphenyl-based di-*o*-carboranyl compounds: Alteration of electronic transition state by terminal phenyl groups. *J. Organomet. Chem.* **2016**, *825–826*, 69–74. [CrossRef]
83. Cossi, M.; Rega, N.; Scalmani, G.; Barone, V. Energies, structures, and electronic properties of molecules in solution with the C-PCM solvation model. *J. Comput. Chem.* **2003**, *24*, 669–681. [CrossRef] [PubMed]
84. Barone, V.; Cossi, M. Quantum Calculation of Molecular Energies and Energy Gradients in Solution by a Conductor Solvent Model. *J. Phys. Chem. A.* **1998**, *102*, 1995–2001. [CrossRef]
85. Binkley, J.S.; Pople, J.A.; Hehre, W.J. Self-consistent molecular orbital methods. 21. Small split-valence basis sets for first-row elements. *J. Am. Chem. Soc.* **1980**, *102*, 939–947. [CrossRef]
86. Runge, E.; Gross, E.K.U. Density-Functional Theory for Time-Dependent Systems. *Phys. Rev. Lett.* **1984**, *52*, 997. [CrossRef]
87. *Gaussian*, version 09 B.01; Gaussian. Inc.: Wallingford, CT, USA, 2016.
88. O'Boyle, N.M.; Tenderholt, A.L.; Langner, K.M. cclib: A library for package-independent computational chemistry algorithms. *J. Comp. Chem.* **2008**, *29*, 839–845. [CrossRef]
89. *GaussView*, version 6; Semichem Inc.: Shawnee Mission, KS, USA, 2016.

Sample Availability: Samples of the *o*-carboranyl compounds are available from the authors.

© 2020 by the authors. Licensee MDPI, Basel, Switzerland. This article is an open access article distributed under the terms and conditions of the Creative Commons Attribution (CC BY) license (http://creativecommons.org/licenses/by/4.0/).

Article

Reactions of Dihaloboranes with Electron-Rich 1,4-Bis(trimethylsilyl)-1,4-diaza-2,5-cyclohexadienes

Li Ma [†], Xiaolin Zhang [†], Wenbo Ming [†], Shengxin Su, Xiaoyong Chang and Qing Ye *

Department of Chemistry, Southern University of Science and Technology, Shenzhen 518055, China; 11849423@mail.sustech.edu.cn (L.M.); zhangxl6003@163.com (X.Z.); wenbo.ming@hotmail.com (W.M.); 11510019@mail.sustech.edu.cn (S.S.); changxy@sustech.edu.cn (X.C.)
* Correspondence: yeq3@sustech.edu.cn; Tel.: +86-0-755-88018354
[†] These authors contributed equally to this work.

Academic Editor: Ashok Kakkar
Received: 31 May 2020; Accepted: 16 June 2020; Published: 22 June 2020

Abstract: The reactions of electron-rich organosilicon compounds 1,4-bis(trimethylsilyl)-1,4-diaza-2,5-cyclohexadiene (**1**), 2,3,5,6-tetramethyl-1,4-bis(trimethylsilyl)-1,4-diaza-2,5-cyclohexadiene (**2**), and 1,1′-bis(trimethylsilyl)-1,1′-dihydro-4,4′-bipyridine (**12**) with B-amino and B-aryl dihaloboranes afforded a series of novel B=N-bond-containing compounds **3–11** and **13**. The B=N rotational barriers of **7** (>71.56 kJ/mol), **10** (58.79 kJ/mol), and **13** (58.65 kJ/mol) were determined by variable-temperature ^{1}H-NMR spectroscopy, thus reflecting different degrees of B=N double bond character in the corresponding compounds. In addition, ring external olefin isomers **11** were obtained by a reaction between **2** and DurBBr$_2$. All obtained B=N-containing products were characterized by multinuclear NMR spectroscopy. Compounds **5**, **9**, **10a**, **11**, and **13a** were also characterized by single-crystal X-ray diffraction analysis.

Keywords: 1,4-bis(trimethylsilyl)-1,4-diaza-2,5-cyclohexadienes; salt-free reduction; rotational barrier; B=N bond

1. Introduction

Low-valent boron compounds are a class of highly reactive species that have been the focus of intense research because of their unique electronic properties [1,2] as well as their diverse and fascinating reactivity patterns such as inert bond activation [3,4], cycloaddition reaction [5–7], and small molecule activation [8]. The progress in this research area is highlighted by the very recent results in terms of borylene-mediated N$_2$ activation [9] and N$_2$ coupling [10]. Nonetheless, the synthetic approach to low-valent boron species is severely limited [11–13]. Almost all of the reported synthetic strategies require a strong metallic reducing agent (e.g., Li, K, Na, KC$_8$) [3,4,14–23], harsh reaction conditions, and a strict moisture- and oxygen-free atmosphere. Therefore, the exploration of metal-free reductants to access low-valent boron species is highly desirable [24–26].

Mashima et al. reported a class of electron-rich organosilicon compounds **1**, **2**, and **12**, which can serve as versatile reducing reagents for the group 4–6 metal chloride complexes. The corresponding low-valent metal species were prepared in a salt-free manner [27–33]. The reducing power mainly derives from the aromatization of the central 1,4-diaza-2,5-cyclohexadiene ring. Deeply inspired by the advantage of the salt-free reduction protocol and easy workup, we decided to examine the ability of the organosilicon compounds **1**, **2**, and **12** for the reduction of trivalent dihaloboranes. Based on the published results, the disubstituted compounds ArXB(N$_2$C$_4$R$_4$)BXAr are proposed as the reduction products. We hypothesized two possible bonding modes (i.e., **A** and **B** in Scheme 1) between the C$_4$N$_2$ ring and the boron atoms. In the first manner, B–N is bound by an electron-precise σ bond (**A**, Scheme 1) and an additional N—B dative π bond. In the second manner, two nitrogen atoms each provide a π-electron for 6π-aromatization, while the remaining two valence electrons form a lone pair on each N atom, donating to the empty sp^2-hybridized orbital of boron, thus leading to the divalent boron radical centers (**B**, Scheme 1).

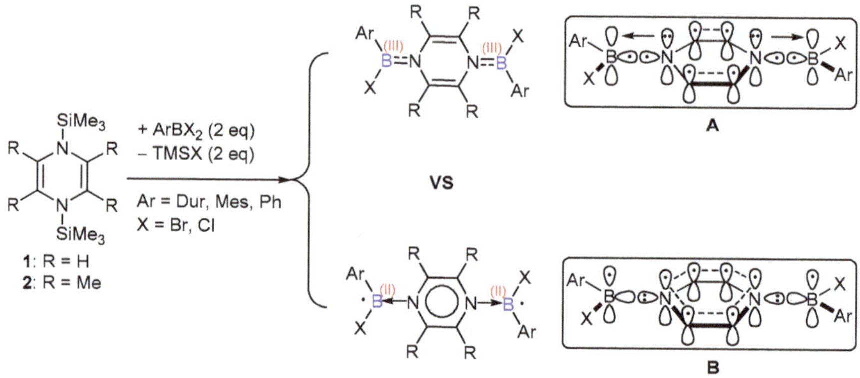

Scheme 1. Proposed products from the reactions of ArBX$_2$ with **1** and **2**, and two possible bonding modes **A** and **B** between the central C$_4$N$_2$ ring and the boron centers.

2. Results and Discussion

First, we examined compound **1** for its ability to reduce ArBX$_2$. The results are summarized in Scheme 2. Compound **1** [34] and ArBX$_2$ (Ar = 2,3,5,6-tetramethylphenyl (Dur), 2,4,6-trimethylphenyl (Mes)) [35] were prepared according to the literature. The reaction of **1** with an equimolar amount of DurBBr$_2$ and MesBCl$_2$ at ambient temperature afforded the expected monosubstituted products **3** and **4**, respectively. Adding the second equiv. of dihaloboranes to the reaction mixture led to the disubstituted products **5** and **6**. In stark contrast, the reaction of **1** with an equimolar amount of the less sterically demanding PhBCl$_2$ caused precipitation, which is insoluble in all ordinary solvents. This is most likely due to the polymerization of PhClBC$_4$N$_2$H$_4$SiMe$_3$ by chlorosilane elimination. Hence, the stepwise synthetic protocol is unsuitable for the synthesis of **7**. Instead, **1** was directly treated with 2 equiv. of PhBCl$_2$ at room temperature (RT), affording **7** in an acceptable yield (48%). Hence, the reaction of the monosubstituted intermediate (i.e., PhClBC$_4$N$_2$H$_4$SiMe$_3$) with PhBCl$_2$ should be much faster than the self-polymerization process. Compounds **3–7** were confirmed by NMR spectroscopic (Figures S1–S15) and HRMS studies. Furthermore, the multinuclear NMR spectroscopic study revealed that the isolated **5–7** all consist of ca. 1:1 cis-trans isomers in the solution phase at ambient temperature due to the nonrotatable B=N double bond (see Electronic Supporting Information (ESI)).

Suitable single crystals of **5** for X-ray diffraction analysis were obtained by slow evaporation of a saturated hexane solution. Two isomers, **5a** and **5b**, co-crystallized in the unit cell. The result is depicted in Figure 1 and Figure S37. The central C$_4$N$_2$ ring is nearly planar. The endocyclic N1–C2 (1.385(7) Å), N2–C3 (1.415(7) Å), C1–C2 (1.311(9) Å), and C3–C3* (1.326(8) Å) distances lie in the expected range for N–C single bonds and C=C double bonds. The bond lengths of B1–N1 (1.423(8) Å) and B2–N2 (1.400(8) Å) are shorter than that of a B–N single bond, which is indicative of a significant B=N double bond character. All these geometric parameters suggest the bonding mode **A** in Scheme 1. Therefore, both boron centers adopt a formal oxidation state of +3.

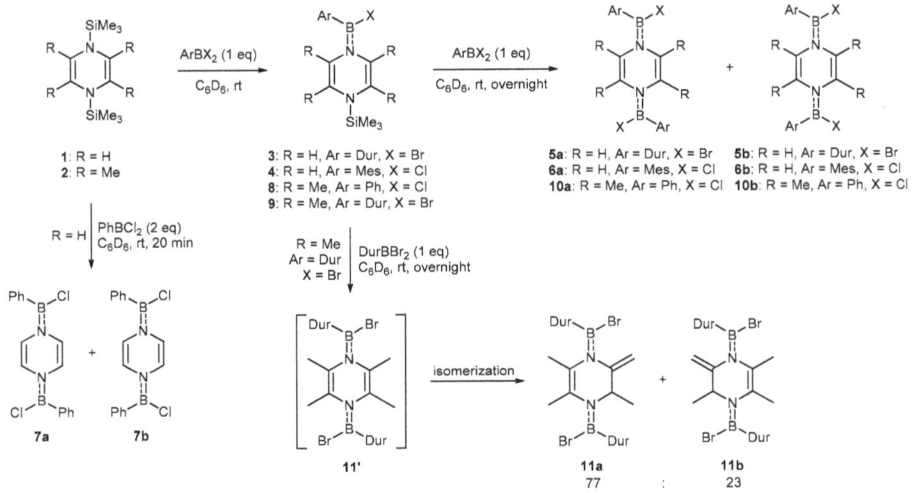

Scheme 2. Synthesis of **3–11**.

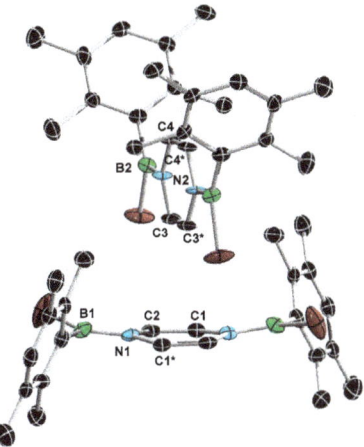

Figure 1. Molecular structures of **5a** (bottom) and **5b** (top) in the solid state (ellipsoids set at 50% probability). Hydrogen atoms are omitted for clarity. Selected bond lengths (Å) and angles (°) for **5a**: B1–N1 1.423(8), N1–C2 1.385(7), C1–C2 1.311(9), C1–C2–N1 123.1(5), C2–N1–C1* 113.9(4); for **5b**: B2–N2 1.400(8), N2–C4 1.411(6), N2–C3 1.415(7), C4–C4* 1.328(8), C3–C3* 1.326(8), C4*–C4–N2 123.7(4), C4–N2–C3 112.6(4), N2–C3–C3* 123.7(5).

Compound **2**, which features a less-negative redox potential (+0.10 V) with respect to **1** (−0.24 V), was further examined to reduce PhBCl$_2$ and DurBBr$_2$ (Scheme 2). Differing from the aforementioned reactions with **1**, both monosubstituted products **8** and **9** could be prepared upon a 1:1 ratio reaction of **2** with PhBCl$_2$ and DurBBr$_2$, respectively. Upon the reaction of **2** with two equiv. of PhBCl$_2$ at RT, the disubstituted compounds **10a** and **10b** were obtained as 1:1 cis-trans isomers. Surprisingly, treatment of **2** with two equiv. of DurBBr$_2$ at ambient temperature led to the formation of **11**, which can be regarded as the product from an isomerization of **11'** [36,37]. Compounds **8–11** were confirmed by NMR spectroscopic (Figures S16–S28) and HRMS studies. There were two sets (intensity ratio of ca. 1:0.3) of ^{1}H signals between 4 and 6 ppm, each consisting of three multiplets with the integration ratio of 1:1:1, which can be assigned to the migrated H and two remaining olefinic protons. These are the most characteristic signs for the formation of the isomerized product. After assigning each peak (with the help of the NOE spectrum, see the ESI Figure S26 for more details), we could determine that the ratio of isomers **11a** and **11b** was 77:23. Furthermore, the isomerization was also observed upon the treatment of the isolated **9** with an equimolar amount of DurBBr$_2$ at RT.

The structures of **9**, **10a**, **11a**, and **11b** were confirmed by single-crystal X-ray diffraction analysis (Figure 2 and Figures S38–S40). All four compounds adopt a boat conformation, which could be explained by the small energy difference between the planar and nonplanar geometry of the C$_4$N$_2$ ring, and the steric congestion between the central exocyclic methyl groups and the bulky boron substituents. Bond lengths (Å) of **9** (B1–N1 1.376(7), N1–C1 1.451(6), C1–C2 1.343(7), N2–C2 1.434(6), N2–Si1 1.759(5)), **10** (B1–N1 1.401(2), N1–C2 1.4494(17), C1–C2 1.328(2), N1–C1* 1.4513(18)) are all as expected. The overall structures of **11a** and **11b** resemble that of **10a**. However, since the C3 position in **11a** and **11b** accepted one H atom from the methyl group at the C2 position, respectively, and thus became sp^3-hybridized, the torsion angles C4–C3–N2–B2 (**11a**: 94.25°; **11b**: 94.69°) are notably greater than those at the other three carbon positions (57–63°) in the central six-membered ring. Due to the disordered nature of the crystal, the bond lengths of **11a** and **11b** cannot be further discussed.

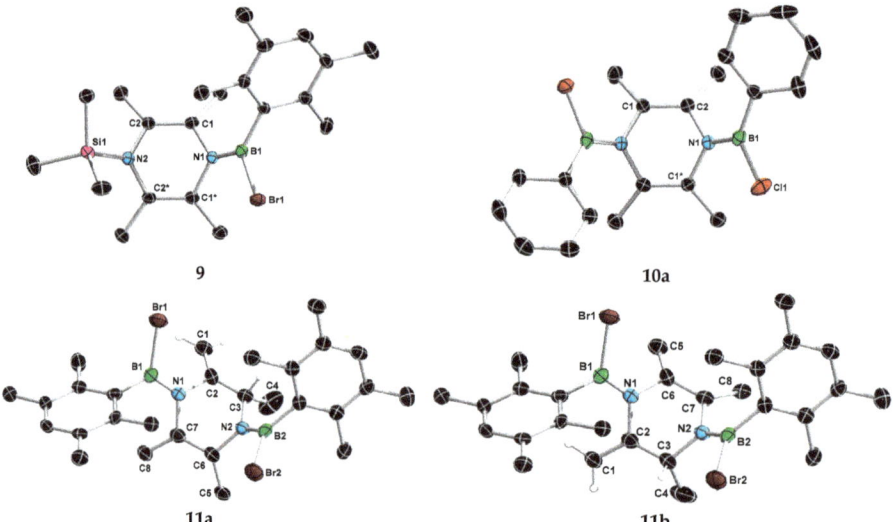

Figure 2. Molecular structures of **9**, **10a**, and **11** in the solid state (ellipsoids set at 50% probability). Hydrogen atoms, except for the C(sp^3)–H and the olefinic H in **11a** and **11b**, are omitted for clarity. Selected bond lengths (Å) and angles (°) for **9**: B1–N1 1.376(7), N1–C1 1.451(6), C1–C2 1.343(7), N2–C2 1.434(6), N2–Si1 1.759(5), C2–C1–N1 115.3(4), C1–N1–C1* 111.1(4); for **10a**: B1–N1 1.401(2), N1–C2 1.4494(17), C1–C2 1.328(2), N1–C1* 1.4513(18), C1–C2–N1 116.25(12), C2–N1–C1* 110.63(11).

The reaction of (SiMe$_3$)$_2$NBCl$_2$ with **12** [38] of greater reducing power (redox potential of −0.40 V) [39] was performed at ambient temperature in C$_6$D$_6$. After the removal of the solvent and extraction with hexane, an NMR spectroscopically pure product **13** was obtained with a yield of 75%. Compound **13** was confirmed by NMR spectroscopic (Figures S29–S31) and HRMS studies. Suitable single crystals of **13a** for X-ray diffraction analysis were obtained upon storage of the reaction mixture overnight at RT (Figure 3 and Figure S41). The N1–C1/N1–C5 (1.402(8)–1.414(8) Å), C1–C2/C5–C4 (1.332(8)–1.347(9) Å), C2–C3/C3–C4 (1.444(9)–1.452(9) Å), C3–C3* (1.376(12) Å) distances are in line with the Lewis structure depicted in Scheme 3.

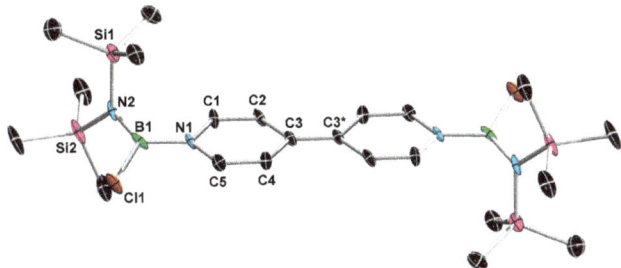

Figure 3. Molecular structure of **13a** in the solid state (ellipsoids set at 50% probability). Selected bond lengths (Å) and angles (°) for **13a**: B1–N1 1.459(8), B1–N2 1.392(10), Si1–N2 1.757(5), Si2–N2 1.770(5), N1–C1 1.402(8), N1–C5 1.414(8), C1–C2 1.332(8), C4–C5 1.347(9), C2–C3 1.452(9), C3–C4 1.444(9), C3–C3* 1.376(12), C1–N1–C5 115.7(5), C2–C1–N1 123.3(6), C1–C2–C3 122.9(6), C4–C3–C2 112.7(5), C5–C4–C3 123.3(6), C4–C5–N1 122.1(6).

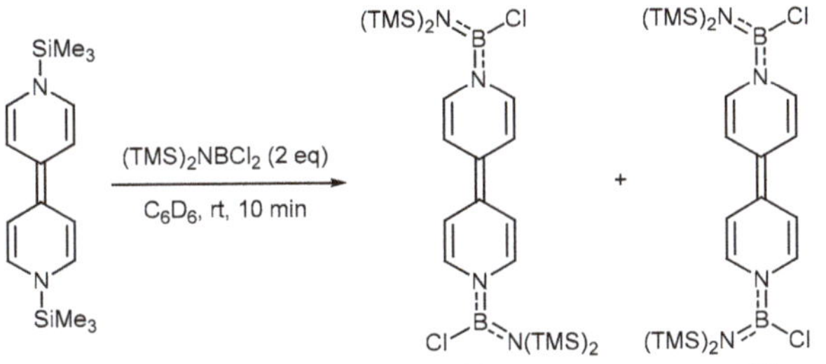

Scheme 3. Synthesis of **13**.

Apparently, both the RT-NMR spectroscopic and crystallographic studies failed to prove any successful reduction of the trivalent borane to divalent boron radical. Since the rotational barrier around an N—B dative bond should be lower than that of a B=N double bond, we assumed that any contribution from the bonding mode **B** (Scheme 1) should slightly lower the rotational barrier around the exocyclic B–N bond. In this context, we conducted a variable-temperature ^{1}H-NMR experiment to provide further insight. Toluene-d^8 was selected as the solvent with a temperature ranging from −60 °C to 80 °C. In general, the exocyclic H or Me as marked in Figure 4 (top right) should display two signals if the B=N bond is nonrotatable. The separated signals will coalesce at an elevated temperature when the B–N bond overcomes the rotational barrier and begins to rotate. Determination of the separation (Hz) of two signals and the coalescent temperature allows calculation of the B–N rotational barrier. The results of the VT-NMR experiments and assignment of the signals of interest are depicted in Figure 4 and Figures S32–S36. The obtained ΔG$^≠$ values are summarized in Table 1. Analysis of the VT-NMR spectra revealed **7** with a strong B=N bond, and **10** and **13** with weak B=N bonds, as reflected by their rotational barriers >71.56 kJ/mol (**7**), 58.79 kJ/mol (**10**), 58.65 kJ/mol (**13**) when compared with ordinary B=N double bonds (71–100 kJ/mol) [40]. When taking the aforementioned assumption into account, the remarkably lower B–N rotational barrier of **10** compared to that of **7** is not in line with the fact that the reducing power of **2** is slightly weaker than that of **1**, according to the CV data. Therefore, the lower B–N rotational barrier in **10** should be mainly due to its boat conformation, which allows for less steric hindrance. Furthermore, although the central C$_4$N$_2$ rings of **7** and **13** both adopt a planar structure, the B–N rotational barrier in **13** is significantly lower than that of **7**. This finding could be explained by the competition in π donation from another *B*-amino function (–N(SiMe$_3$)$_2$) in **13**.

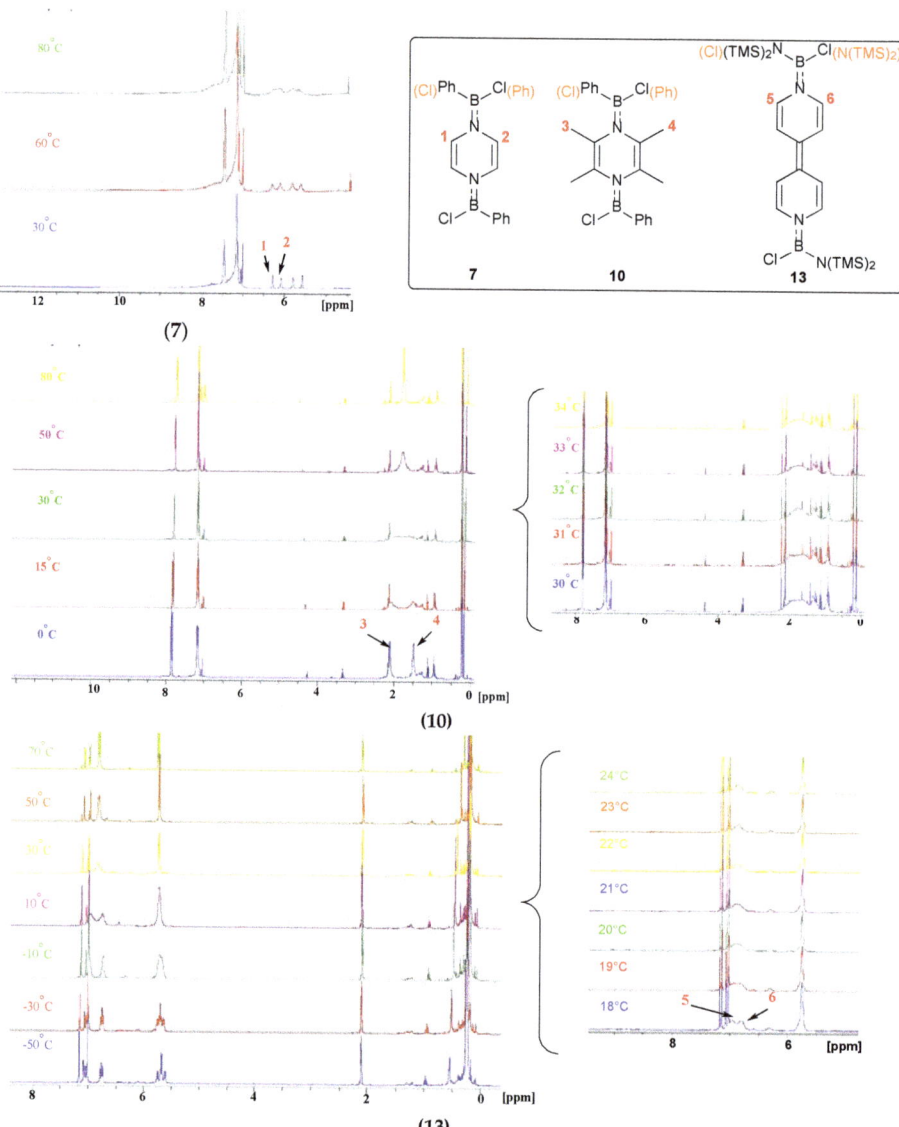

Figure 4. Variable-temperature ^{1}H-NMR (400 MHz, toluene-d^8) spectra of **7**, **10**, and **13**.

Table 1. Rotational barrier of **7**, **10**, and **13**.

Compound	Tc	Δν	ΔG$^{\neq}$
7	>80 °C (353 K)	85.0 Hz	>71.56 kJ/mol
10	32 °C (305 K)	243.5 Hz	58.79 kJ/mol
13	20 °C (293 K)	96.0 Hz	58.65 kJ/mol

Tc = coalescence temperature; Δν = the separation in hertz between the two singlets in the absence of exchange; ΔG$^{\neq}$ = rotational barrier.

3. Materials and Methods

3.1. General Information

All manipulations were performed under dry argon using standard Schlenk line or glovebox techniques. Solvents were purified by distillation from Na under dry argon. C_6D_6 was dried over an Na/K alloy and then degassed by freeze–pump–thaw cycles. PhBCl$_2$ was purchased from Beijing MREDA Technologie Co., Ltd., without any special treatment before use. The NMR spectra were acquired on a Bruker AVANCE 400 (^{1}H: 400 MHz, ^{13}C{^{1}H}: 101 MHz, ^{11}B: 128 MHz) NMR spectrometer at 298 K. Variable-temperature NMR experiments were conducted on a Bruker AVANCE 400 NMR spectrometer (^{1}H: 400 MHz, 213–353 K). Chemical shifts are given in ppm. ^{1}H and ^{13}C{^{1}H} NMR spectra were referenced to an external tetramethylsilane (TMS) via the residual protons of the solvent (^{1}H) or the solvent itself (^{13}C{^{1}H}). ^{11}B NMR spectra were referenced to the external BF$_3$·OEt$_2$. High-resolution mass spectrometry (HMRS) was performed with a Thermo Fisher Scientific Q Exactive Mass Spectrometer (MS) system.

3.2. Synthesis of 3 and 4:

In the glove box, DurBBr$_2$ (30.3 mg, 0.1 mmol, 1.0 equiv.) and 1,4-bis(trimethylsilyl)-1,4-diaza-2,5-cyclohexadiene (**1**) (22.6 mg, 0.1 mmol, 1.0 equiv.) were added into C_6D_6 (0.6 mL) in a J. Young NMR tube. The mixture was rested for 10 min prior to the removal of the volatiles under vacuum to get **3** as a pale yellow solid (23.6 mg, 72 mmol, 63%). Compound **4** was synthesized in a similar manner, with a yield of 64%.

3: ^{1}**H-NMR** (400 MHz, C_6D_6): δ = 6.86 (s, 1H, H of Dur), 6.35 (d, *J* = 6.6 Hz, 1H, H of C_4N_2), 5.13 (d, *J* = 6.5 Hz, 1H, H of C_4N_2), 4.91 (d, *J* = 6.6 Hz, 1H, H of C_4N_2), 4.64 (d, *J* = 6.5 Hz, 1H, H of C_4N_2), 2.29 (s, 6H, Me of Dur), 2.06 (s, 6H, Me of Dur), −0.23 (s, 9H, Me of TMS). ^{13}**C{^{1}H}-NMR** (101 MHz, C_6D_6): δ = 134.5, 133.4, 132.0, 119.8, 118.7, 113.5, 112.8, 19.4, 18.4, −2.3 (9C, C of TMS). The carbon atom directly attached to boron was not detected, likely due to quadrupolar broadening. ^{11}**B-NMR** (128 MHz, C_6D_6): δ = 34.1. **HRMS**: calc. for [M]$^+$ $C_{17}H_{26}BBrN_2Si^+$ 376.11362; found: 376.11308.

4: ^{1}**H-NMR** (400 MHz, C_6D_6): δ = 6.75 (s, 2H, H of Mes), 6.17 (d, *J* = 6.6 Hz, 1H, H of C_4N_2), 5.07 (d, *J* = 6.5 Hz, 1H, H of C_4N_2), 4.91 (d, *J* = 6.6 Hz, 1H, H of C_4N_2), 4.65 (d, *J* = 6.5 Hz, 1H, H of C_4N_2), 2.37 (s, 6H, Me of Mes), 2.15 (s, 3H, Me of Mes), 0.21 (s, 9H, Me of TMS). ^{13}**C{^{1}H}-NMR** (101 MHz, C_6D_6): δ = 139.2, 137.8, 127.5, 119.4, 118.5, 112.4 (1C, C of C_4N_2), 112.8 (1C, C of C_4N_2), 21.3 (2C, *o*-CH$_3$ of Mes), 20.9 (C, *p*-CH$_3$ of Mes), −2.3 (9C, C of TMS). The carbon atom directly attached to boron was not detected, likely due to quadrupolar broadening. ^{11}**B-NMR** (128 MHz, C_6D_6): δ = 34.2. **HRMS**: calc. for [M]$^+$ $C_{16}H_{24}BClN_2Si^+$ 318.14848; found: 318.14877.

3.3. Synthesis of 5–7

In the glove box, DurBBr$_2$ (60.4 mg, 0.2 mmol, 2.0 equiv.) and 1,4-bis(trimethylsilyl)-1,4-diaza-2,5-cyclohexadiene (**1**) (22.6 mg, 0.1 mmol, 1.0 equiv.) were added into C_6D_6 (0.6 mL) in a J. Young NMR tube. The mixture was rested overnight prior to the removal of the volatiles under vacuum to get **5** as a yellow oil with a 45% yield. Mixture **5** contains the *cis*-structure **5a** and *trans*-structure **5b**, and the ratio of the *cis-trans* isomers was about 1:1. Compounds **6–7** were synthesized in a similar manner, with a *cis-trans* isomers ratio of about 1:1 (yield: 52% (**6**) and 48% (**7**)].

5a + 5b: ^{1}**H-NMR** (400 MHz, C_6D_6): δ = 6.86 (s, 2H), 6.82 (s, 2H), 6.52 (s, 2H), 6.26 (d, $^3J_{H\text{-}H}$ = 1.6 Hz, 1H), 6.24 (d, $^3J_{H\text{-}H}$ =1.6 Hz, 1H), 5.36 (d, $^3J_{H\text{-}H}$ = 1.6 Hz, 1H), 5.34 (d, $^3J_{H\text{-}H}$ = 1.6 Hz, 1H), 4.99 (s, 2H), 2.09 (s, 12 H), 2.07 (s, 12 H), 2.03 (s, 12 H), 2.01 (s, 12H). ^{13}**C{^{1}H}-NMR** (101 MHz, C_6D_6): δ = 134.0, 133.9, 133.6, 133.5, 132.5, 132.3, 118.2, 117.7, 117.2, 116.6, 19.2, 19.1, 18.5, 18.4. The carbon atom directly attached to boron was not detected, likely due to quadrupolar broadening. ^{11}**B-NMR** (128 MHz, C_6D_6): δ = 38.9. **HRMS**: calc. for [M]$^+$ $C_{24}H_{30}N_2B_2Br_2^+$ 526.09564; found: 526.09549.

6a + 6b: ^{1}**H-NMR** (400 MHz, C_6D_6): δ = 6.69 (s, 4H), 6.67 (s, 4H), 6.34 (s, 2H), 6.09 (d, $^3J_{H\text{-}H}$ = 1.6 Hz, 1H), 6.08 (d, $^3J_{H\text{-}H}$ = 1.6 Hz, 1H), 5.32 (d, $^3J_{H\text{-}H}$ = 1.6 Hz, 1H), 5.30 (d, $^3J_{H\text{-}H}$ = 1.6 Hz, 1H), 4.94 (s,

2H), 2.19 (s, 12H), 2.16 (s, 12H), 2.13 (s, 6H), 2.12 (s, 6H). ^{13}C{^{1}H}-NMR (101 MHz, C$_6$D$_6$): δ = 138.9, 138.8, 138.6, 138.5, 127.6, 127.6, 117.3, 116.4, 116.3, 115.5, 21.2, 21.1, 20.9, 20.8. The carbon atom directly attached to boron was not detected, likely due to quadrupolar broadening. ^{11}B-NMR (128 MHz, C$_6$D$_6$): δ = 38.5. **HRMS**: calc. for [M]$^+$ C$_{22}$H$_{26}$N$_2$B$_2$Cl$^+$ 410.16537; found: 410.16492.

7a + 7b: ^{1}H-NMR (400 MHz, C$_6$D$_6$): δ = 7.86 (s, 2 H), 7.51–7.46 (m, 8 H), 7.19–7.15 (m, 10 H), 6.29 (s, 2H), 6.08 (d, $^3J_{H-H}$ = 1.68 Hz, 1H), 6.06 (d, $^3J_{H-H}$ = 1.64 Hz, 1H), 5.78 (d, $^3J_{H-H}$ = 1.60 Hz, 1H), 5.76 (d, $^3J_{H-H}$ = 1.70 Hz, 1H), 5.54 (s, 2H). ^{13}C{^{1}H}-NMR (101 MHz, C$_6$D$_6$): δ = 133.3, 133.2, 130.2, 130.1, 127.9, 127.8, 118.0, 117.5, 116.8, 116.5. The carbon atom directly attached to boron was not detected, likely due to quadrupolar broadening. ^{11}B-NMR (128 MHz, C$_6$D$_6$): δ = 36.5. **HRMS**: calc. for [M]$^+$ C$_{16}$H$_{14}$N$_2$B$_2$Cl$_2$$^+$ 326.07147; found: 326.07069.

3.4. Synthesis of 8 and 9

Compounds **8** and **9** were synthesized in a similar manner as **3** and **4**, with yields of 65% and 75%, respectively.

8: ^{1}H-NMR (400 MHz, C$_6$D$_6$): δ = 7.93–7.90 (m, 2H), 7.85–7.83 (m, 1H), 7.24–7.13 (m, 2H), 2.14 (s, 3H), 1.68 (s, 3H), 1.62 (s, 3H), 1.52 (s, 3H), 0.19 (s, 9H). ^{13}C{^{1}H}-NMR (101 MHz, C$_6$D$_6$): δ = 132.5, 132.2, 131.9, 131.9, 131.6, 128.6, 128.0, 122.8, 121.8, 16.8, 16.7, 16.7, 16.5, 0.2. The carbon atom directly attached to boron was not detected, likely due to quadrupolar broadening. ^{11}B-NMR (128 MHz, C$_6$D$_6$): δ = 35.9. **HRMS**: calc. for [M+H]$^+$ C$_{17}$H$_{27}$N$_2$BClSi$^+$ 333.17196; found: 333.17233.

9: ^{1}H-NMR (400MHz, C$_6$D$_6$): δ = 6.88 (s, 1H), 2.43 (s, 3H), 2.29 (s, 3H), 2.24 (s, 3H, Me of Dur), 2.10 (s, 3H), 2.06 (s, 3H), 1.71 (s, 3H), 1.48 (s, 3H), 1.44 (s, 3H), 0.25 (s, 9H). ^{13}C{^{1}H}-NMR (101 MHz, C$_6$D$_6$): δ = 133.8, 133.2, 133.1, 133.0, 132.8, 131.6, 131.4, 122.5, 122.0, 19.6, 19.4, 19.3, 19.3, 19.2, 18.1, 18.0, 16.1, 1.8. The carbon atom directly attached to boron was not detected, likely due to quadrupolar broadening. ^{11}B-NMR (128 MHz, C$_6$D$_6$): δ = 37.7. **HRMS**: calc. for [M+H]$^+$ C$_{21}$H$_{35}$N$_2$BBrSi$^+$ 433.18405; found: 433.18483.

3.5. Synthesis of 10

In the glove box, PhBCl$_2$ (31.6 mg, 0.2 mmol, 2.0 equiv.) and 2,3,5,6-tetramethyl-1,4-bis(trimethylsilyl)-1,4-diaza-2,5-cyclohexadiene (**2**) (28.2 mg, 0.1 mmol, 1.0 equiv.) were added into C$_6$D$_6$ (0.6 mL) in a J. Young NMR tube. The mixture was rested overnight prior to the removal of the volatiles under vacuum to get **10** as a pale yellow solid (24.9 mg, 0.53 mmol, 53%).

10a + 10b: ^{1}H-NMR (400 MHz, toluene-d^8): δ = 7.72–7.20 (m, 8H), 7.11–7.06 (m, 12H), 1.95 (br, 12H), 1.45 (br, 12H). ^{13}C{^{1}H}-NMR (101 MHz, toluene-d^8): δ = 133.2, 130.2, 129.1, 128.6, 128.2, 128.0, 127.9, 127.7, 125.4, 124.9, 20.8, 20.7, 20.3, 20.1. The carbon atom directly attached to boron was not detected, likely due to quadrupolar broadening. ^{11}B-NMR (128 MHz, toluene-d^8): δ = 36.9. **HRMS**: calc. for [M+H]$^+$ C$_{20}$H$_{23}$N$_2$B$_2$Cl$_2$$^+$ 383.14189; found: 383.14083.

3.6. Synthesis of 11a and 11b

In the glove box, **9** (37.6 mg, 0.1 mmol, 1 equiv.) and DurBBr$_2$ (30.3 mg, 1 mmol, 1.0 equiv.) were added into C$_6$D$_6$ (0.6 mL) in a J. Young NMR tube. The mixture was rested overnight prior to the removal of the volatiles under vacuum to get **11** as a yellow oil with a yield of 76%. Mixture **11** contains two olefin isomers, **11a** and **11b**, with a ratio of about 1:0.3.

11a: ^{1}H-NMR (400 MHz, C$_6$D$_6$): δ = 6.91 (s, 1H), 6.89 (s, 1H), 5.14–5.13 (m, 1H), 4.72–4.71 (m, 1H), 4.38 (s, 1H), 2.58 (s, 3H), 2.55 (s, 3H), 2.35 (s, 3H), 2.32 (s, 3H), 2.11 (s, 3H), 2.09 (s, 6H), 2.07 (s, 3H), 1.99 (d, $^3J_{H-H}$ = 0.8 Hz, 3H), 1.41 (d, $^3J_{H-H}$ = 0.8 Hz, 3H), 0.91 (d, $^3J_{H-H}$ = 6.6 Hz, 3H).

11b: ^{1}H-NMR (400 MHz, C$_6$D$_6$): δ = 6.89 (s, 1H), 6.88 (s, 1H), 5.75–5.70 (m, 1H), 4.47–4.46 (m, 1H), 4.07–4.06 (m, 1H), 2.54 (s, 3H), 2.52 (s, 3H), 2.36 (s, 3H), 2.33 (s, 3H), 2.10 (s, 3H), 2.09 (s, 9H), 2.07 (s, 3H), 1.98 (d, $^3J_{H-H}$ = 1.1 Hz, 3H), 1.45 (d, $^3J_{H-H}$ = 1.1 Hz, 3H), 1.13 (d, $^3J_{H-H}$ = 8.0 Hz, 3H).

11a + 11b: ^{13}C{^{1}H}-NMR (101 MHz, C$_6$D$_6$): δ = 152.0, 151.4, 133.7, 133.6, 133.6, 133.5, 133.4, 133.3, 133.2, 133.2, 132.6, 132.2, 132.1, 132.0, 131.7, 131.4, 131.4, 131.1, 130.8, 129.9, 105.6, 101.7, 61.2,

60.8, 22.8, 22.6, 20.2, 20.1, 20.0, 19.7, 19.7, 19.6, 19.5, 19.4, 19.3, 19.3, 19.2, 19.2, 19.2, 19.1, 18.9, 18.1, 17.9, 15.4. The carbon atom directly attached to boron was not detected, likely due to quadrupolar broadening. ^{11}**B-NMR** (128 MHz, C_6D_6): δ = 39.1. **HRMS**: calc. for [M+H]$^+$ $C_{28}H_{39}N_2B_2Br_2{}^+$ 585.16402; found: 585.16454.

3.7. Synthesis of 13a and 13b

In the glove box, (TMS)$_2$NBCl$_2$ (86.4 mg, 0.2 mmol, 2 equiv.) and 1,1'-bis(trimethylsilyl)-1H,1'H-4,4'-bipyridinylidene (**12**) (30.2 mg, 0.1 mmol, 1 equiv.) were added into C_6D_6 (0.6 mL). The mixture was rested for 10 min prior to the removal of the volatiles under vacuum to get a yellowish green powder. The yellowish green powder was extracted with hexane, filtered, and the solvent was again removed under reduced pressure to yield **13** as a yellow powder (60.0 mg, 0.84 mmol, 84%). The ratio of **13a:13b** is ca. 1:1.

13a + 13b: ^{1}**H-NMR** (400 MHz, C_6D_6): δ = 6.87 (br, 8H), 5.79 (d,$^3J_{H-H}$ = 8.1 Hz, 8H), 0.21 (s, 72H). ^{13}C{^{1}H}-**NMR** (101 MHz, C_6D_6): δ = 114.3, 111.3, 2.25 (C of TMS). The carbon atom directly attached to boron was not detected, likely due to quadrupolar broadening. ^{11}**B-NMR** (128 MHz, C_6D_6): δ = 34.4. **HRMS**: calc. for [M]$^+$ $C_{22}H_{44}N_4B_2Cl_2Si_4{}^+$ 568.22007; found: 568.21887.

4. Conclusions

In summary, the reactions of electron-rich organosilicon compounds **1**, **2**, and **12** with various B-amino and B-aryl dihaloboranes were comprehensively studied. No direct evidence for the presence of divalent boron radical character could be obtained from NMR spectra and single-crystal structures. The rotational barrier around the exocyclic B–N bonds was studied by VT ^{1}H-NMR spectroscopy, which revealed relatively small barriers for **10** and **13**. The steric hindrance as well as the competition from additional B-amino functions were the main factors affecting the B–N rotational barrier. In addition, the reaction between **2** and DurBBr$_2$ resulted in **11** via an isomerization process. Although this study does not access the desired biradial species, we believe that the novel B=N-containing products could act as an RXB• source upon the liberation of the aromatic linker (i.e., pyrazine and 4,4'-bipyridine). Studies of the mechanism of the isomerization reaction, as well as the application of **10** and **13** as RXB• transfer reagents to unsaturated organic substrates, are currently underway in our laboratory, and will be reported in due course.

Supplementary Materials: Supplementary materials are available online. Figures S1–S31: NMR spectra for **3–11**, **13**. Figures S32–S36: Variable-temperature ^{1}H-NMR spectra for **7**, **10**, and **13**. Figures S37–S41, Single crystal structure for **5**, **9–11**, and **13**. Table S1: Crystal data for **5**, **9–11**, and **13**.

Author Contributions: Q.Y. conceived and designed the experiments; L.M., X.Z., S.S., and W.M. performed the experiments; L.M., W.M., and Q.Y. analyzed the data; L.M., X.Z., and X.C. tested and refined single crystals. L.M., W.M., and Q.Y. wrote the paper. All authors have read and agreed to the published version of the manuscript.

Funding: This research was funded by the start-up fund of SUSTech.

Acknowledgments: The authors thank Zhanying Ren, Yinhua Yang, and Jianfei Qu (SUSTech) for the VT NMR measurement, and Hua Li (SUSTech) for the HRMS measurement.

Conflicts of Interest: The authors declare no conflicts of interest.

References

1. Krasowska, M.; Bettinger, H.F. Reactivity of borylenes toward ethyne, ethene, and methane. *J. Am. Chem. Soc.* **2012**, *134*, 17094–17103. [CrossRef] [PubMed]
2. Krasowska, M.; Edelmann, M.; Bettinger, H.F. Electronically excited states of borylenes. *J. Phys. Chem. A* **2016**, *120*, 6332–6341. [CrossRef] [PubMed]
3. Wang, Y.; Robinson, G.H. Carbene stabilization of highly reactive main-group molecules. *Inorg. Chem.* **2011**, *50*, 12326–12337. [CrossRef] [PubMed]

4. Bissinger, P.; Braunschweig, H.; Damme, A.; Dewhurst, R.D.; Kupfer, T.; Radacki, K.; Wagner, K. Generation of a carbene-stabilized bora-borylene and its insertion into a C–H bond. *J. Am. Chem. Soc.* **2011**, *133*, 19044–19047. [CrossRef] [PubMed]
5. Bissinger, P.; Braunschweig, H.; Kraft, K.; Kupfer, T. Trapping the elusive parent borylene. *Angew. Chem. Int. Ed.* **2011**, *50*, 4704–4707. [CrossRef] [PubMed]
6. Curran, D.P.; Boussonnière, A.; Geib, S.J.; Lacôte, E. The parent borylene: Betwixt and between. *Angew. Chem. Int. Ed.* **2012**, *51*, 1602–1605. [CrossRef]
7. Braunschweig, H.; Claes, C.; Damme, A.; Deißenberger, A.; Dewhurst, R.D.; Hörl, C.; Kramer, T. A facile and selective route to remarkably inert monocyclic NHC-stabilized boriranes. *Chem. Commun.* **2015**, *51*, 1627–1630. [CrossRef]
8. Frey, G.D.; Lavallo, V.; Donnadieu, B.; Schoeller, W.W.; Bertrand, G. Facile splitting of hydrogen and ammonia by nucleophilic activation at a single carbon center. *Science* **2007**, *316*, 439–441. [CrossRef]
9. Légaré, M.-A.; Bélanger-Chabot, G.; Dewhurst, R.D.; Welz, E.; Krummenacher, I.; Engels, B.; Braunschweig, H. Nitrogen fixation and reduction at boron. *Science* **2018**, *359*, 896–900. [CrossRef]
10. Légaré, M.-A.; Rang, M.; Bélanger-Chabot, G.; Schweizer, J.I.; Krummenacher, I.; Bertermann, R.; Arrowsmith, M.; Holthausen, M.C.; Braunschweig, H. The reductive coupling of dinitrogen. *Science* **2019**, *363*, 1329–1332. [CrossRef]
11. Krasowska, M.; Bettinger, H.F. Ring enlargement of three-membered boron heterocycles upon reaction with organic π systems: Implications for the trapping of borylenes. *Chem. Eur. J.* **2016**, *22*, 10661–10670. [CrossRef] [PubMed]
12. Timms, P.L. Boron-fluorine chemistry. I. Boron monofluoride and some derivatives. *J. Am. Chem. Soc.* **1967**, *89*, 1629–1632. [CrossRef]
13. Timms, P.L. Chemistry of boron and silicon subhalides. *Acc. Chem. Res.* **1973**, *6*, 118–123. [CrossRef]
14. Wang, Y.; Quillian, B.; Wei, P.; Wannere, C.S.; Xie, Y.; King, R.B.; Schaefer, H.F.; Schleyer, P.v.R.; Robinson, G.H. A stable neutral diborene containing a B=B double bond. *J. Am. Chem. Soc.* **2007**, *129*, 12412–12413. [CrossRef] [PubMed]
15. Kinjo, R.; Donnadieu, B.; Celik, M.A.; Frenking, G.; Bertrand, G. Synthesis and characterization of a neutral tricoordinate organoboron isoelectronic with amines. *Science* **2011**, *333*, 610–613. [CrossRef]
16. Woon, E.C.Y.; Tumber, A.; Kawamura, A.; Hillringhaus, L.; Ge, W.; Rose, N.R.; Ma, J.H.Y.; Chan, M.C.; Walport, L.J.; Che, K.H.; et al. Linking of 2-oxoglutarate and substrate binding sites enables potent and highly selective inhibition of Jmjc histone demethylases. *Angew. Chem. Int. Ed.* **2012**, *51*, 1631–1634. [CrossRef]
17. Kong, L.; Li, Y.; Ganguly, R.; Vidovic, D.; Kinjo, R. Isolation of a bis(oxazol-2-ylidene)–phenylborylene adduct and its reactivity as a boron-centered nucleophile. *Angew. Chem. Int. Ed.* **2014**, *53*, 9280–9283. [CrossRef]
18. Wang, H.; Zhang, J.; Lin, Z.; Xie, Z. Synthesis and structural characterization of carbene-stabilized carborane-fused azaborolyl radical cation and dicarbollyl-fused azaborole. *Organometallics* **2016**, *35*, 2579–2582. [CrossRef]
19. Lu, W.; Li, Y.; Ganguly, R.; Kinjo, R. Alkene-carbene isomerization induced by borane: Access to an asymmetrical diborene. *J. Am. Chem. Soc.* **2017**, *139*, 5047–5050. [CrossRef]
20. Wang, H.; Zhang, J.; Lin, Z.; Xie, Z. The synthesis and structure of a carbene-stabilized iminocarboranyl-boron(I) compound. *Chem. Commun.* **2015**, *51*, 16817–16820. [CrossRef]
21. Ruiz, D.A.; Melaimi, M.; Bertrand, G. An efficient synthetic route to stable bis(carbene)borylenes [(L$_1$)(L$_2$)BH]. *Chem. Commun.* **2014**, *50*, 7837–7839. [CrossRef] [PubMed]
22. Hahn, F.E.; Wittenbecher, L.; Boese, R.; Bläser, D. N,N'-bis(2,2-dimethylpropyl)benzimidazolin-2-ylidene: A stable nucleophilic carbene derived from benzimidazole. *Chem. Eur. J.* **1999**, *5*, 1931–1935. [CrossRef]
23. Lavallo, V.; Canac, Y.; Donnadieu, B.; Schoeller, W.W.; Bertrand, G. Cyclopropenylidenes: From interstellar space to an isolated derivative in the laboratory. *Science* **2006**, *312*, 722–724. [CrossRef]
24. Connelly, N.G.; Geiger, W.E. Chemical redox agents for organometallic chemistry. *Chem. Rev.* **1996**, *96*, 877–910. [CrossRef] [PubMed]
25. Broggi, J.; Terme, T.; Vanelle, P. Organic electron donors as powerful single-electron reducing agents in organic synthesis. *Angew. Chem. Int. Ed.* **2014**, *53*, 384–413. [CrossRef] [PubMed]
26. Eberle, B.; Hubner, O.; Ziesak, A.; Kaifer, E.; Himmel, H.J. What makes a strong organic electron donor (or acceptor)? *Chem. Eur. J.* **2015**, *21*, 8578–8590. [CrossRef]

27. Tsurugi, H.; Saito, T.; Tanahashi, H.; Arnold, J.; Mashima, K. Carbon radical generation by d^0 tantalum complexes with α-diimine ligands through ligand-centered redox processes. *J. Am. Chem. Soc.* **2011**, *133*, 18673–18683. [CrossRef]
28. Tsurugi, H.; Tanahashi, H.; Nishiyama, H.; Fegler, W.; Saito, T.; Sauer, A.; Okuda, J.; Mashima, K. Salt-free reducing reagent of bis(trimethylsilyl)cyclohexadiene mediates multielectron reduction of chloride complexes of W(VI) and W(IV). *J. Am. Chem. Soc.* **2013**, *135*, 5986–5989. [CrossRef]
29. Saito, T.; Nishiyama, H.; Tanahashi, H.; Kawakita, K.; Tsurugi, H.; Mashima, K. 1,4-bis(trimethylsilyl)-1,4-diaza-2,5-cyclohexadienes as strong salt-free reductants for generating low-valent early transition metals with electron-donating ligands. *J. Am. Chem. Soc.* **2014**, *136*, 5161–5170. [CrossRef]
30. Tsurugi, H.; Mashima, K. A new protocol to generate catalytically active species of group 4–6 metals by organosilicon-based salt-free reductants. *Chem. Eur. J.* **2019**, *25*, 913–919. [CrossRef]
31. Arteaga-Muller, R.; Tsurugi, H.; Saito, T.; Yanagawa, M.; Oda, S.; Mashima, K. New tantalum ligand-free catalyst system for highly selective trimerization of ethylene affording 1-hexene: New evidence of a metallacycle mechanism. *J. Am. Chem. Soc.* **2009**, *131*, 5370–5371. [CrossRef] [PubMed]
32. Saito, T.; Nishiyama, H.; Kawakita, K.; Nechayev, M.; Kriegel, B.; Tsurugi, H.; Arnold, J.; Mashima, K. Reduction of (^{t}BuN=)NbCl$_3$(py)$_2$ in a salt-free manner for generating Nb(IV) dinuclear complexes and their reactivity toward benzo[c]cinnoline. *Inorg. Chem.* **2015**, *54*, 6004–6009. [CrossRef] [PubMed]
33. Tanahashi, H.; Ikeda, H.; Tsurugi, H.; Mashima, K. Synthesis and characterization of paramagnetic tungsten imido complexes bearing α-diimine ligands. *Inorg. Chem.* **2016**, *55*, 1446–1452. [CrossRef] [PubMed]
34. Weyenberg, D.R.; Toporcer, L.H. The synthesis of 3,6-disilyl-1,4-cyclohexadienes by the trapping of benzene anion-radicals. *J. Am. Chem. Soc.* **1962**, *84*, 2843–2844. [CrossRef]
35. Braunschweig, H.; Ye, Q.; Radacki, K. High yield synthesis of a neutral and carbonyl-rich terminal arylborylene complex. *Chem. Commun.* **2012**, *48*, 2701–2703. [CrossRef]
36. Cowherd, F.G.; Von Rosenberg, J.L. Mechanism of iron pentacarbonyl-catalyzed 1,3-hydrogen shifts. *J. Am. Chem. Soc.* **1969**, *91*, 2157–2158. [CrossRef]
37. Hvistendahl, G.; Williams, D.H. Energy barrier to symmetry-forbidden 1,3-hydrogen shifts in simple oxonium ions. Metastable peaks from fast dissociations. *J. Am. Chem. Soc.* **1975**, *97*, 3097–3101. [CrossRef]
38. Laguerre, M.; Dunogues, J.; Calas, R.; Duffaut, N. Silylation d'hydrocarbures mono-aromatiques mono- ou disubstitues. *J. Organomet. Chem.* **1976**, *112*, 49–59. [CrossRef]
39. Lis, A.V.; Gostevskii, B.A.; Albanov, A.I.; Yarosh, N.O.; Rakhlin, V.I. Synthesis of volatile bis[bis(trimethylsilyl)amide]-substituted boron derivatives. *Russ. J. Gen. Chem.* **2017**, *87*, 353–356. [CrossRef]
40. Brown, C.; Cragg, R.H.; Miller, T.J.; Smith, D.O.N. Organoboron compounds: Xxii. A ^{13}C NMR study of some dialkylaminophenylboranes. *J. Organomet. Chem.* **1983**, *244*, 209–215. [CrossRef]

Sample Availability: Samples of the compounds are not available from the authors.

© 2020 by the authors. Licensee MDPI, Basel, Switzerland. This article is an open access article distributed under the terms and conditions of the Creative Commons Attribution (CC BY) license (http://creativecommons.org/licenses/by/4.0/).

Article

A SF$_5$ Derivative of Triphenylphosphine as an Electron-Poor Ligand Precursor for Rh and Ir Complexes

Maria Talavera, Silke Hinze, Thomas Braun *, Reik Laubenstein and Roy Herrmann

Department of Chemistry, Humboldt–Universität zu Berlin, Brook-Taylor-Str. 2, 12489 Berlin, Germany; talaverm@hu-berlin.de (M.T.); silkehinze@gmx.de (S.H.); reik.laubenstein@gmx.de (R.L.); roy.herrmann@gmx.de (R.H.)
* Correspondence: thomas.braun@cms.hu-berlin.de

Academic Editor: Ashok Kakkar
Received: 11 August 2020; Accepted: 26 August 2020; Published: 1 September 2020

Abstract: The synthesis of the triarylphosphine, P(p-C$_6$H$_4$SF$_5$)$_3$ containing a SF$_5$ group, has been achieved. The experimental and theoretical studies showed that P(p-C$_6$H$_4$SF$_5$)$_3$ is a weaker σ-donor when compared with other substituted triarylphosphines, which is consistent with the electron-withdrawing effect of the SF$_5$ moiety. The studies also revealed a moderate air stability of the phosphine. The σ-donor capabilities of P(p-C$_6$H$_4$SF$_5$)$_3$ were estimated from the phosphorus-selenium coupling constant in SeP(p-C$_6$H$_4$SF$_5$)$_3$ and by DFT calculations. The behavior of P(p-C$_6$H$_4$SF$_5$)$_3$ as ligand has been investigated by the synthesis of the iridium and rhodium complexes [MCl(COD){P(p-C$_6$H$_4$SF$_5$)$_3$}], [MCl(CO)$_2${P(p-C$_6$H$_4$SF$_5$)$_3$}$_2$] (M = Ir, Rh), or [Rh(μ-Cl)(COE){P(p-C$_6$H$_4$SF$_5$)$_3$}]$_2$, and the molecular structures of [IrCl(COD){P(p-C$_6$H$_4$SF$_5$)$_3$}] and [Rh(μ-Cl)(COE){P(p-C$_6$H$_4$SF$_5$)$_3$}]$_2$ were determined by single X-ray diffraction. The structures revealed a slightly larger cone angle for P(p-C$_6$H$_4$SF$_5$)$_3$ when compared to other *para*-substituted triarylphosphines.

Keywords: fluorosulfanyl group; fluorinated ligands; phosphines; rhodium; iridium

1. Introduction

Phosphines are one of the most important and widely used ligands in homogeneous catalysis and organometallic chemistry due to the extensive options to modify their electronic and steric properties by varying their substitution pattern [1–4]. Phosphines are σ-donor and π-acceptor ligands and the choice of the phosphorus-bound entities can effectively tune their electronic characteristics. Thus, the electronic, but also steric properties of triarylphosphines are highly dependent on the employed aryl groups [1–4].

Triarylphosphines bearing fluorine atoms or fluorinated substituents are characterized by a higher steric demand and a lower basicity than the non-halogenated counterparts which might lead to changes in the reactivity [5]. Thus, complexes bearing fluorinated triarylphosphines have been widely investigated over the last years [5–13]. In particular, arylphosphines containing a CF$_3$ group have been studied for catalytic processes such as hydroformylation [14], C-C coupling [13,15–17] as well as hydroalkylation reactions [18]. They present an alternative to fluoroarylphosphines with respect to the increased electron-withdrawing properties of the phosphine.

Another alternative and fascinating chemically and thermally rather stable group is the SF$_5$ group [19,20], which displays a large σ-withdrawing nature ($F = 0.56$) and π-withdrawing ability ($R = 0.12$) based on the Swann-Lupton constants. In addition, phenyl groups with SF$_5$ moieties at both *meta* and *para* positions exhibit a stronger electron-withdrawing effect according to Hammett constants ($\sigma_m = 0.61$, $\sigma_p = 0.68$), when compared with a fluorine substituent ($F = 0.45$, $R = -0.39$, $\sigma_m = 0.34$,

σ_p = 0.06) or the trifluoromethyl group (F = 0.38, R = 0.16, σ_m = 0.43, σ_p = 0.54) [21]. Despite the increasing interest on SF_5-containing compounds described in literature and their applications as building blocks in many fields such as agrochemical, medicinal, or materials chemistry [22–30], SF_5 derivatized ligands in transition metal complexes are still rare [31–41]. Examples mainly focus on C^N cyclometalated ligands which contain the SF_5 moiety in their phenyl rings in order to stabilize iridium(III) [34–37] or platinum(II) [38] complexes with optoelectronic properties.

Herein, we present the synthesis and characterization of a SF_5 derivative of triphenylphosphine $P(p-C_6H_4SF_5)_3$ (**1**). Its air stability and electronic properties have been estimated and the derivatives $O=P(p-C_6H_4SF_5)_3$ (**2**) and $Se=P(p-C_6H_4SF_5)_3$ (**3**) were prepared. In addition, the rhodium and iridium complexes [MCl(COD){$P(p-C_6H_4SF_5)_3$}] (M = Ir (**4**), Rh (**5**)), [MCl(CO)$_2${$P(p-C_6H_4SF_5)_3$}$_2$] (M = Ir (**8**), Rh (**9**)) or [Rh(μ-Cl)(COE){$P(p-C_6H_4SF_5)_3$}]$_2$ (**6**) were synthesized.

2. Results and Discussion

2.1. Synthesis and Air-Stability of $P(p-C_6H_4SF_5)_3$ (**1**)

Treatment of 4-iodophenylsulfur pentafluoride with an excess of tBuLi and a subsequent reaction with triethylphosphite afforded the tris(p-pentafluorosulfanylphenyl)phosphine (**1**) in 39% isolated yield (Scheme 1). Compound **1** shows in the ^{19}F NMR spectrum the characteristic signal pattern for the SF_5 moiety [42], a doublet at 62.6 ppm corresponding to the four equatorial equivalent fluorine atoms and a pentet at 83.7 ppm for the axial fluorine with coupling constants of 150 Hz. The ^{31}P{^{1}H} NMR spectrum shows a singlet at −7.8 ppm, whereas the GC/MS gave a mass peak of m/z 640.

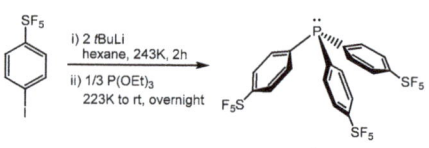

Scheme 1. Synthesis of the phosphine **1**.

Single crystals of compound **1**, which were suitable for X-ray crystallography were obtained from concentrated solutions in benzene. The molecular structure is shown in Figure 1. Compound **1** shows a trigonal pyramidal arrangement at the phosphorus atom when taking the electron lone-pair into account. The C–P–C angles are comparable to the ones reported for triphenylphosphine, where only one angle is slightly larger (101.72(1)° for PPh$_3$ vs. 99.18(10)° for **1**) than the others [43].

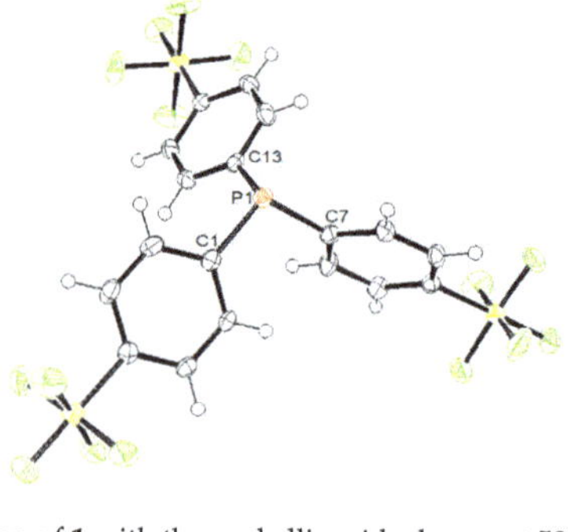

Figure 1. ORTEP representation of **1** with thermal ellipsoids drawn at 50% probability level. The C_6H_6 molecule in the asymmetric unit has been omitted for clarity. Selected bond lengths [Å] and bond angles [deg]: P1–C1 1.829(2), P1–C7 1.833(2), P1–C13 1.831(2), C1–P1–C7 103.21(10), C7–P1–C13 99.18(10), C13–P1–C1 103.08(10).

In order to test the air-stability of the phosphine **1**, a solution of **1** in toluene-d_8 was left under air for 2 weeks. Seventy-eight percent of the phosphine **1** was then converted into phosphine oxide **2** (Scheme 2). In the solid state 41% of conversion was achieved after a month of air exposure. These data indicate a higher stability of **1** towards air when compared with the CF_3 analogue, which is fully oxidized in the solid state after one month [44]. $^{31}P\{^1H\}$ NMR data of the solution as well as liquid injection field desorption/ionization mass spectrometry (LIFDI-MS) confirm the formation of the oxide with a shift of the signal in the $^{31}P\{^1H\}$ NMR to lower field (δ = 21.09 ppm) and a mass peak of m/z 656.

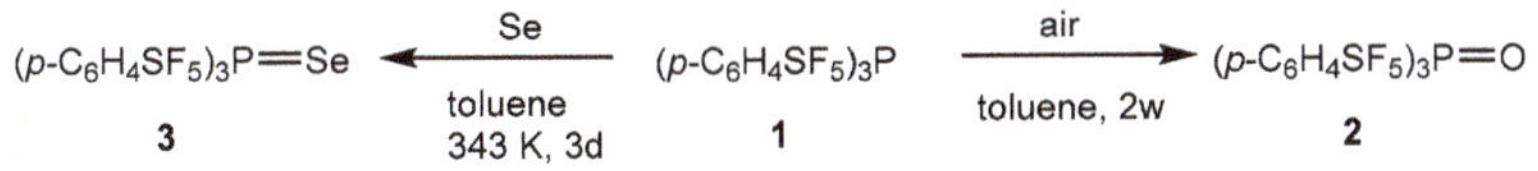

$$(p\text{-}C_6H_4SF_5)_3P=Se \xleftarrow[\text{toluene} \atop 343\,K,\,3d]{Se} (p\text{-}C_6H_4SF_5)_3P \xrightarrow[\text{toluene, 2w}]{air} (p\text{-}C_6H_4SF_5)_3P=O$$
$$\quad\quad\quad\quad\quad\text{3}\quad\quad\quad\quad\quad\quad\quad\quad\quad\quad\text{1}\quad\quad\quad\quad\quad\quad\quad\quad\quad\text{2}$$

Scheme 2. Formation of the phosphine oxide **2** and the phosphine selenide **3**.

Crystals of **2** suitable for X-ray diffraction analysis were obtained from the reaction solution (Figure 2). The compound shows the expected tetrahedral arrangement with a C–P–C mean angle of 106.6°, a O–P–C mean angle of 112.2° and the P–O bond length of 1.4867(14) Å. All of the data are consistent with these of other reported triarylphosphine oxides derivatives [45].

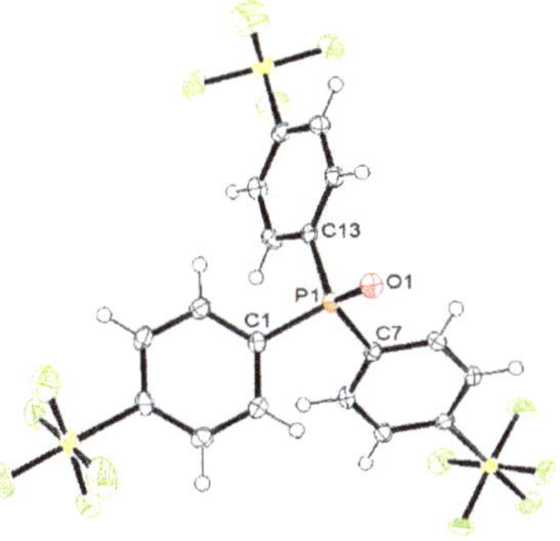

Figure 2. ORTEP representation of **2** with thermal ellipsoids drawn at 50% probability level. The two C_6H_6 molecules contained in the asymmetric unit have been omitted for clarity. Selected bond lengths [Å] and bond angles [deg]: P1–O1 1.4867(14), P1–C1 1.8086(19), P1–C7 1.8086(19), P1–C13 1.8019(19), C1–P1–C7 107.19(8), C7–P1–C13 106.28(8), C13–P1–C1 106.35(9), O1–P1–C1 113.23(8), O1–P1–C7 111.35(8), O1–P1–C13 112.02(8).

In order to further compare the air-stability of phosphine **1** with other triarylphosphines, DFT calculations were performed. It has been suggested previously that the steric demand of the phosphines, but also the SOMO energy of radical cations of phosphines can be correlated with their air-stability [46]. Thus, a radical cation with a SOMO at lower energy would be more prone to react with dioxygen generating the corresponding phosphine oxide [46]. Therefore, the energy of the SOMO of different triarylphosphines radical cations was calculated using the CAM-B3LYP functional (Table 1). The phosphine **1** has a SOMO energy of −12.22 eV, which is lower than the energy of other electron-withdrawing triarylphosphines, and lower than the one of the air-stable triphenylphoshine (−11.11 eV). The data are in accordance with the observed moderate air sensitivity of phosphine **1**.

Table 1. DFT calculated SOMO energies (eV) of triarylphosphines radical cations [PAr₃]•⁺ (CAM-B3LYP/6-311G(d,p)).

Ar Group	SOMO	Ar Group	SOMO
p-C$_6$H$_4$SF$_5$	−12.22	3,4,5-C$_6$H$_2$F$_3$	−11.93
p-C$_6$H$_4$CF$_3$	−11.82	p-C$_6$H$_4$Me	−10.62
m-C$_6$H$_3$(CF$_3$)$_2$	−12.36	p-C$_6$H$_4$OMe	−10.01
p-C$_6$H$_4$F	−11.21	C$_6$H$_5$	−11.11

2.2. Estimation of the Donor Properties

Different methods have been reported to determine the electronic properties of phosphines. The σ-donor ability increases when the s-character of the lone pair of the phosphine decreases, which is associated with a higher energy level of the HOMO [47,48]. Thus, DFT studies have been performed in order to calculate the energy level of the HOMO of compound **1** and compare it with other triarylphosphines which were also calculated. The data indicate that compound **1** (−8.69 eV) is a less pronounced σ-donor than most of the calculated phosphines (Table 2). It is worth noting that P(p-C$_6$H$_4$CF$_3$)$_3$ (−8.17 eV) seems, according to this data, to be a better σ-donor than **1**, whereas P(m-C$_6$H$_3$(CF$_3$)$_2$)$_3$ is a weaker σ-donor (−8.78 eV).

The HOMO energy level and, therefore, the s-character of the lone pair is also related to the phosphorus-selenium coupling constant of the corresponding selenide. This has been commonly used to experimentally estimate the σ-donor abilities of a broad range of phosphines [49–51]. Thus, for more electron-withdrawing substituents a larger coupling is expected [51]. Taking this into account, phosphine **1** was reacted with selenium and after 3d, a full conversion to SeP(p-C$_6$H$_4$SF$_5$)$_3$ (**3**) was observed (Scheme 2). The ^{31}P{^{1}H} NMR spectrum shows a singlet at δ 32.5 ppm with selenium satellites and a phosphorus-selenium coupling constant of 792 Hz. Correspondingly, the resonance in the ^{77}Se NMR spectrum at −273.3 ppm appears as a doublet with the same coupling constant. Among the data for the phosphines shown in Table 2 only P(m-C$_6$H$_3$(CF$_3$)$_2$)$_3$ shows a larger coupling constant of 802 Hz, which is consistent with the lower HOMO energy values.

Another common method to determine the electronic properties of ligands is the calculation of the Tolman's electronic parameter TEP [52]. This method consists in the analysis of the frequency of the A$_1$ carbonyl vibration mode of [Ni(CO)$_3$L] complexes, which will decrease due to the back-donation into the CO π* orbitals when the ligand L is a better donor. While Tolman experimentally determined the parameter for a broad range of phosphines [52], DFT studies have demonstrated the correlation between the calculated and experimental values for different ligands L [53,54]. Thus, the calculated values determined in this work correlate well with the experimentally obtained for PPh$_3$ and P(p-C$_6$H$_4$Me)$_3$ ($\Delta_{exp\text{-}calc} \approx 5$ cm^{-1}) (Table 2) [52]. The data also indicate that the phosphine **1** might be a slightly more electron-withdrawing phosphine than P(p-C$_6$H$_4$CF$_3$)$_3$, but somewhat weaker than P(m-C$_6$H$_3$(CF$_3$)$_2$)$_3$, although the calculated values are very close.

Table 2. Calculated HOMO energies and TEP values of triarylphosphines (PAr₃) and experimental ^{77}Se NMR data of the corresponding selenides.

Ar Group	HOMO (eV)	TEP (cm^{-1}) [a]	$^1J_{Se\text{-}P}$ (Hz)
p-C$_6$H$_4$SF$_5$	−8.69	2072.6	792
p-C$_6$H$_4$CF$_3$	−8.17	2069.3	765 [b]
m-C$_6$H$_3$(CF$_3$)$_2$	−8.78	2075.2	802 [b]
p-C$_6$H$_4$F	−7.58	n.d.	741 [b]
3,4,5-C$_6$H$_2$F$_3$	−8.36	n.d.	792 [c]
p-C$_6$H$_4$Me	−7.10	2061.7 (2166.7)	715 [b]
p-C$_6$H$_4$OMe	−6.78	n.d.	710 [b]
C$_6$H$_5$	−7.34	2063.5 (2068.9)	733 [b]

n.d. = not determined; [a] Calculated values from calculated ν$_{CO}$ (A$_1$ band) of Ni(CO)$_3$L complexes using the relation TEP = 0.9572ν$_{CO}$ + 4.081 [54]. In brackets experimentally determined values given in ref. [52] [b] Values given in ref. [50]; [c] Value given in ref. [55].

2.3. Synthesis of Iridium and Rhodium Complexes

Treatment of the binuclear iridium complex [Ir(μ-Cl)(COD)]$_2$ (COD = 1,5-cyclooctadiene) with two equivalents of the phosphine **1** in toluene yielded the iridium(I) complex [IrCl(COD){P(p-C$_6$H$_4$SF$_5$)$_3$}] (**4**) (Scheme 3). The same reactivity was reported for other phosphines such as P(p-C$_6$H$_4$CF$_3$)$_3$ [56]. The structure of **4** is supported by the ^{31}P{^{1}H} NMR spectrum, which shows a resonance at δ 21.7 ppm and the ^{1}H NMR spectrum with two resonances at δ 5.64 and 2.45 ppm corresponding to the olefinic protons of the COD ligand in a *trans* arrangement to the phosphine and the chlorido ligands. The LIFDI mass spectrometry reveals a mass peak of *m/z* 976.

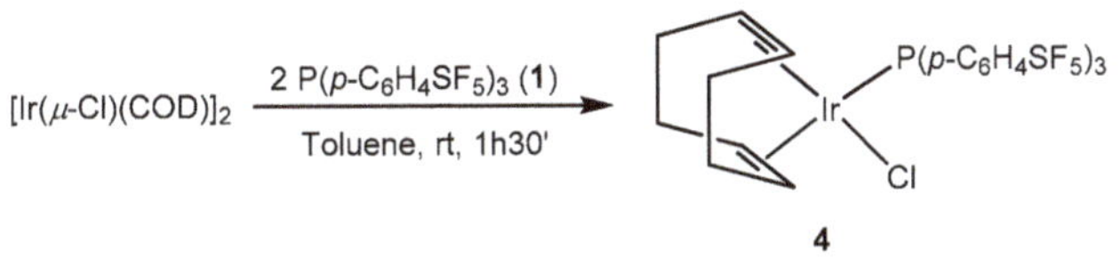

Scheme 3. Synthesis of the iridium(I) complex **4**.

Analogously, [Rh(μ-Cl)(COD)]$_2$ reacted with two equivalents of the phosphine **1** affording the rhodium(I) complex [RhCl(COD){P(p-C$_6$H$_4$SF$_5$)$_3$}] (**5**) (Scheme 4). The ^{31}P{^{1}H} NMR spectrum of **5** depicts a doublet at δ 31.5 ppm with a rhodium-phosphorus coupling constant of 155.9 Hz. The coupling constant of **5** is 1.8 Hz higher than the one for [RhCl(COD){P(p-C$_6$H$_4$CF$_3$)$_3$}] ($^1J_{\text{P-Rh}}$ = 154.1 Hz) [57]. As for the iridium complex **4**, ^{1}H NMR spectrum of complex **5** shows two resonances for the olefinic protons of the cyclooctadiene ligand at δ 5.67 and 3.12 ppm which are also consistent with the CF$_3$ derivative [57].

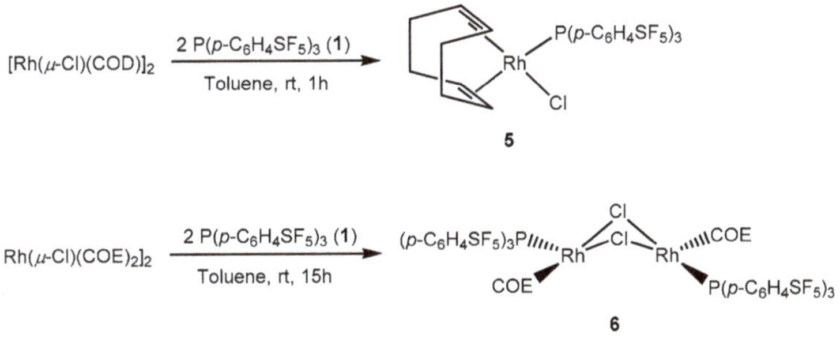

Scheme 4. Synthesis of the rhodium(I) complexes **5** and **6**.

In contrast, when [Rh(μ-Cl)(COE)$_2$]$_2$ (COE = *cis*-cyclooctene) was used as binuclear starting compound, the reaction with two equivalents of phosphine **1** selectively provided *trans*-[Rh(μ-Cl)(COE){P(p-C$_6$H$_4$SF$_5$)$_3$}]$_2$ (**6**) (Scheme 5). The dimeric nature of the product is supported by the ^{31}P{^{1}H} NMR data revealing a signal at δ 55.5 ppm with an increased value of the rhodium-phosphorus coupling constant ($^1J_{\text{P-Rh}}$ = 194.5 Hz). This coupling constant value is slightly larger than for other *trans*-[Rh(μ-Cl)(COE)(L)]$_2$ complexes, for which L is a σ-donor phosphine ($^1J_{\text{P-Rh}}$ = 183–188Hz) [58,59].

2.4. Steric Properties of 1

Steric properties of phosphines is another important feature when studying a ligand. The Tolman cone angle [52] of a phosphine is a widely used parameter to describe the steric effects in phosphines and can be estimated not only from molecular models but also from molecular structures determined by X-ray crystallography [60].

The molecular structures of the complexes **4** and **6** were obtained by X-ray diffraction analysis from concentrated C_6H_6 solutions (Figures 3 and 4). In the case of complex **4**, the iridium center displays a square planar coordination geometry when considering COD as a bidentate ligand with an Ir–P bond distance of 2.2888(12) Å, which is slightly shorter than the corresponding separation in [IrCl(COD){P(C_6H_5)$_3$}] (2.3172(9) Å) [61].

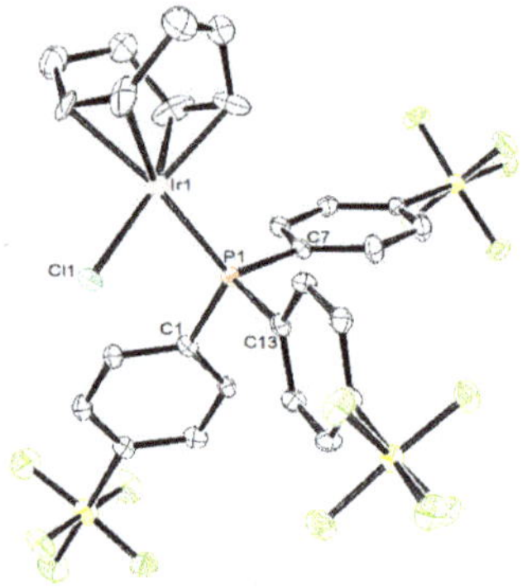

Figure 3. ORTEP representation of **4** with thermal ellipsoids drawn at 50% probability level. Hydrogen atoms and the additional C_6H_6 molecule in the unit cell have been omitted for clarity. Note that some disorder is observed in COD ligand. Selected bond lengths [Å] and bond angles [deg]: Ir1–P1 2.2888(12), Ir1–Cl1 2.4070(10), P1–C1 1.831(5), P1–C7 1.834(5), P1–C13 1.838(5), Cl1–Ir1–P1 90.27(4), Ir1–P1–C1 118.23(15), Ir1–P1–C7 118.61(15), Ir1–P1–C13 108.27(16), C1–P1–C7 100.7(2), C7–P1–C13 104.2(2), C13–P1–C1 105.3(2).

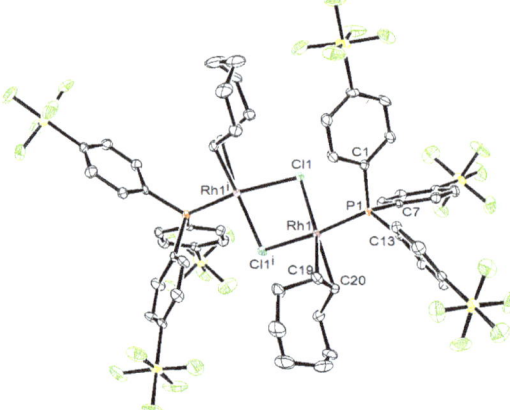

Figure 4. ORTEP representation of **6** with thermal ellipsoids drawn at 50% probability level. Hydrogen atoms and the C_6H_6 molecule in the unit cell have been omitted for clarity. Selected bond lengths [Å] and bond angles [deg]: Rh1–P1 2.1930(11), Rh1–Cl1 2.3911(11), Rh1–Cl1^i 2.4368(11), Rh1–C19 2.120(4), Rh1–C20 2.131(4), P1–C1 1.827(4), P1–C7 1.842(4), P1–C13 1.825(4), Cl1–Rh1–Cl1^i 80.57(4), Cl1–Rh1–P1 89.66(4), Cl1^i–Rh1–P1 166.80(4), P1–Rh1–C19 94.48(12), P1–Rh1–C20 93.38(12), Rh1–P1–C1 113.38(14), Rh1–P1–C7 111.20(14), Rh1–P1–C13 120.94(14), C1–P1–C7 103.73(19), C7–P1–C13 103.51(19), C13–P1–C1 102.26(19).

The Tolman cone angle of compound **1** can be estimated to be 150.8° as the average value for the complexes **4** and **6**, and by considering van der Waals radii of 1.20 Å and 1.47 Å for the H and F nuclei, respectively [62]. For the calculations, the algorithm reported by Müller and Mingos was used and the metal-phosphorus distances were fixed to 2.28 Å [60]. The obtained value is slightly larger than for other triarylphosphines with a substituent at the *para* position P(p-C$_6$H$_4$X)$_3$ (X = NMe$_2$, Me, OMe, F, CF$_3$), all of which have a cone angle of 145° [52,63].

2.5. Reactivity of Complexes 4 and 5 towards CO

Bubbling of CO into a dichloromethane solution of [IrCl(COD){P(p-C$_6$H$_4$SF$_5$)$_3$}] (4) resulted in a partial conversion of 4 to yield a complex, for which we suggest the structure [IrCl(CO)(COD){P(p-C$_6$H$_4$SF$_5$)$_3$}] (7). However, when a degassed solution was treated with CO gas, full conversion of complex 4 was observed and a dark unidentified precipitate was formed. In the solution, the formation of a complex containing both COD and CO ligands without the presence of phosphine 1 was detected. Unfortunately, no further identification of this complex was possible. In addition, two products bearing the phosphine ligand 1 in a 1:1 ratio were obtained. Thus complex [IrCl(CO)(COD){P(p-C$_6$H$_4$SF$_5$)$_3$}] (7) was formed together with a second complex, the analytical data of which are consistent with the structure [IrCl(CO)$_2${P(p-C$_6$H$_4$SF$_5$)$_3$}$_2$] (8). To support the structural assignments further, ^{13}C labeled carbon monoxide was reacted with a solution of complex 4 to give 7' and 8'. However, the formation of the unknown complex as well as the mixture of products was avoided and a full conversion of 4 into complexes 8 or 8' was achieved, when the reaction was performed in presence of one equivalent of phosphine 1 (Scheme 5).

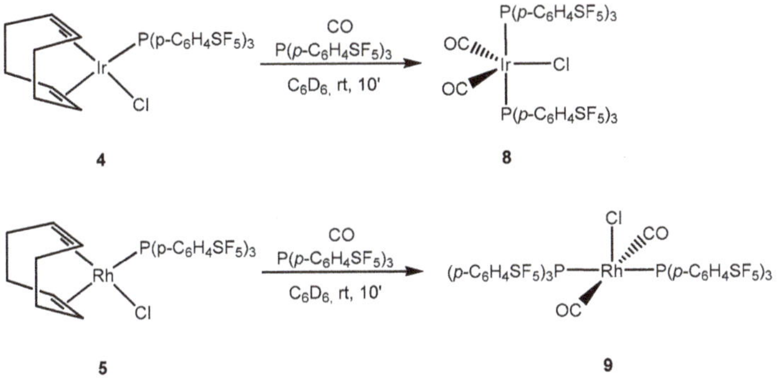

Scheme 5. Reactivity of the complexes 4 and 5 towards CO in presence of phosphine 1.

Complexes 7 and 8 show singlet resonances for the phosphine ligands at δ 3.65 and 0.84 ppm, respectively, in the ^{31}P{^{1}H} NMR spectrum. In addition, the signals for the olefinic protons of COD in complex 7 appear at δ 4.25 and 3.91 ppm, which would correspond to the protons *trans* to the CO and Cl ligands, respectively, suggesting that the phosphine ligand is in an apical position. The ^{31}P{^{1}H} NMR spectrum for the mixture of complexes 7' and 8' showed the two resonances at δ 3.52 and 0.84 ppm as a doublet ($^2J_{P-C}$ = 13.7 Hz) and a triplet ($^2J_{P-C}$ = 13.5 Hz), respectively. Note that the values of the carbon-phosphorus coupling constants are consistent with *cis* arrangements [64,65]. On the other hand, in the ^{13}C{^{1}H} NMR spectrum a doublet (δ = 175.6 ppm, complex 7') and a triplet (δ = 179.6 ppm, complex 8') with similar coupling constants are observed for the carbonyl ligands.

The trigonal bipyramid proposed for complex 8 is supported by the IR data in the solid state. The IR spectrum of complex 8 shows two bands at 1940 and 1986 cm^{-1} for the symmetric and the asymmetric stretching bands of the CO ligands, which shift to 1896 and 1933 cm^{-1} for 8' suggesting that the CO ligands are in an equatorial position (see Figure S39). This data for 8 are in the same range as the ones observed for [IrCl(CO)$_2$(PPh$_3$)$_2$] [66].

For comparison, [RhCl(COD){P(p-C$_6$H$_4$SF$_5$)$_3$}] (5) was also reacted with CO or ^{13}CO in the presence of one equivalent of phosphine 1 to give complexes, for which we suggest the structures *trans,trans*-[RhCl(CO)$_2${P(p-C$_6$H$_4$SF$_5$)$_3$}$_2$] (9) and *trans,trans*-[RhCl(^{13}CO)$_2${P(p-C$_6$H$_4$SF$_5$)$_3$}$_2$]

(**9'**) (Scheme 5). The NMR data of **9** showed a broad band in the $^{31}P\{^1H\}$ NMR spectrum at room temperature at δ 28.6 ppm. When the sample was measured at −70 °C, the coupling to rhodium was observed ($^1J_{P-Rh}$ = 130 Hz). As also found for the PPh$_3$ analogue [67], the coupling to carbon in the ^{13}C labeled complex was not observed even not at −70 °C. The $^{13}C\{^1H\}$ NMR spectrum revealed a Rh-C coupling constant of 72 Hz in the doublet at 186.5 ppm. Finally, in the IR spectrum, one unique band was observed for the CO ligands at 1992 and 1945 cm^{-1} for the complexes **9** and **9'**, respectively (see Figure S40). This is consistent with data for [RhCl(CO)$_2$(PPh$_3$)$_2$], where only one stretching band was observed for CO at 1990 cm^{-1} and a square pyramidal structure with the chlorido ligand in the apical position as the most probable structure was proposed [67].

3. Materials and Methods

3.1. General Procedures, Methods and Materials

All experiments were carried out under an atmosphere of argon by Schlenk techniques. Solvents were dried by the usual procedures [68] and, prior to use, distilled under argon. The rhodium and iridium complexes [Rh(μ-Cl)(COE)$_2$]$_2$ and [Ir(μ-Cl)(COD)]$_2$ were prepared as described in the literature [69,70]. All reagents were obtained from commercial sources. Unless stated, NMR spectra were recorded at room temperature on a Bruker DPX 300 (Bruker BioSpin, Rheinstetten, Germany,) or a Bruker Avance 300 spectrometer (Bruker BioSpin, Rheinstetten, Germany). ^1H and $^{13}C\{^1H\}$ signals are referred to residual solvent signals, those of $^{31}P\{^1H\}$ to external 85% H$_3$PO$_4$, the ^{19}F NMR spectra to external CFCl$_3$ and the ^{77}Se NMR spectra to external SePh$_2$ (δ = 414 ppm). ^1H and $^{13}C\{^1H\}$ NMR signal assignments were confirmed by $^1H\{^{31}P\}$, $^1H,^1H$ COSY, $^1H,^{13}C$ HMQC and $^1H,^{13}C$ HMBC NMR experiments. Mass spectra were measured with a Micromass Q–Tof–2 instrument equipped with a Linden LIFDI source (Linden CMS GmbH, Weyhe, Germany). GC/MS analyses were performed with an Agilent 6890N gas–phase chromatograph (Shimadzu, Berlin, Germany) equipped with an Agilent 5973 Network mass selective detector at 70eV. Infrared spectra were recorded with the Platinum ATR module of a Bruker FT-IR Alpha II spectrometer (Bruker Optics, Leipzig, Germany) equipped with an ATR unit (diamond). NMR spectra are included as Supplementary Material (Figures S1–S38).

3.2. Synthesis of Tris-(p-pentafluorosulfanylphenyl)phosphine (1)

4-iodophenylsulfur pentafluoride (250 mg, 0.76 mmol) was dissolved in 10 mL of hexane at 243 K. Then, two equivalents of *tert*-buthyllithium (1.7 M in pentanes, 1.52 mmol, 0.9 mL) was added dropwise to the solution and the reaction mixture was stirred for 2 h at 243 K. Afterwards, the mixture was cooled down to 223 K and triethylphosphite (0.25 mmol, 45 μL) was added slowly. The mixture was stirred while warming up overnight. The volatiles were removed under vacuum, toluene (2 × 10 mL) was added and the product extracted. The solvent was removed from the extract and the beige solid obtained dried under vacuum. Yield: 190 mg (39%).

GC–MS (toluene): Calculated (m/z) for [M]: 640.43; found: 640. $^{31}P\{^1H\}$ NMR (121.5 MHz, C$_6$D$_6$): δ = −7.8 (s) ppm. ^1H NMR (300.1 MHz, C$_6$D$_6$): δ = 7.26 (dm, ^3J(H,H) = 8.7 Hz, 6H, *m*–CH); 6.74 (dd, ^3J(H,H) = 8.6, ^2J(H,P) = 6.8 Hz, 6H, *o*–CH) ppm. ^{19}F NMR (282.4 MHz, C$_6$D$_6$): δ = 83.7 (p, ^2J(F,F) = 150 Hz, 1F, SF$_5$); 62.6 (d, ^2J(F,F) = 150 Hz, 4F, SF$_5$) ppm. $^{13}C\{^1H\}$ NMR (75.4 MHz, CD$_2$Cl$_2$): 155.5–154.1 (m, C$_q$–SF$_5$); 140.6 (d, ^1J(C,P) = 15.4 Hz, P–C$_q$); 134.5 (d, ^2J(C,P) = 20.9 Hz, *o*–CH); 126.8 (dp, ^3J(C,P) = 6.8 Hz, ^3J(C,F$_{eq}$) = 4.3 Hz, *m*–CH) ppm.

3.3. Formation of Tris-(p-pentafluorosulfanylphenyl)phosphine oxide (2)

Method a: Tris-(*p*-pentafluorosulfanylphenyl)phosphine **1** (20 mg, 0.03 mmol) was dissolved in toluene-d^8 (0.4 mL) and the solution was exposed under air. The conversion was followed by NMR spectroscopy, and after 2 weeks 78% conversion was observed.

Method b: Tris-(*p*-pentafluorosulfanylphenyl)phosphine **1** (20 mg, 0.03 mmol) was exposed under air in an open vial. After 1 month, a 41% conversion to compound **2** was found.

LIFDI (toluene-d^8): Calculated (m/z) for [M]: 656.4; found: 656. ^{31}P{^{1}H} NMR (121.5 MHz, toluene-d_8): δ = 21.1 (s) ppm. ^{1}H NMR (300.1 MHz, toluene–d^8): δ = 7.30 (dd, ^3J(H,H) = 8.3; ^3J(H,P) = 2.2 Hz, 6H, *m*-CH); 7.21 (dd, ^2J(H,P) = 10.9, ^3J(H,H) = 8.3 Hz, 6H, *o*-CH) ppm. ^{19}F NMR (282.4 MHz, toluene-d^8): δ = 82.7 (p, ^2J(F,F) = 150 Hz, 1F, SF$_5$); 62.3 (d, ^2J(F,F) = 150 Hz, 4F, SF$_5$) ppm.

3.4. Synthesis of Tris-(p-pentafluorosulfanylphenyl)phosphine Selenide (3)

Tris-(*p*-pentafluorosulfanylphenyl)phosphine **1** (50 mg, 0.08 mmol) was dissolved in toluene (10 mL) and one equivalent of selenium (6 mg, 0.08 mmol) was added. Then, the reaction mixture was stirred at 343 K for 3 days. The reaction solution was filtered and the volatiles were removed from the filtrate. The dark solid was dried in vacuum. Yield: 54 mg (94%).

LIFDI (toluene-d^8): Calculated (m/z) for [M]: 719.39; found: 719. ^{31}P{^{1}H} NMR (202.4 MHz, toluene-d_8): δ = 32.5 (s + sat, ^1J(P,Se) = 791.8 Hz) ppm. ^{77}Se NMR (95.4 MHz, toluene-d^8): δ = −273.3 (d, ^1J(Se,P) = 792.3 Hz) ppm. ^{1}H NMR (500.1 MHz, toluene-d^8): δ = 7.27 (dd, ^2J(H,P) = 12.8; ^3J(H,H) = 8.7 Hz, 6H, *o*-CH); 7.20 (dd, ^3J(H,H) = 8.8, ^3J(H,P) = 2.2 Hz, 6H, *m*-CH) ppm. ^{19}F NMR (470.6 MHz, toluene-d_8): δ = 82.4 (p, ^2J(F,F) = 150 Hz, 1F, SF$_5$); 62.3 (d, ^2J(F,F) = 150 Hz, 4F, SF$_5$) ppm.

3.5. Synthesis of [IrCl(COD){P(p-C$_6$H$_4$SF$_5$)$_3$}] (4)

[Ir(μ-Cl)(COD)]$_2$ (100 mg, 0.15 mmol) (COD = cyclooctadiene) was dissolved in toluene (7 mL) and a solution of P(*p*-C$_6$H$_4$SF$_5$)$_3$ (**1**) (192 mg, 0.30 mmol) in 5 mL of toluene was added slowly. After stirring for 1 h 30′, the volatiles were removed under vacuum and a dark red solid was obtained. The solid was washed with hexane (3 × 5 mL) and dried in vacuum. Yield: 117 mg (80%).

LIFDI-TOF-MS (toluene): Calculated (m/z) for [M]$^+$: 976.28; found: 976. ^{31}P{^{1}H} NMR (121.5 MHz, C$_6$D$_6$): δ = 21.7 (s) ppm. ^{1}H NMR (300.1 MHz, C$_6$D$_6$): δ = 7.41–7.23 (m, 12H, *Ph*); 5.64 (s br, 2H, =CH *trans* to P); 2.45 (s br, 2H, =CH *trans* to Cl); 2.16–1.84 (m, 2H, CH$_2$); 1.82–1.57 (m, 2H, CH$_2$); 1.54–1.29 (m, 2H, CH$_2$); 1.17–0.85 (m, 2H, CH$_2$) ppm. ^{19}F NMR (282.4 MHz, C$_6$D$_6$): δ = 83.0 (p, ^2J(F,F) = 150 Hz, 1F, SF$_5$); 62.4 (d, ^2J(F,F) = 150 Hz, 4F, SF$_5$) ppm.

3.6. Synthesis of [RhCl(COD){P(p-C$_6$H$_4$SF$_5$)$_3$}] (5)

[Rh(μ-Cl)(COD)]$_2$ (50 mg, 0.10 mmol) was dissolved in toluene (5 mL) and P(*p*-C$_6$H$_4$SF$_5$)$_3$ (**1**) (130 mg, 0.20 mmol) was added to the solution. After stirring for 1h15′, the volatiles were removed under vacuum and a yellow solid was obtained. The solid was washed with cold hexane (2 × 4 mL) and finally dried in vacuum. The NMR spectra of the yellow solid confirmed the formation of complex **5**. Yield: 162 mg (92%).

^{31}P{^{1}H} NMR (121.5 MHz, C$_6$D$_6$): δ = 31.5 (d, ^1J(P,Rh) = 155.9 Hz) ppm. ^{1}H NMR (300.1 MHz, C$_6$D$_6$): δ = 7.89–7.78 (m, 12H, *Ph*); 5.67 (s br, 2H, =CH *trans* to P); 3.12 (s br, 2H, =CH *trans* to Cl); 2.55–2.35 (m, 4H, CH$_2$); 2.25–1.94 (m, 4H, CH$_2$) ppm. ^{19}F NMR (282.4 MHz, C$_6$D$_6$): δ = 82.33 (p, ^2J(F,F) = 150 Hz, 1F, SF$_5$); 61.89 (d, ^2J(F,F) = 150 Hz, 4F, SF$_5$) ppm. ^{13}C{^{1}H} NMR (75.4 MHz, CD$_2$Cl$_2$): 156.4–155.0 (m, C$_q$–SF$_5$); 135.7 (m, overlapped with CH signals, P–C$_q$); 135.6 (d, ^2J(C,P) = 12.8 Hz, *o*–CH); 126.4 (dp, ^3J(C,P) = 9.6 Hz, ^3J(C,F$_{eq}$) = 4.8 Hz, *m*-CH); 108.7 (dd, ^1J(C,Rh) = 12.0 Hz; ^2J(C,P) = 7.2 Hz, =CH *trans* to P); 72.2 (d, ^1J(C,Rh) = 12.8 Hz, =CH *trans* to Cl); 33.44, 33.43 and 29.2 (all s, CH$_2$) ppm.

3.7. Synthesis of [Rh(μ-Cl)(COE){P(p-C$_6$H$_4$SF$_5$)$_3$}]$_2$ (6)

[Rh(μ-Cl)(COE)$_2$]$_2$ (50 mg, 0.07 mmol) (COE = cyclooctene) was dissolved in toluene (7 mL) and a solution of P(*p*-C$_6$H$_4$SF$_5$)$_3$ (**1**) (90 mg, 0.14 mmol) in 5 mL of toluene was added slowly. Instantly, the solution turned red and after stirring for 1 day, the volatiles were removed under vacuum. The red solid obtained was dried in vacuum. Yield: 123 mg (99%)

^{31}P{^{1}H} NMR (121.5 MHz, C$_6$D$_6$): δ = 54.4 (d, ^1J(P,Rh) = 194.5 Hz) ppm. ^{1}H NMR (300.1 MHz, C$_6$D$_6$): δ = 7.39–7.28 (m, 6H, *m*-CH); 7.54–7.40 (m, 6H, *o*-CH); 2.66–2.20 (m, 6H, COE); 1.55–1.15 (m, 8H, COE) ppm. ^{19}F NMR (282.4 MHz, C$_6$D$_6$): δ = 83.2 (p, ^2J(F,F) = 150 Hz, 1F, SF$_5$); 62.4 (d, ^2J(F,F) = 150 Hz, 4F, SF$_5$) ppm.

3.8. Reaction of [IrCl(COD){P(p-C₆H₄SF₅)₃}] (4) with CO. Formation of [IrCl(CO)(COD){P(p-C₆H₄SF₅)₃}] (7) and [IrCl(CO)₂{P(p-C₆H₄SF₅)₃}₂] (8)

In a Young NMR tube, a solution of [IrCl(COD){P(p-C₆H₄SF₅)₃}] (4) (25 mg, 0.03 mmol) in CD_2Cl_2 (0.4 mL) was cooled to 77 K, degassed and treated with CO. After 10' the solution turned dark brown and a black precipitate was formed. The NMR analysis showed full conversion to yield the complexes 7 and 8 in a 1:1 ratio together with an unknown iridium complex bearing no phosphine ligand.

NMR data for 7: $^{31}P\{^{1}H\}$ NMR (121.5 MHz, CD_2Cl_2): δ = 3.65 (s) ppm. ^{1}H NMR (300.1 MHz, CD_2Cl_2): δ 7.90–7.82 (m, 12H, *Ph*); 4.25 (s br, 2H, =C*H* trans to CO); 3.91 (s br, 2H, =C*H* trans to Cl); 2.95 (s br, 4H, C*H₂*); 2.79–2.53 (m, 2H, C*H₂*); 2.06-1.83 (m, 2H, C*H₂*) ppm. ^{19}F NMR (282.4 MHz, CD_2Cl_2): δ = 82.0 (p, $^{2}J(F,F)$ = 150 Hz, 1F, S*F₅*); 61.9 (d, $^{2}J(F,F)$ = 150 Hz, 4F, S*F₅*) ppm.

When the reaction was performed with ^{13}CO, the complexes 7' and 8' were obtained.

Selected NMR data for 7': $^{31}P\{^{1}H\}$ NMR (121.5 MHz, CD_2Cl_2): δ = 3.65 (d, $^{2}J(P,C)$ = 13.7 Hz) ppm. $^{13}C\{^{1}H\}$ NMR (75.4 MHz, CD_2Cl_2): 176.0 (t, $^{2}J(C,P)$ = 13.6 Hz, ^{13}CO); 135.6-135.2 (m, P–C_q and *o*–C*H*);127.1–126.6 (m, *m*–C*H*) ppm.

NMR data for the unknown complex: ^{1}H NMR (300.1 MHz, CD_2Cl_2): δ 4.79 (s br, 4H, =C*H*); 2.79–2.53 (m, 8H, C*H₂*) ppm. $^{13}C\{^{1}H\}$ NMR (75.4 MHz, CD_2Cl_2): 169.7 (s br, CO); 80.4 (s, =CH); 33.7 (s, CH₂) ppm.

3.9. Independent Formation of [IrCl(CO)₂{P(p-C₆H₄SF₅)₃}₂] (8)

In a Young NMR tube, complex [IrCl(COD){P(p-C₆H₄SF₅)₃}] (4) (25 mg, 0.03 mmol) was dissolved in C_6D_6 (0.4 mL) and the phosphine 1 (19 mg, 0.03 mmol) was added. Then, the solution was cooled to 77 K, degassed and treated with CO. After 10 min, the solution turned orange and a yellow precipitate was formed. The solid was filtered off, washed with C_6D_6 (2 × 0.25 mL) and dried under vacuum. Yield: 34 mg (85%).

$^{31}P\{^{1}H\}$ NMR (121.5 MHz, CD_2Cl_2): δ = 0.84 (s) ppm. ^{1}H NMR (300.1 MHz, CD_2Cl_2): δ 7.92 (dm, $^{3}J(H,H)$ = 8.6 Hz, *m*–C*H*); 7.79 (dd, $^{3}J(H,H)$ = 8.6, $^{2}J(H,P)$ = 5.8 Hz, *o*–C*H*) ppm. ^{19}F NMR (282.4 MHz, CD_2Cl_2): δ = 81.7 (p, $^{2}J(F,F)$ = 150 Hz, 1F, S*F₅*); 61.8 (d, $^{2}J(F,F)$ = 150 Hz, 4F, S*F₅*) ppm. $^{13}C\{^{1}H\}$ NMR (75.4 MHz, CD_2Cl_2): 156.1 (p, $^{1}J(C,P)$ = 18.5 Hz, C_q–SF₅); 135.0 (m, overlapped with CH signals, P–C_q); 134.9 (d, $^{2}J(C,P)$ = 12.8 Hz, *o*–CH); 126.5 (dp, $^{3}J(C,P)$ = 10.0 Hz, $^{3}J(C,F_{eq})$ = 4.9 Hz, *m*–CH) ppm. The resonance for the CO ligand was not observed. IR (ATR): 1986 (C≡O), 1940 (C≡O), 824 (S–F) cm^{-1}.

When the reaction was performed with ^{13}CO, the complex 8' was obtained.

Selected NMR data: $^{31}P\{^{1}H\}$ NMR (121.5 MHz, CD_2Cl_2): δ = 0.84 (t, $^{2}J(P,C)$ = 13.5 Hz) ppm. $^{13}C\{^{1}H\}$ NMR (75.4 MHz, CD_2Cl_2): 179.6 (t, $^{2}J(C,P)$ = 13.6 Hz, ^{13}CO) ppm. IR (ATR): 1933 (C≡O), 1896 (C≡O), 824 (S–F) cm^{-1}.

3.10. Formation of Trans,trans-[RhCl(CO)₂{P(p-C₆H₄SF₅)₃}₂] (9)

Complex [RhCl(COD){P(p-C₆H₄SF₅)₃}] (5) (20 mg, 0.02 mmol) was dissolved in C_6D_6 (0.4 mL) and the phosphine 1 (14 mg, 0.02 mmol) was added. Then, the solution was cooled to 77 K, degassed and treated with CO. After 10 min, the solution turned clear and a yellow precipitate was formed. The solid was filtered off and washed with C_6D_6 (2 × 0.25 mL) and dried under vacuum Yield: 29 mg (89%).

$^{31}P\{^{1}H\}$ NMR (121.5 MHz, acetone-d_6): δ = 27.6 (s, br) ppm. $^{31}P\{^{1}H\}$ NMR (121.5 MHz, 243 K, acetone-d_6): δ = 29.15 (d, br, $^{2}J(P,Rh)$ ≈ 130 Hz) ppm. ^{1}H NMR (300.1 MHz, acetone-d^6): δ 8.22–8.00 (m, *m*–C*H* + *o*–C*H*) ppm. ^{19}F NMR (282.4 MHz, 243 K acetone-d^6): δ = 82.9 (p, $^{2}J(F,F)$ = 148 Hz, 1F, S*F₅*); 61.6 (d, $^{2}J(F,F)$ = 148 Hz, 4F, S*F₅*) ppm. $^{13}C\{^{1}H\}$ NMR (75.4 MHz, acetone-d^6): 156.1–154.7 (m, C_q–SF₅); 136.3 (s br, P–C_q + *o*–CH); 127.0 (s br, *m*–CH) ppm. The resonance for the CO ligand was not observed. IR (ATR): 1992 (C≡O), 828 (S–F) cm^{-1}.

When the reaction was performed with ^{13}CO, complex 9' was obtained.

Selected NMR data: $^{31}P\{^{1}H\}$ NMR (121.5 MHz, acetone-d_6): δ = 28.62 (s, br) ppm. $^{31}P\{^{1}H\}$ NMR (121.5 MHz, 203 K, acetone-d^6): δ = 29.49 (d, br, $^{2}J(P,Rh)$ ≈ 130 Hz) ppm. $^{13}C\{^{1}H\}$ NMR (75.4 MHz, acetone-d^6): 187.4 (t, ^1J(C,Rh) = 73.0 Hz, ^{13}CO) ppm. ^{13}C{^{1}H} NMR (75.4 MHz, 203 K, acetone-d^6): 187.1 (d, ^1J(C,Rh) = 72.6 Hz, ^{13}CO) ppm. IR (ATR): 1945 (C≡O), 824 (S–F) cm^{-1}.

3.11. X-ray Diffraction Analysis

For the structure determination, the data collection was performed with a BRUKER APEX-II CCD diffractometer (Bruker AXS, Karlsruhe, Germany) using Mo-K$_\alpha$ radiation (λ = 0.71073 Å). Multi-scan absorption corrections implemented in SADABS [71] were applied to the data. The structures were solved by intrinsic phasing method (SHELXT-2014) [72] and refined by full–matrix least square procedures based on F2 with all measured reflections (SHELXL-2018) [73], with anisotropic temperature factors for all non-hydrogen atoms. In complex **4·0.5C$_6$H$_6$**, C(sp^3) atoms of COD ligand were treated with ISOR while in complex **6·C$_6$H$_6$**, all the carbon atoms of the solvent molecule were treated with EADP. All carbon bound hydrogen atoms were added geometrically and refined by using a riding model. CCDC 2021751–2021754 contain the supplementary crystallographic data. These data can be obtained free of charge via http://www.ccdc.cam.ac.uk/conts/retrieving.html (or from the CCDC, 12 Union Road, Cambridge CB2 1EZ, UK; Fax: +44 1223 336033; E–mail: deposit@ccdc.cam.ac.uk).

Crystal Data for **1·C$_6$H$_6$**. C$_{18}$H$_{12}$F$_{15}$PS$_3$·C$_6$H$_6$, M_W 718.53, triclinic, space group $P\bar{1}$, a: 10.0772(6) Å, b: 10.2132(6) Å, c: 15.1197(12) Å, α: 70.446(3)°, β: 70.770(4)°, γ = 85.849(3)°, V = 1383.31(16) Å^3, Z = 2, D_{calc}: 1.725 mg m^{-3}, T = 103(2) K, μ = 0.443 mm^{-1}. 28,056 measured reflections (2θ: 2.142–26.416°), 5658 unique (R_{int} = 0.0533). Final agreement factors were R^1 = 0.0402 ($I > 2\sigma(I)$) and wR^2 = 0.0862.

Crystal Data for **2·2C$_6$H$_6$**. C$_{30}$H$_{24}$F$_{15}$OPS$_3$·2C$_6$H$_6$, M_W 812.64, triclinic, space group $P\bar{1}$, a: 7.7689(7) Å, b: 12.3481(10) Å, c: 17.6929(17) Å, α: 106.915(3)°, β: 90.217(4)°, γ = 92.514(3)°, V = 1622.1(3) Å^3, Z = 2, D_{calc}: 1.664 mg m^{-3}, T = 102(2) K, μ = 0.391 mm^{-1}. 65,957 measured reflections (2θ: 2.374–26.846°), 6935 unique (R_{int} = 0.0577). Final agreement factors were R^1 = 0.0388 ($I > 2\sigma(I)$) and wR^2 = 0.0890.

Crystal Data for **4·0.5C$_6$H$_6$**. C$_{26}$H$_{24}$ClF$_{15}$IrPS$_3$·0.5C$_6$H$_6$, M_W 1015.30, monoclinic, space group $P2_1/c$, a: 16.8589(11) Å, b: 17.1921(11) Å, c: 11.8495(8) Å, β: 101.744(2)°, V = 3362.6(4) Å^3, Z = 2, D_{calc}: 2.006 mg m^{-3}, T = 100(2) K, μ = 4.390 mm^{-1}. 41,026 measured reflections (2θ: 2.264–26.436°), 6901 unique (R_{int} = 0.0505). Final agreement factors were R^1 = 0.0340 ($I > 2\sigma(I)$) and wR^2 = 0.0763.

Crystal Data for **6·C$_6$H$_6$**. C$_{52}$H$_{52}$Cl$_2$F$_{30}$Rh$_2$P$_2$S$_6$·C$_6$H$_6$, M_W 1856.06, monoclinic, space group $C2/c$, a: 19.8800(13) Å, b: 11.6762(7) Å, c: 29.2074(19) Å, β: 96.426(2)°, V = 6737.1(7) Å^3, Z = 4, D_{calc}: 1.830 mg m^{-3}, T = 103(2) K, μ = 0.926 mm^{-1}. 62,530 measured reflections (2θ: 2.182–24.840°), 5816 unique (R_{int} = 0.1005). Final agreement factors were R^1 = 0.0431 ($I > 2\sigma(I)$) and wR^2 = 0.0910.

3.12. Computational Details

Calculations were run using the Gaussian 09 (Revision D.01, Gaussian, Inc., Wallingford, CT, USA) program package [74]. In the case of phosphines and phosphine radical cations the CAM-B3LYP functional was used and 6-311G(d,p) basis set were employed for all atoms. For the nickel complexes, the B3LYP functional was chosen. Nickel was described with RECPs and the associated LANL2DZ basis sets [75] while the ligands were described with 6-31G(d,p). All calculated structures were identified as minima (no negative eigenvalues). All xyz coordinates are included as Supplementary Material (Tables S1–S3).

4. Conclusions

An unprecedented phosphine bearing SF$_5$ groups, P(p-C$_6$H$_4$SF$_5$)$_3$ (**1**), was successfully synthesized. It exhibits a moderate air stability, which is corroborated experimentally and by theoretical methods. The HOMO energy level and the phosphorus-selenium coupling constant of the synthesized phosphine selenide indicate that **1** is a weaker donor than P(p-C$_6$H$_4$CF$_3$)$_3$ and very close to P(m-C$_6$H$_4$(CF$_3$)$_2$)$_3$. The calculated Tolman electronic parameter are consistent with an electron-withdrawing character. The molecular structures of iridium and rhodium complexes allowed for an estimation of the cone angle, which is slightly larger than the one for P(p-C$_6$H$_4$CF$_3$)$_3$. Finally, reactivity studies of [MCl(COD){P(p-C$_6$H$_4$SF$_5$)$_3$}] (M = Ir (**4**), Rh (**5**)) towards CO revealed a preference for the formation of the diphosphine complexes [MCl(CO)$_2${P(p-C$_6$H$_4$SF$_5$)$_3$}$_2$] (M = Ir (**8**), Rh (**9**)) instead of [MCl(CO)$_2${P(p-C$_6$H$_4$SF$_5$)$_3$}]. Overall, the presence of SF$_5$ groups in the phosphine **1** qualifies it as an electron-poor and sterically demanding phosphine. This might provide new opportunities for applications in catalysis.

Supplementary Materials: The Supplementary Materials are available online.

Author Contributions: Conceptualization, T.B.; methodology, S.H. and M.T.; investigation, S.H., M.T., R.H. and R.L.; writing—original draft preparation, M.T.; writing—review and editing, M.T. and T.B.; supervision, T.B.; funding acquisition, T.B. All authors have read and agreed to the published version of the manuscript.

Funding: We acknowledge the CRC 1349 funded by the Deutsche Forschungsgemeinschaft (DFG, German Research Foundation; Gefördert durch die Deutsche Forschungsgemeinschaft–Projektnummer 387284271–SFB 1349). The APC was funded by MDPI.

Acknowledgments: The authors want to thank Philipp Wittwer and Mike Ahrens for helpful discussions.

Conflicts of Interest: The authors declare no conflict of interest.

References

1. Crabtree, R.H. *The Organometallic Chemistry of the Transition Metals*, 6th ed.; John Wiley & Sons, Inc: Hoboken, NJ, USA, 2014.
2. Hartwig, J.F. *Organotransition Metal Chemistry: From Bonding to Catalysis*; University Science Books: Sausalito, CA, USA, 2010.
3. McAuliffe, C.A.; Levason, W. *Phosphine, Arsine and Stibine Complexes of the Transition Elements*; Elsevier Scientific Pub. Co.: Amsterdam, The Netherlands, 1999.
4. Pignolet, L.H. *Homogeneous Catalysis with Metal Phosphine Complexes*; Plenum Press: New York, NY, USA, 1983.
5. Pollock, C.L.; Saunders, G.C.; Smyth, E.C.M.S.; Sorokin, V.I. Fluoroarylphosphines as ligands. *J. Fluor. Chem.* **2008**, *129*, 142–166. [CrossRef]
6. Clarke, M.L.; Ellis, D.; Mason, K.L.; Orpen, A.G.; Pringle, P.G.; Wingad, R.L.; Zaher, D.A.; Baker, R.T. The electron-poor phosphines P{C$_6$H$_3$(CF$_3$)$_2$-3,5}$_3$ and P(C$_6$F$_5$)$_3$ do not mimic phosphites as ligands for hydroformylation. A comparison of the coordination chemistry of P{C$_6$H$_3$(CF$_3$)$_2$-3,5}$_3$ and P(C$_6$F$_5$)$_3$ and the unexpectedly low hydroformylation activity of their rhodium complexes. *Dalton Trans.* **2005**, *7*, 1294–1300.
7. Fawcett, J.; Hope, E.G.; Kemmitt, R.D.; Paige, D.R.; Russell, D.R.; Stuart, A.M.; Stuart, M. Platinum group metal complexes of arylphosphine ligands containing perfluoroalkyl ponytails; crystal structures of [RhCl$_2$(η^5-C$_5$Me$_5$){P(C$_6$H$_4$C$_6$F$_{13}$-4)$_3$}] and cis- and trans-[PtCl$_2${P(C$_6$H$_4$C$_6$F$_{13}$-4)$_3$}$_2$]. *J. Chem. Soc. Dalton Trans.* **1998**, *22*, 3751–3764. [CrossRef]
8. Hope, E.G.; Kemmitt, R.D.W.; Paige, D.R.; Stuart, A.M.; Wood, D.R.W. Synthesis and coordination chemistry of meta-perfluoroalkyl-derivatised triarylphosphines. *Polyhedron* **1999**, *18*, 2913–2917. [CrossRef]
9. Corcoran, C.; Fawcett, J.; Friedrichs, S.; Holloway, J.H.; Hope, E.G.; Russell, D.R.; Saunders, G.C.; Stuart, A.M. Structural and electronic impact of fluorine in the ortho positions of triphenylphosphine and 1,2-bis(diphenylphosphino)ethane; a comparison of 2,6-difluorophenyl- with pentafluorophenyl-phosphines. *J. Chem. Soc. Dalton Trans.* **2000**, *2*, 161–172. [CrossRef]
10. Croxtall, B.; Fawcett, J.; Hope, E.G.; Stuart, A.M. Synthesis and coordination chemistry of ortho-perfluoroalkyl-derivatised triarylphosphines. *J. Chem. Soc. Dalton Trans.* **2002**, *4*, 491–499. [CrossRef]
11. Saunders, G.C. Structural and electronic properties of tris(4-trifluoromethyltetrafluorophenyl)phosphine. *J. Fluor. Chem.* **2015**, *180*, 15–20. [CrossRef]
12. Uson, R.; Oro, L.A.; Fernandez, M.J. Preparation, reactions and catalytic activity of complexes of the type [Ir(COD){P(p-RC$_6$H$_4$)$_3$}$_2$]A (R = Cl, F, H, CH$_3$ or CH$_3$O.; A = ClO$_4^-$ or B(C$_6$H$_5$)$_4^-$). *J. Organomet. Chem.* **1980**, *193*, 127–133. [CrossRef]
13. Matsubara, K.; Fujii, T.; Hosokawa, R.; Inatomi, T.; Yamada, Y.; Koga, Y. Fluorine-Substituted Arylphosphine for an NHC-Ni(I) System, Air-Stable in a Solid State but Catalytically Active in Solution. *Molecules* **2019**, *24*, 3222. [CrossRef]
14. Moser, W.R.; Papile, C.J.; Brannon, D.A.; Duwell, R.A.; Weininger, S.J. The mechanism of phosphine-modified rhodium-catalyzed hydroformylation studied by CIR-FTIR. *J. Mol. Catal.* **1987**, *41*, 271–292. [CrossRef]

15. Chen, C.; Wang, Y.; Shi, X.; Sun, W.; Zhao, J.; Zhu, Y.-P.; Liu, L.; Zhu, B. Palladium-Catalyzed C-2 and C-3 Dual C–H Functionalization of Indoles: Synthesis of Fluorinated Isocryptolepine Analogues. *Org. Lett.* **2020**, *22*, 4097–4102. [CrossRef] [PubMed]
16. Paterson, A.J.; Dunås, P.; Rahm, M.; Norrby, P.-O.; Kociok-Köhn, G.; Lewis, S.E.; Kann, N. Palladium Catalyzed Stereoselective Arylation of Biocatalytically Derived Cyclic 1,3-Dienes: Chirality Transfer via a Heck-Type Mechanism. *Org. Lett.* **2020**, *22*, 2464–2469. [CrossRef] [PubMed]
17. Jakab, A.; Dalicsek, Z.; Holczbauer, T.; Hamza, A.; Pápai, I.; Finta, Z.; Timári, G.; Soós, T. Superstable Palladium(0) Complex as an Air- and Thermostable Catalyst for Suzuki Coupling Reactions. *Eur. J. Org. Chem.* **2015**, *2015*, 60–66. [CrossRef]
18. Cheng, L.; Li, M.-M.; Wang, B.; Xiao, L.-J.; Xie, J.-H.; Zhou, Q.-L. Nickel-catalyzed hydroalkylation and hydroalkenylation of 1,3-dienes with hydrazones. *Chem. Sci.* **2019**, *10*, 10417–10421. [CrossRef] [PubMed]
19. Sheppard, W.A. The Electrical Effect of the Sulfur Pentafluoride Group. *J. Am. Chem. Soc.* **1962**, *84*, 3072–3076. [CrossRef]
20. Hansch, C.; Muir, R.M.; Fujita, T.; Maloney, P.P.; Geiger, F.; Streich, M. The Correlation of Biological Activity of Plant Growth Regulators and Chloromycetin Derivatives with Hammett Constants and Partition Coefficients. *J. Am. Chem. Soc.* **1963**, *85*, 2817–2824. [CrossRef]
21. Hansch, C.; Leo, A.; Taft, R.W. A survey of Hammett substituent constants and resonance and field parameters. *Chem. Rev.* **1991**, *91*, 165–195. [CrossRef]
22. Savoie, P.R.; Welch, J.T. Preparation and Utility of Organic Pentafluorosulfanyl-Containing Compounds. *Chem. Rev.* **2015**, *115*, 1130–1190. [CrossRef]
23. Lentz, D.; Seppelt, K. The -SF$_5$, -SeF$_5$, and -TeF$_5$ Groups in Organic Chemistry. In *Chemistry of Hypervalent Compounds*; Akiba, K.-Y., Ed.; Wiley-VCH: New York, NY, USA, 1999; pp. 295–325.
24. Gard, G.L. Recent Milestones in SF$_5$-Chemistry. *Chim. Oggi* **2009**, *27*, 10–13.
25. Altomonte, S.; Zanda, M. Synthetic chemistry and biological activity of pentafluorosulphanyl (SF$_5$) organic molecules. *J. Fluor. Chem.* **2012**, *143*, 57–93. [CrossRef]
26. Kanishchev, O.S.; Dolbier, W.R. Chapter One—SF$_5$-Substituted Aromatic Heterocycles. In *Advances in Heterocyclic Chemistry*; Scriven, E.F.V., Ramsden, C.A., Eds.; Academic Press: Cambridge, MA, USA, 2016; Volume 120, pp. 1–42.
27. Chan, J.M.W. Pentafluorosulfanyl group: An emerging tool in optoelectronic materials. *J. Mater. Chem. C* **2019**, *7*, 12822–12834. [CrossRef]
28. Beier, P. Synthesis and reactivity of novel sulfur pentafluorides—Effect of the SF$_5$ group on reactivity of nitrobenzenes in nucleophilic substitution. *Phosphorus Sulfur Silicon Relat. Elem.* **2017**, *192*, 212–215. [CrossRef]
29. Beier, P. Pentafluorosulfanylation of Aromatics and Heteroaromatics. In *Emerging Fluorinated Motifs*; Ma, J.-A., Cahard, D., Eds.; Wiley-VCH: Weinheim, Germany, 2020; Volume 2, pp. 551–570.
30. Haufe, G. Pentafluorosulfanylation of Aliphatic Substrates. In *Emerging Fluorinated Motifs*; Ma, J.-A., Cahard, D., Eds.; Wiley-VCH: Weinheim, Germany, 2020; Volume 2, pp. 571–609.
31. Damerius, R.; Leopold, D.; Schulze, W.; Seppelt, K. Strukturen von SF$_5$-substituierten Metallkomplexen. *Z. Anorg. Allg. Chem.* **1989**, *578*, 110–118. [CrossRef]
32. Henkel, T.; Klauck, A.; Seppelt, K. Pentafluoro-λ^6-sulfanylacetylene complexes of cobalt. *J. Organomet. Chem.* **1995**, *501*, 1–6. [CrossRef]
33. Preugschat, D.; Thrasher, J.S. Pentacarbonylchrom-Komplexe SF$_5$-substituierter Isocyanide. *Z. Anorg. Allg. Chem.* **1996**, *622*, 1411–1414. [CrossRef]
34. Shavaleev, N.M.; Xie, G.; Varghese, S.; Cordes, D.B.; Slawin, A.M.Z.; Momblona, C.; Ortí, E.; Bolink, H.J.; Samuel, I.D.W.; Zysman-Colman, E. Green Phosphorescence and Electroluminescence of Sulfur Pentafluoride-Functionalized Cationic Iridium(III) Complexes. *Inorg. Chem.* **2015**, *54*, 5907–5914. [CrossRef]
35. Ma, X.-F.; Luo, X.-F.; Yan, Z.-P.; Wu, Z.-G.; Zhao, Y.; Zheng, Y.-X.; Zuo, J.-L. Syntheses, Crystal Structures, and Photoluminescence of a Series of Iridium(III) Complexes Containing the Pentafluorosulfanyl Group. *Organometallics* **2019**, *38*, 3553–3559. [CrossRef]
36. Pal, A.K.; Henwood, A.F.; Cordes, D.B.; Slawin, A.M.Z.; Samuel, I.D.W.; Zysman-Colman, E. Blue-to-Green Emitting Neutral Ir(III) Complexes Bearing Pentafluorosulfanyl Groups: A Combined Experimental and Theoretical Study. *Inorg. Chem.* **2017**, *56*, 7533–7544. [CrossRef]

37. Groves, L.M.; Schotten, C.; Beames, J.; Platts, J.A.; Coles, S.J.; Horton, P.N.; Browne, D.L.; Pope, S.J.A. From Ligand to Phosphor: Rapid, Machine-Assisted Synthesis of Substituted Iridium(III) Pyrazolate Complexes with Tuneable Luminescence. *Chem. A Eur. J.* **2017**, *23*, 9407–9418. [CrossRef]
38. Henwood, A.F.; Webster, J.; Cordes, D.; Slawin, A.M.Z.; Jacquemin, D.; Zysman-Colman, E. Phosphorescent platinum(ii) complexes bearing pentafluorosulfanyl substituted cyclometalating ligands. *RSC Adv.* **2017**, *7*, 25566–25574. [CrossRef]
39. Berg, C.; Braun, T.; Laubenstein, R.; Braun, B. Palladium-mediated borylation of pentafluorosulfanyl functionalized compounds: The crucial role of metal fluorido complexes. *Chem. Commun.* **2016**, *52*, 3931–3934. [CrossRef] [PubMed]
40. Golf, H.R.A.; Reissig, H.-U.; Wiehe, A. Synthesis of SF_5-Substituted Tetrapyrroles, Metalloporphyrins, BODIPYs, and Their Dipyrrane Precursors. *J. Org. Chem.* **2015**, *80*, 5133–5143. [CrossRef] [PubMed]
41. Berry, A.D.; De Marco, R.A. Reaction of pentafluoro[(trifluoromethyl)acetylenyl]sulfur with nickel tetracarbonyl. *Inorg. Chem.* **1982**, *21*, 457–458. [CrossRef]
42. Sergeeva, T.A.; Dolbier, W.R. A New Synthesis of Pentafluorosulfanylbenzene. *Org. Lett.* **2004**, *6*, 2417–2419. [CrossRef]
43. Dunne, B.J.; Orpen, A.G. Triphenylphosphine: A redetermination. *Acta Cryst.* **1991**, *C47*, 345–347. [CrossRef]
44. Eapen, K.C.; Tamborski, C. The synthesis of tris-(trifluoromethylphenyl)phosphines and phosphine oxides. *J. Fluor. Chem.* **1980**, *15*, 239–243. [CrossRef]
45. See, R.F.; Dutoi, A.D.; Fettinger, J.C.; Nicastro, P.J.; Ziller, J.W. The crystal structures of (*p*-ClPh)$_3$PO and (*p*-OMePh)$_3$PO, including an analysis of the P-O bond in triarylphosphine oxides. *J. Chem. Crystallogr.* **1998**, *28*, 893–898. [CrossRef]
46. Stewart, B.; Harriman, A.; Higham, L.J. Predicting the Air Stability of Phosphines. *Organometallics* **2011**, *30*, 5338–5343. [CrossRef]
47. Dunne, B.J.; Morris, R.B.; Orpen, A.G. Structural systematics. Part 3. Geometry deformations in triphenylphosphine fragments: A test of bonding theories in phosphine complexes. *J. Chem. Soc. Dalton Trans.* **1991**, 653–661. [CrossRef]
48. Palau, C.; Berchadsky, Y.; Chalier, F.; Finet, J.-P.; Gronchi, G.; Tordo, P. Tris(monochlorophenyl)- and Tris(dichlorophenyl)phosphines: Molecular Geometry, Anodic Behavior, and ESR Studies. *J. Phys. Chem.* **1995**, *99*, 158–163. [CrossRef]
49. Chevykalova, M.N.; Manzhukova, L.F.; Artemova, N.V.; Luzikov, Y.N.; Nifant'ev, I.E.; Nifant'ev, E.E. Electron-donating ability of triarylphosphines and related compounds studied by 31P NMR spectroscopy. *Russ. Chem. Bull.* **2003**, *52*, 78–84. [CrossRef]
50. Howell, J.S.; Lovatt, J.; Yates, P.; Gottlieb, H.; Hursthouse, M.; Light, M. Effect of fluorine and trifluoromethyl substitution on the donor properties and stereodynamical behaviour of triarylphosphines. *J. Chem. Soc. Dalton Trans.* **1999**, *17*, 3015–3028. [CrossRef]
51. Allen, D.W.; Taylor, B.F. The chemistry of heteroarylphosphorus compounds. Part 15. Phosphorus-31 nuclear magnetic resonance studies of the donor properties of heteroarylphosphines towards selenium and platinum(II). *J. Chem. Soc. Dalton Trans.* **1982**, *1*, 51–54. [CrossRef]
52. Tolman, C.A. Steric effects of phosphorus ligands in organometallic chemistry and homogeneous catalysis. *Chem. Rev.* **1977**, *77*, 313–348. [CrossRef]
53. Gusev, D.G. Donor Properties of a Series of Two-Electron Ligands. *Organometallics* **2009**, *28*, 763–770. [CrossRef]
54. Perrin, L.; Clot, E.; Eisenstein, O.; Loch, J.; Crabtree, R.H. Computed Ligand Electronic Parameters from Quantum Chemistry and Their Relation to Tolman Parameters, Lever Parameters, and Hammett Constants. *Inorg. Chem.* **2001**, *40*, 5806–5811. [CrossRef]
55. Kawaguchi, S.-I.; Minamida, Y.; Okuda, T.; Sato, Y.; Saeki, T.; Yoshimura, A.; Nomoto, A.; Ogawa, A. Photoinduced Synthesis of P-Perfluoroalkylated Phosphines from Triarylphosphines and Their Application in the Copper-Free Cross-Coupling of Acid Chlorides and Terminal Alkynes. *Adv. Synth. Catal.* **2015**, *357*, 2509–2519. [CrossRef]
56. Ma, X.-Y.; Wang, K.; Zhang, L.; Li, X.-J.; Li, R.-X. Selective Hydrogenation of Avermectin Catalyzed by Iridium-Phosphine Complexes. *Chin. J. Chem.* **2007**, *25*, 1503–1507. [CrossRef]

57. Tiburcio, J.; Bernès, S.; Torrens, H. Electronic and steric effects of triarylphosphines on the synthesis, structure and spectroscopical properties of mononuclear rhodium(I)–chloride complexes. *Polyhedron* **2006**, *25*, 1549–1554. [CrossRef]
58. Naaktgeboren, A.J.; Nolte, R.J.M.; Drenth, W. Phosphorus-31 nuclear magnetic resonance studies of polymer-anchored rhodium(I) complexes. *J. Am. Chem. Soc.* **1980**, *102*, 3350–3354. [CrossRef]
59. Canepa, G.; Brandt, C.D.; Werner, H. Mono- and Dinuclear Rhodium(I) and Rhodium(III) Complexes with the Bulky Phosphine 2,6-Me$_2$C$_6$H$_3$CH$_2$CH$_2$PtBu$_2$, Including the First Structurally Characterized Cis-Configurated Dicarbonyl Compound, *cis*-[RhCl(CO)$_2$(PR$_3$)]. *Organometallics* **2004**, *23*, 1140–1152. [CrossRef]
60. Müller, T.E.; Mingos, D.M.P. Determination of the Tolman cone angle from crystallographic parameters and a statistical analysis using the crystallographic data base. *Transit. Met. Chem.* **1995**, *20*, 533–539. [CrossRef]
61. Reyna-Madrigal, A.; Ortiz-Pastrana, N.; Paz-Sandoval, M.A. Cyclooctadiene iridium complexes with phosphine and pentadienyl ligands. *J. Organomet. Chem.* **2019**, *886*, 13–26. [CrossRef]
62. Bondi, A. van der Waals Volumes and Radii. *J. Phys. Chem.* **1964**, *68*, 441–451. [CrossRef]
63. Joerg, S.; Drago, R.S.; Sales, J. Reactivity of Phosphorus Donors. *Organometallics* **1998**, *17*, 589–599. [CrossRef]
64. Ortega-Moreno, L.; Fernández-Espada, M.; Moreno, J.J.; Navarro-Gilabert, C.; Campos, J.; Conejero, S.; López-Serrano, J.; Maya, C.; Peloso, R.; Carmona, E. Synthesis, properties, and some rhodium, iridium, and platinum complexes of a series of bulky m-terphenylphosphine ligands. *Polyhedron* **2016**, *116*, 170–181. [CrossRef]
65. Von Hahmann, C.N.; Talavera, M.; Xu, C.; Braun, T. Reactivity of 3,3,3-Trifluoropropyne at Rhodium Complexes: Development of Hydroboration Reactions. *Chem. A Eur. J.* **2018**, *24*, 11131–11138. [CrossRef]
66. Vaska, L. Reversible Combination of Carbon Monoxide with a Synthetic Oxygen Carrier Complex. *Science* **1966**, *152*, 769–771. [CrossRef]
67. Sanger, A.R. Five-coordinate dicarbonyl complexes of rhodium(I): [RhX(CO)$_2$(PPh$_3$)$_2$] (X = Cl, Br, I). *Can. J. Chem.* **1985**, *63*, 571–575. [CrossRef]
68. Perrin, D.D.; Armarego, W.L.F. *Purification of Laboratory Chemicals*, 3rd ed.; Butterworth/Heinemann: London/Oxford, UK, 1988.
69. Herde, J.-L.; Lambert, J.C.; Senoff, C.V. Cyclooctene and 1,5-Cyclooctadiene Complexes of Iridium. *Inorg. Synth.* **1974**, *15*, 18–20.
70. Van der Ent, A.; Onderdelinden, A.L. Chlorobis(cyclooctene)rhodium(I) and -iridium(I) Complexes. *Inorg. Synth.* **1973**, *14*, 92–95.
71. Sheldrick, G.M. *SADABS, Program for Empirical Absorption Correction of Area Detector Data*, May 2014; University of Göttingen: Göttingen, Germany, 1996.
72. Sheldrick, G.M. *SHELXT-2014, Program for the Solution of Crystal Structures from X-ray Data*; University of Göttingen: Göttingen, Germany, 2013.
73. Sheldrick, G.M. *SHELXL-2018, Program for the Refinement of Crystal Structures from X-ray Data*; University of Göttingen: Göttingen, Germany, 2018.
74. Frisch, M.J.; Schlegel, H.B.; Scuseria, G.E.; Robb, M.A.; Cheeseman, J.R.; Scalmani, G.; Barone, V.; Petersson, G.A.; Nakatsuji, H.; Li, X.; et al. *Gaussian 09, Revision D.01*; Gaussian, Inc.: Wallingford, CT, USA, 2016.
75. Hay, P.J.; Wadt, W.R. Ab initio effective core potentials for molecular calculations. Potentials for K to Au including the outermost core orbitals. *J. Chem. Phys.* **1985**, *82*, 299–310. [CrossRef]

Sample Availability: Samples of the compounds are not available from the authors.

© 2020 by the authors. Licensee MDPI, Basel, Switzerland. This article is an open access article distributed under the terms and conditions of the Creative Commons Attribution (CC BY) license (http://creativecommons.org/licenses/by/4.0/).

Article

Magnetic Nanoparticles Fishing for Biomarkers in Artificial Saliva

Arpita Saha [1], Hamdi Ben Halima [2], Abhishek Saini [1], Juan Gallardo-Gonzalez [2], Nadia Zine [2], Clara Viñas [1], Abdelhamid Elaissari [3], Abdelhamid Errachid [2,*] and Francesc Teixidor [1,*]

1. Inorganic Materials Laboratory, Institut de Ciencia de Materials de Barcelona (ICMAB-CSIC), Campus de la UAB, 08193 Bellaterra, Spain; arpitasaha666@gmail.com (A.S.); ab76779@gmail.com (A.S.); clara@icmab.es (C.V.)
2. Micro & Nanobiotechnology Laboratory, Université de Lyon, CNRS, University Claude Bernard Lyon 1, Institut des Sciences Analytiques, UMR 5280, 5 rue de la Doua, F-69100 Villeurbanne, France; ben.halima.hamdi@hotmail.com (H.B.H.); juagalgon@gmail.com (J.G.-G.); Nadia.zine@univ-lyon1.fr (N.Z.)
3. LAGEPP-UMR 5007, CNRS, University Claude Bernard Lyon-1, University of Lyon, F-69622 Lyon, France; abdelhamid.elaissari@univ-lyon1.fr
* Correspondence: abdelhamid.errachid-el-salhi@univ-lyon1.fr (A.E.); teixidor@icmab.es (F.T.); Tel.: +34-935-801-853 (F.T.)

Academic Editor: Ashok Kakkar
Received: 31 July 2020; Accepted: 29 August 2020; Published: 31 August 2020

Abstract: Magnetic nanoparticles (MNPs) were synthesized using the colloidal co-precipitation method and further coated with silica using the Stöber process. These were functionalized with carboxylic and amine functionalities for further covalent immobilization of antibodies on these MNPs. The procedure for covalent immobilization of antibodies on MNPs was developed using 1-ethyl-3-(dimethylaminopropyl)carbodiimide (EDC) and N-hydroxysuccinimide (NHS). The evaluation of the efficiency of the coupling reaction was carried out by UV-vis spectrophotometry. The developed antibodies coupled to MNPs were tested for the pre-concentration of two biomarkers tumor necrosis factor alpha (TNF-α) and Interleukin-10 (IL-10). Both biomarkers were assessed in the matrix based on phosphate-buffered saline solution (PBS) and artificial saliva (AS) to carry out the demonstration of the format assay. Supernatants were used to determine the number of free biomarkers for both studies. Reduction of the nonspecific saliva protein adsorption on the surface of the complex antibodies-MNPs to levels low enough to allow the detection of biomarkers in complex media has been achieved.

Keywords: immobilization of antibodies; IL-10; magnetic nanoparticles; pre-concentration of antigens; saliva matrix; TNF-α

1. Introduction

Superparamagnetic iron oxide nanoparticles (SPIONs) are of great importance when grafted with biomarkers for applications in modern biological and biotechnology areas. The surface-modified magnetic nanoparticles (MNPs) can be used both in in-vitro and in-vivo systems effectively. The size of the MNPs needs to be controlled and innovative functionalization techniques need to be utilized for the effective implementation of these modified MNPs in medical applications [1–6].

The small size of MNPs (a few nanometers) is essential for them to be able to interact, bind, or penetrate biological entities. This is because in the nano range, their dimensions are comparable to those of proteins, cells, or viruses, and this facilitates their movement through biological structures [7]. These MNPs are highly attractive for use in biomedical applications such as magnetic resonance imaging (MRI), targeted drug delivery, and treatment of hyperthermia. These are due to certain unique properties that arise due to a combination of their small dimension, enhanced sensing, and nanoscale-dependent magnetism, as well as physiological properties [8]. MNPs are also very commonly used these days for in-vivo applications as drug carriers in a magnetic 'tag drag-release' process, termed as targeted drug delivery. The MNPs are usually loaded with special drug molecules or chemotherapy agents, which are directly vectorized to tumor cells by targeting ligands on their surfaces, or by the application of an external magnetic field. In recent years, substantial interest has been focused on multifunctional MNPs in which diagnostic (MRI) and therapeutic (hyperthermia treatment and drug delivery) capabilities are combined [9–12].

The successful design of MNPs for biological applications needs a careful selection of magnetic core and surface coating material, where the former mainly determines the MNPs' heating and sensing capabilities with regards to application efficiency and the latter specifies the interaction of these MNPs with a physiological environment. To enable the direct use of MNPs in biomedical applications, the MNPs should be further functionalized by conjugating them with functional groups. The surface coating provides a suitable base for the attachment of these functional groups on MNPs. These groups such as antibodies, peptides, polysaccharides, etc., permit specific recognition of cell types and direct the MNPs to a specific tissue or cell type by binding to a cell surface receptor. The silica coating facilitates the functionalization of the surface of the MNPs with either amine groups or carboxylic groups, which help in bonding with biological entities for medical applications. Also, the silica coating does not affect the magnetic property of the Fe_3O_4 core in a substantial way. Among the different types of MNPs, iron oxides (magnetite (Fe_3O_4), maghemite (γ-Fe_2O_3), and hematite (α-Fe_2O_3)) are by far the most commonly employed ones for in-vivo applications since iron is physiologically well tolerated. Silica is inorganic but bio-friendly and it is known for its chemical stability and ease of formation. The biggest advantage of having a surface enriched in silica is the presence of silanol groups, which can easily react with coupling agents, providing strong attachment of surface ligands on MNPs [13]. There are several successful methods available for the formation of silica coating, amongst which the most commonly used is the Stöber method. Here, a hydrolysis reaction of tetraethyl orthosilicate (TEOS) is governed in alcohol media under the catalytic action of ammonia [14]. In this paper, a variation of this method is applied to produce more homogeneous coatings of silica.

The coupling strategy used for linking the functionalizing agent to the particle is a pivotal component for success, and will largely depend both on the coating of the MNPs and the available functional groups on the target moiety [15–26]. To improve the sensitivity of sensing devices, signal amplification has been attempted using MNPs. The role of MNPs in our experiment is to pre-concentrate the analyte and, following magnetic separation, remove the undesired effect of the matrix, so that a clean measurable signal can be obtained. In the last decade, MNPs-based EC biosensors [27–37], for salivary diagnostics, have received increasing attention because many substances enter saliva from the blood through transcellular or paracellular diffusion [38]. Consequently, most substances found in blood are also present in saliva. Therefore, saliva is functionally equivalent to serum in reflecting the physiological state of the body but has several advantages over blood [39], as saliva collection is easy, stress free, requires painless sampling, has a simple matrix (less complex than blood), is non-invasive, and offers the possibility of performing real-time monitoring. Saliva has recently been the prominent body fluid for the study of SARS-CoV-2 during the COVID-19 pandemic. Saliva is not only a reliable tool to detect the virus, but can also help in studying the evolution of the virus [40–42]. Hence, there are compelling reasons for exploring saliva as a diagnostic and prognostic fluid in heart failure research [43,44] as proposed here. However, saliva is not as constant as blood, as we must keep in mind the change in salivating nature of humans on the sight and smell of food and also depending on the hunger quotient of a particular individual at the specific time. In any case, saliva is there as a potential diagnostic tool due to its ease and non-invasive accessibility along with its abundance of biomarkers, such as genetic material and proteins [45].

To learn if salivary diagnosis is applicable in the conditions defined earlier on pre-concentration in saliva we tested tumor necrosis factor alpha (TNF alpha), responsible for a diverse range of signaling events within cells, and Interleukin 10 (IL-10) that plays a central role in limiting host immune response to pathogens, thereby preventing damage to the host. To study the feasibility of saliva and, in these preliminary tests, avoid the variability in concentration of the saliva, we considered adequate to test the immunoreagents in artificial saliva and ensure that there is no cross-reactivity between the immunoreagents used. In this research, we have immobilized anti-TNF-α and anti-IL-10 antibodies, which are present in saliva on MNPs as a proof of concept. The 3D nano-collectors generated were evaluated using UV-vis spectrophotometry (JENWAY 7205) for their ability to pre-concentrate the antigens TNF-α and IL-10, respectively, by employing an external magnetic field. Tests were performed both in phosphate buffered saline (PBS) and artificial saliva (AS) as the matrix. The obtained results have proven that our MNPs can represent a promising tool for rapid pre-concentration of heart failure biomarkers (TNF-α/IL-10) in saliva. Further investigation will be carried out to integrate our MNPs with sensors to increase the sensitivity, selectivity, and decrease matrix effect.

On the other hand, the colloidal stability is a big issue with the MNPs or magnetic nanocomposites, since their advent there remains a big issue surrounding the stability of the MNPs in colloids. The more stable they are in a colloidal state, the less easy it is to use them for magnetic separation, while on the other hand, if they are less stable then they are easier to isolate by external magnetic field application. So, for this a compromise has to be made regarding them and based on their application they are synthesized as needed. In order to improve magnetization, larger particles can be synthesized; these show higher magnetization but much less stability in colloidal state as, due to their large sizes, they have a greater tendency to aggregate. MNPs are more prone to aggregation compared to other nanoparticles and agglomeration is even more significant for MNPs as there exists magnetic dipole–dipole attraction among themselves [46]. In particular, the magnitude of this magnetic dipole–dipole attraction is directly proportional to power 2 of the particle saturation magnetization and power 3 of the particle size. Sedimentation of the agglomerates from its suspension is a challenge [47] that could limit mobility and applicability of MNPs [48] especially in biosensing applications.

In order to improve stability and also preserve the magnetic properties, the following few things can be done: Sonication, which is a mechanical way and very effective in disaggregating the particles and dispersing them in colloidal state. One of the other methods is using inorganic shell coating like Silica or carbon coating to improve the stability. Also, surface coating with macromolecules like polymers or surfactants can be used for improving stability. Increasing viscosity of the medium can also prevent aggregation as the particles cannot get dispersed easily and travel through the viscous medium, thus preventing aggregation. This can be done by gum gelation or by emulsion formation.

In this paper, one of the first studies done using saliva as a diagnostic fluid for heart failure research is being reported. There lies a novelty in the synthesis procedure in the use of a non-magnetic mechanical stirrer for the coating of SiO_2 layer on the iron oxide core. This strategy coupled with the concave nature of the reaction flask used results in more homogenized coating of the magnetic core with the Silica layer. The traditional method of using a magnetic bead in this step results in inefficient coating as a major part of the magnetic core becomes attracted to the bead and does not fully participate in the reaction. Using our strategy of a specifically designed mechanical stirrer in the concave flask with Teflon coating on top leads to an increase in the chemical kinetics of the reaction by creating a vortex that leads to homogenized coating with almost negligible reactant losses. Usually, when magnetic nanoparticles tend to aggregate, this causes difficulty in magnetic separation as it is well established that the smaller the size of the particles, the faster is the separation time, while phenomenon like aggregation and flocculation slow down the process [49].

Furthermore, besides a modification in the synthesis procedure of the MNPs and artificial saliva as the fluid for heart failure detection, colloidal stability of MNPs has been studied here with different surfactants and by increasing the viscosity of the medium by using a surfactant. Relevant issues like colloidal stability of MNPs in long-term use for varied applications and how the medium affects their stability are avoided in most studies dealing with MNPs. In this paper, preliminary results of how the stability varies with the change in viscosity of the medium and how difficult it is to extract with an external magnet is reported. Certain common issues with MNPs, which are not reported when its synthesis and application is stated like its long-term usage, stability, and extraction time using external magnets and stirrer systems, are dealt with here.

2. Results

The syntheses of $Fe_3O_4@SiO_2$-NH_2 and $Fe_3O4@SiO_2$-NH_2-$COOH$ were done using co-precipitation method, followed by Stober process and functionalization [50,51], though the critical point in the synthesis procedure is to use the mechanical stirrer and not the magnetic bead stirring procedure as is commonly used in other colloidal synthesis routes. This is necessary to avoid the MNPs from sticking to the bead, hence the mechanical stirring is necessary to disperse the MNPs in the solution for the reaction to take place. This is used only in the silica coating procedure; after that, magnetic stirring can be used in these MNPs as, after the silica coating, the direct contact to the magnetite core is prevented by the silica shell. After the synthesis of the magnetite core, it is important to first disperse the MNPs using an ultrasonicator before any further reactions. There is a slight increase in the size of the MNPs after silica coating, which is further increased by functionalizing with the NH_2 and $COOH$ groups. The $Fe_3O_4@SiO_2$ MNPs are shown in TEM images (see Figure 1a) with the characteristic spherical shape of these MNPs with a mean size of 9.3 ± 1.6 nm (see Figure 1b). When the $Fe_3O_4@SiO_2$ MNPs are coated with amine and a carboxylic acid, the spherical shape is maintained as Figure 2 shows but the average size increases to 10.7 + 2.1 nm and 11.6 ± 1.7 nm (see Figure 3).

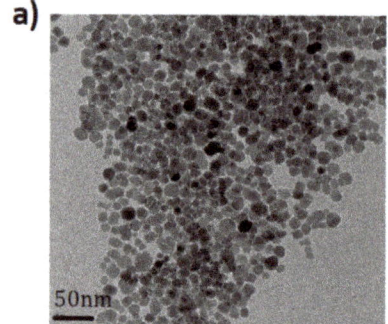

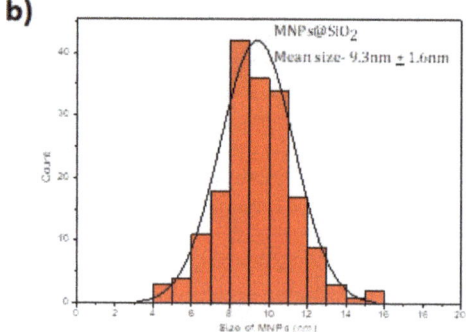

Figure 1. (a) The $Fe_3O_4@SiO_2$ modified magnetic nanoparticles (MNPs) under TEM at 120 kV. They were measured in a copper grid. (b) The average size of $Fe_3O_4@SiO_2$ MNPs is 9.3 ± 1.6 nm.

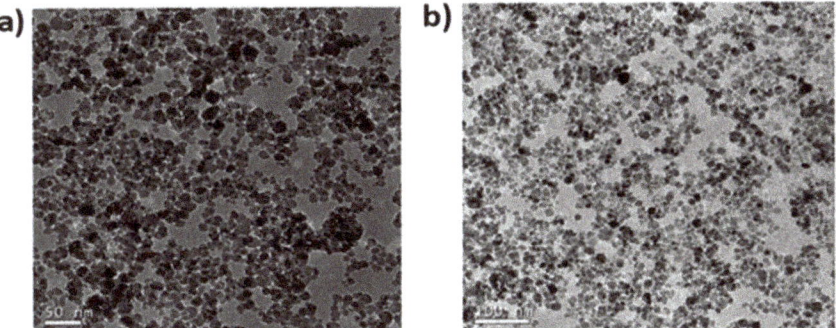

Figure 2. (a) Fe$_3$O$_4$@SiO$_2$-NH$_2$ and (b) The Fe$_3$O$_4$@SiO$_2$-NH$_2$-COOH MNPs under TEM at 120 kV. They were measured in a copper grid. These MNPs were made by using maleic anhydride.

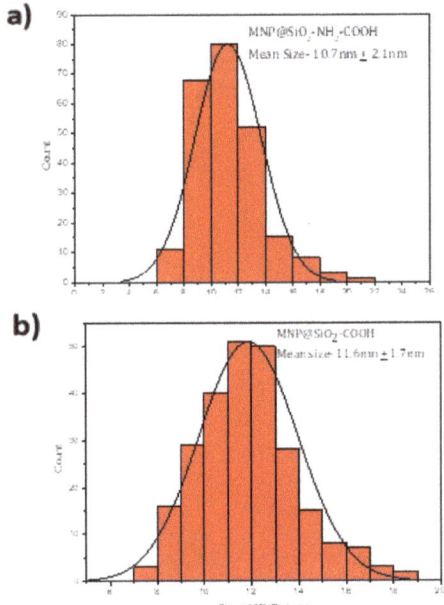

Figure 3. The average size of (a) Fe$_3$O$_4$@SiO$_2$-NH$_2$ MNPs is 10.7 ± 2.1 nm and (b) Fe$_3$O$_4$@SiO$_2$-NH$_2$-COOH MNPs is 11.6 ± 1.7 nm.

The SEM and EDS techniques provide information about the morphology of the MNPs (Figure S1a) (indicating that the sample is homogenous) and the composition (Figure S2), respectively. The sample showed uniform-sized spheres along with a consistent presence of Fe, Si, and O throughout the sample. The composition was studied using EDS and the analysis was done across the whole sample and it was consistent throughout. Using the EDS technique, it is possible to know the percentage coated by SiO$_2$. In this sample, the coating is 1/3 of Si/Fe that is confirmed with the size we get by TEM (Fe$_3$O$_4$ has a size near 7 nm and Fe$_3$O$_4$@SiO$_2$ has 9.3 nm). Figure S1b shows that the electron diffraction (ED) confirms the cubic spinel structure of the MNPs. This ensures that the magnetization of these MNPs results from the Fe^{2+} ions of the MNPs. Also, further chemical composition was analyzed using the IR spectrum. Fe$_3$O$_4$@SiO$_2$ and Fe$_3$O$_4$@SiO$_2$-COOH was measured with IR and the characteristic peaks of Si-O and C=O was observed in the respective samples (Figure S3) at 1058 cm^{-1} and 1739 cm^{-1}, respectively.

Further, these MNPs were also characterized using IR before and after being dispersed in PBS, and the peaks of Si-O and C=O were still visible, thus confirming the presence of the functionalized coating of the MNPs (Figure S4).

Magnetic characterization of the MNPs was carried out in a superconductive quantum interference device (SQUID) magnetometer (Quantum Design MPMS5XL). Magnetization vs. magnetic field measurements were performed at 300K in a 6T field. The samples were prepared using a polycarbonate capsule each filled with 1 mg of $Fe_3O_4@SiO_2$-NH_2 or $Fe_3O_4@SiO_2$-NH_2-COOH and compacted cotton. We already know that the MNPs show the cubic spinel structure by ED. The magnetic property of iron oxide (Fe_3O_4) nanoparticles is dependent on the distribution of Fe ions in octahedral and tetrahedral sites of the spinel structure [52]. The magnetic spins of the ions in the octahedral sites are ferromagnetically coupled to each other and antiferromagnetically coupled with tetrahedral sites. Since the number of Fe^{3+} ions in the octahedral sites and the tetrahedral sites are the same, their magnetic spins cancel out each other. Consequently, the magnetic spins of only Fe^{2+} ions in the octahedral sites contribute to the net magnetic moment in a spinel structure. Figure 4 shows a typical magnetization curve at 300K for superparamagnetic nanoparticles in which neither remnant magnetization (magnetization at zero field, M_R) nor coercivity (hysteresis loop, H_c) was observed. The saturation magnetization value of MNPs at 300K was 38.66 emu/g for $Fe_3O_4@SiO_2$-NH_2 and 40.76 emu/g for $Fe_3O_4@SiO_2$-NH_2-COOH. Also, the magnetization curve for $Fe_3O_4@SiO_2$ is shown where the saturated magnetization is 41.2 emu/g. (Figure 4)

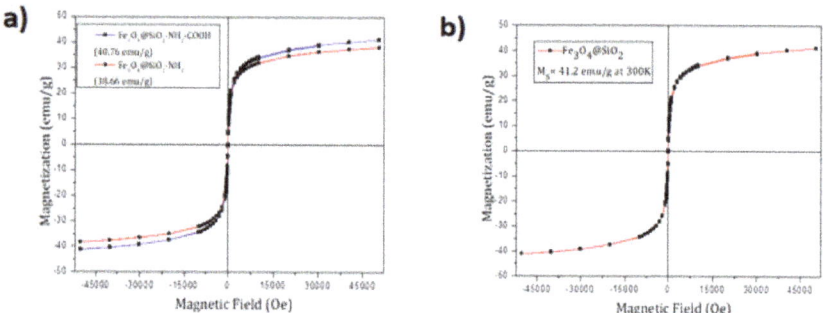

Figure 4. The hysteresis curve of $Fe_3O_4@SiO_2$-NH_2 MNPs: 38.66 emu/g, $Fe_3O_4@SiO_2$-NH_2-COOH: 40.76 emu/g (**a**) and $Fe_3O_4@SiO_2$: 41.2 emu/g (**b**).

The colloidal stability as mentioned at the start is an issue with the MNPs. The MNPs we synthesized were extremely small in size and hence had a tendency to aggregate due to the magnetic dipole–dipole interaction and precipitate over the course of time. This also aided in easily extracting them with the help of an external magnet. To be able to use these MNPs in biosensors, it was important to test their stability and, if possible, increase it without affecting their facile extraction using an external magnet. Five different surfactants were used for this purpose and they are as follows: Citric Acid, Tricaprylylmethyl ammonium chloride, Cetyl trimethyl ammonium chloride (CTAC), Tetrabutyl ammonium chloride, and Dimethyl di-octadecyl ammonium chloride. Out of these five, the last one was successful in providing long-term stability. This was so because it increased the viscosity of the medium and, hence, stopped the MNPs from aggregating. The surfactant prevented the diffusion of the MNPs through the solvent by increasing the viscosity and hence hindering their aggregation. However, this came with a major drawback, as these MNPs were harder to extract using an external magnet. Different concentrations of the surfactant were studied but the trend observed was the same; the more stable they became, the harder it was to extract them. The zeta-potential values obtained for each surfactant is given in Table 1 as well as the different zeta-potential values obtained for different concentrations of dimethyl di-n-octadecyl ammonium chloride in Table 2.

Table 1. Stability of $Fe_3O_4@SiO_2-NH_2$ with different surfactants.

Surfactant	Zeta-Potential (mV)
Citric acid	15.6
Tricaprylyl methyl ammonium chloride	26.8
CTAC	22.6
Tetrabutyl ammonium chloride	36.9
Dimethyl di-n-octadecyl ammonium chloride	54.2

Table 2. Stability of $Fe_3O_4@SiO_2-NH_2$ with different amounts of dimethyl di-n-octadecyl ammonium chloride.

MNPs (wt) (mg/mL)	Dimethyl di-n-octadecyl Ammonium Chloride (wt) (mg)	Zeta-Potential (mV)
0.5	2.5	15.4
0.5	5.0	18.9
0.5	10	28.4
0.5	15	30.2
0.5	20	38.7
0.5	25	44.6
0.5	30	50.2

Up until 15 mg of the surfactant, the dispersion was stable, and with time, a small number of MNPs could be extracted by the magnet. However, to obtain them completely dry and free of the surfactant is difficult as it makes the suspension extremely viscous. Figure S10 shows the different time periods needed to extract the MNPs with increasing stability and viscosity of the medium.

The edges of toroidal external magnets have been used in all the cases to extract the magnets as they comprise of the maximum magnetic fields passing through them. The magnets were used on the sidewalls of the vials or reaction flasks to extract the MNPs. Figure S11 shows a photo of the external magnet and the vials used for extraction of the MNPs.

To learn about the bio-functionalization efficacy of MNPs with anti-TNF-α antibody, the following experiments were performed: A known amount of MNP@SiO$_2$-NH$_2$-COOH activated with EDC/NHS was incubated for a fixed period of time with a known concentration of anti-TNF-α antibody. Following a magnetic separation, the supernatant liquid was analyzed by UV-vis spectrophotometry. The bio-functionalization efficacy was evaluated by comparing the absorbance reading to that obtained for the supernatants of the same concentration of anti-TNF-α antibody after incubation with non-activated MNPs. Consequently, the physisorption of antibodies onto MNPs was taken into account and differences in absorbance can only be attributed to the concentration of unreacted antibody remaining in the supernatants. An illustration of the steps involved in the bio-functionalization of MNPs is presented in Figure 5. Firstly, the –COOH groups present on the MNPs surface were activated using a mixture of EDC/NHS followed by the incubation of the activated MNPs with a fixed concentration of antibody to create the complex Antibody-MNPs. Afterwards, the unreacted sites R–COO-NHS were deactivated (BSA was used in the illustration). Finally, the 3D nano-collector MNP@SiO$_2$-NH$_2$-CO-anti-TNF-α was incubated with the antigen at different concentrations.

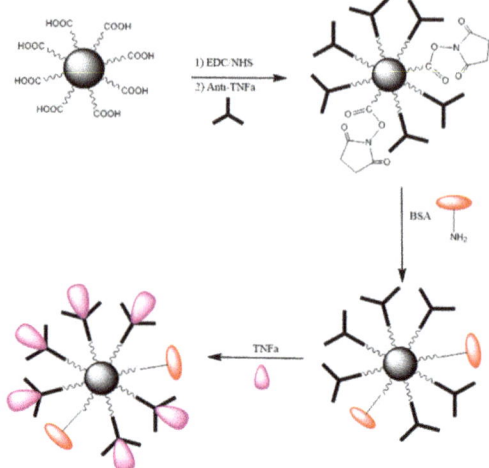

Figure 5. MNPs@SiO$_2$-NH$_2$-COOH bio-functionalization with anti-TNF-α-antibody. Red: BSA Purple: TNF-α.

Figure 6 shows that the absorbance of anti-TNF-α antibody solutions after incubating it with activated MNPs is lower than the absorbance of the anti-TNF-α antibody solutions after incubating with non-activated MNPs. The small difference in the absorbance seen between the Activated MNPs and non-activated MNPs confirms that the activated MNPs binds anti-TNF-α antibody on their surface by the reaction between the amine group of antibody TNF-α and R–COO-NHS groups obtained on the MNPs using EDC/NHS.

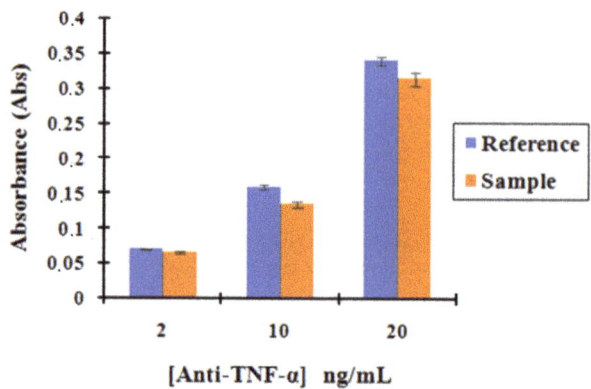

Figure 6. Bar graph showing (blue) the measurement of the absorbance solutions obtained after the immobilization of anti-TNF-α with activated MNPs (with 1-ethyl-3-(dimethylaminopropyl) carbod (EDC)/N-hydroxysuccinimide (NHS)) at different concentrations: 2, 10, and 20 ng/mL; (Orange) the measurement of the absorbance solutions obtained after the immobilization of anti-TNF-α with non-activated MNPs (without EDC/NHS) at different concentration: 2, 10 and 20 ng/mL. Error bars correspond to three replicates per sample.

This was confirmed by the reproducibility of our three replication measurements. In addition, our bio-functionalized MNPs are vortexed before the removal of supernatant so that the weakly bonded by physical interaction are removed and only the strongly bonded by covalent interaction remain, enabling us to evaluate the effectiveness of our chemistry. These results confirm that the bio-functionalization of MNPs with the antibodies was successfully achieved. Moreover, the concentration of anti-TNF-α in the supernatant decreased by 7%, 16%, and 8% for an initial concentration of 2, 10, and 20 ng/mL, respectively, using the "MNP@SiO$_2$-NH$_2$-COOH". Therefore, based on these results, the concentration of anti-TNF-α antibody to run the incubation was fixed at 10 ng/mL to ensure the successful bio-functionalization of MNPs.

The bio-functionalization efficacy was calculated using Beer–Lambert law. Beer–Lambert Law is difficult for quantitative measurement for wavelengths near 200 nm though it has been used here and has provided satisfactory results.

$$A = \varepsilon c l \quad (1)$$

A is the absorbance, ε is the molar attenuation coefficient, l is the optical path length, and c is the concentration of the attenuation species (Equation (1)).

Concentrations (ng/mL)	2	10	20
Absorbance Sample	0.066	0.135	0.315
Absorbance Reference	0.071	0.160	0.341

The methodology that has been used is the following:

$$A_{sample} = \varepsilon\ C_{sample}\ l \quad (2)$$

$$A_{reference} = \varepsilon\ C_{reference}\ l \quad (3)$$

Equation (2)/Equation (3): $A_{sample}/A_{reference} = C_{sample}/C_{reference}$
$C_{sample} = A_{sample} \times C_{reference}/A_{reference}$
2 ng/mL
$C_{sample} = A_{sample} \times C_{reference}/A_{reference} = 0.066 \times 2/0.071 = 1.85$ ng/mL
This is means that the concentration of Antibodies fixed on MNPs is 0.15 ng/mL and the percentage of antibodies fixed in the MNPs is 7%.
10 ng/mL
$C_{sample} = A_{sample} \times C_{reference}/A_{reference} = 0.135 \times 10/0.160 = 8.43$ ng/mL
This is means that the concentration of Antibodies fixed on MNPs is 1.57 ng/mL and the percentage of antibodies fixed in the MNPs is 16%.
20 ng/mL
$C_{sample} = A_{sample} \times C_{reference}/A_{reference} = 0.315 \times 20/0.341 = 18.47$ ng/mL
This means that the concentration of Antibodies fixed on MNPs is 1.53 ng/mL and the percentage of antibodies fixed in the MNPs is 8%.

The complex MNP@SiO$_2$-NH$_2$-CO-anti-TNF-α described in the previous section was used to pre-concentrate TNF-α in PBS. This section details the experiments performed to incubate TNF-α at different concentrations with the complex MNP@SiO$_2$-NH$_2$-CO-anti-TNF-α and afterwards to measure the supernatants containing the unreacted TNF-α by UV-vis spectrophotometry. Thus, a comparison with the initial concentration of the antigen could be carried out and the percentage of TNF-α, which has been bonded to the complex MNP@SiO$_2$-NH$_2$-CO-anti-TNF-α, could be assessed.

Figure 7 shows the calibration curve obtained for TNF-α after incubation with the complex MNP@SiO$_2$-NH$_2$-CO-anti-TNF-α at three different concentrations of TNF-α: 2, 5 and 10 ng/L. The curves, corresponding to reference (reference 1 and 2) to which the former plot was compared, are also shown. Reference (1) depicts supernatants containing unreacted TNF-α after incubation with MNPs that were non-activated with EDC/NHS during the functionalization with anti-TNF-α. Since -COOH groups present on the MNPs were not activated, we did not expect any covalent bonding between anti-TNF-α-MNPs. Therefore, we only expected interaction of TNF-α with those antibodies physisorbed on the MNPs. Reference (2) depicts Supernatants containing unreacted TNF-α after incubation with the initial MNPs that have only been washed four times with 1 mL of 10 mM PBS. In this case, the only physisorption of TNF-α on the MNPs was expected.

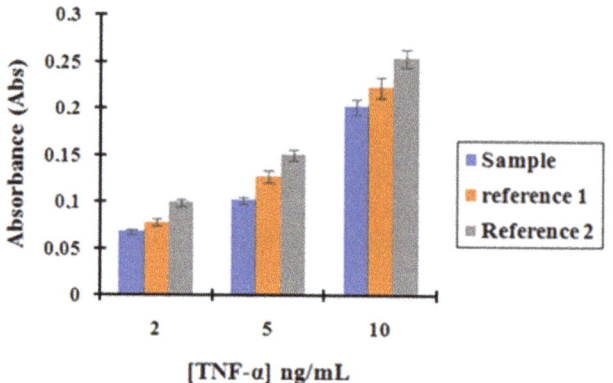

Figure 7. Bar graph showing (blue) the measurement of the absorbance after TNF-α incubation with the complex MNP@SiO$_2$-NH$_2$-CO-anti-TNF-α at three different concentrations of TNF-α: 2, 5, and 10 ng/L; (Orange) reference 1 represent the measurement of the absorbance of TNF-α after incubation with non-activated MNPs; (grey) Reference 2 represent the measurement of the absorbance of TNF-α after incubation with the initial MNPs.

Based on the results, it can be asserted that the concentration of unreacted TNF-α in the supernatants after incubation with the complex MNP@SiO$_2$-NH$_2$-CO-anti-TNF-α is lower than the concentration of TNF-α after incubation with the other two references.

As shown in Figure 7, the absorbance decreased following the trend Reference 2 > Reference 1 > Sample. This is consistent with the results expected in Reference 2. The decrease in the amount of TNF-α in the supernatants when compared with its initial concentration can only be attributed to physisorption on the MNPs surfaces. However, in Reference 1, TNF-α can interact with anti-TNF-α antibodies that have been adsorbed to the surface of the MNPs. This effect is shown by a decrease in the absorbance read when compared to Reference 2. Finally, in "Sample", the complex MNP@SiO$_2$-NH$_2$-CO-anti-TNF-α binds the antigen TNF-α more effectively as antibodies are both covalently bonded and physisorbed to the MNPs. Therefore, the pre-concentration effect when incubating TNF-α with "Sample" is much more important.

Thus, we can confirm these two hypotheses:

(1) There is an effective coupling between MNPs and anti-TNF-α antibody when the MNPs were activated using EDC/NHS.
(2) The complex MNP@SiO$_2$-NH$_2$-CO-Anti-TNF-α can bind the antigen TNF-α and it can be followed by UV-vis spectrophotometry.

The previous experiment as described above was tested using artificial saliva as a matrix. First, a stock solution of artificial saliva (AS) was prepared by dissolving 0.6 g/L Na$_2$HPO$_4$, 0.6 g/L anhydrous CaCl$_2$, 0.4 g/L KCl, 0.4 g/L NaCl, 4 g/L mucin, and 4 g/L urea (purchased from Sigma-Aldrich, France) in deionized water. The pH was adjusted to 7.2 by adding NaOH and it was stored at 4 °C until use.

A dilution profile was performed by UV-vis spectrophotometry to find out the optimal dilution range of artificial saliva at which the matrix effect is removed. For this purpose, AS was diluted using 10 mM PBS to obtain saliva/PBS ratios of 1/1, 1/10, 1/100, 1/500, and 1/1000. The results are shown in

Figure S5. Based on the results obtained, AS diluted with PBS at a ratio 1/500 was chosen as a matrix. Afterwards, a new calibration curve of TNF-α using artificial saliva diluted 1/500 as the matrix was carried out at different concentrations of TNF-α: 2, 5, and 10 ng/L. The results are shown in Figure S6 and confirmed the dependency of the absorbance with the concentration of TNF-α.

In the first stage, the complex MNP@SiO$_2$-NH$_2$-CO-anti-TNF-α was obtained as described above. Subsequently, TNF-α was prepared in AS diluted in PBS 1/500 at three different concentrations: 2, 5, and 10 ng/mL. The solutions prepared were incubated with the complex MNP@SiO$_2$-NH$_2$-CO-Anti-TNF-α as described above and the supernatants containing the unreacted TNF-α were measured by UV-vis spectrophotometry.

Once again, the solution used for the reference was based on the supernatants containing unreacted TNF-α after incubation with MNPs that were non-activated with EDC/NHS during the functionalization with anti-TNF-α antibody.

Based on these results shown in Figure 8, we could confirm that the concentration obtained for the supernatant containing the unreacted amount of TNF-α after incubation with the nano-collector MNP@SiO$_2$-NH$_2$-CO-anti-TNF-α was lower than the supernatant of TNF-α after incubation with non-activated MNP (without activation with EDC/NHS) used as a reference. Therefore, it can be concluded that the pre-concentration effect when incubating TNF-α with the complex MNP@SiO$_2$-NH$_2$-CO-anti-TNF-α is much more important. The concentration of TNF-α decreased by 9%, 12%, and 9% when compared to the initial concentration of 2, 5, and 10 ng/mL, respectively, using the MNP@SiO$_2$-NH$_2$-COOH.

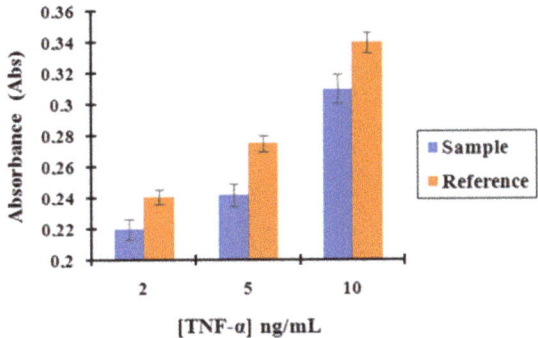

Figure 8. Bar graph showing (blue) the measurement of the absorbance after TNF-α incubation with the complex MNP@SiO$_2$-NH$_2$-CO-anti-TNF-α at three different concentrations of TNF-α: 2, 5, and 10 ng/L; (Orange) the measurement of the absorbance of TNF-α after incubation with no activated MNPs. Background fixed in artificial saliva at 1/500. Error bars correspond to three replicates per sample.

The same experiment described with TNF-α was repeated for the cytokine IL-10 and its corresponding anti-IL-10 antibody. First, the concentration range at which IL-10 can be measurable by UV-vis spectrophotometry in a fixed background of AS diluted 1/500 with PBS was obtained. In the UV region, IL-10 was shown to absorb 10 times less than TNF-α for the same concentration. Consequently, the results showed a working range of IL-10 at a concentration between 10 and 100 ng/mL (Figure S7).

The concentration of IL-10 decreased by 7%, 6%, and 4% when compared to the initial concentration of 10, 40, and 100 ng/mL, respectively, using MNP@SiO$_2$-NH$_2$-COOH. Once again, the results demonstrated the ability of the complex antibody-MNPs to pre-concentrate the antigen.

A comparative study of different deactivating compounds was carried out. The aim was to assess the adsorption effect of the TNF-α antigen when using different chemicals/proteins for the deactivation step performed after bio-functionalization of MNPs with an antibody. For this purpose, four different deactivation mixtures were used: 0.1% bovine serum albumin (BSA) in 10 mM PBS, ethanolamine 0.1% in 10mM PBS, poly (ethylene glycol) methyl ether amine (PEG-NH$_2$) 0.1% in 10 mM PBS, and ethanolamine cyanoborohydride (NaBH$_3$CN) 0.1% in 10 mM PBS.

The comparative study was performed using activated MNPs with EDC/NHS: MNPs@SiO$_2$-NH$_2$-COOH. They were reacted with anti-TNF-α antibody at a concentration of 10 ng/mL, which was chosen as the antibody for the bio-functionalization of MNPs. The experimental procedure concerning the MNPs bio-functionalization was described above.

The deactivation protocol was the same as described above. The MNPs were activated using a mixture of EDC/NHS. Then, anti-TNF-α antibody was added to the activate MNPs. Afterwards, the deactivating solution was changed consecutively as aforementioned. As an illustration, the deactivation of remaining R-CO-NHS groups using BSA/Ethanolamine/PEG-NH$_2$/Ethanolamine + NaBH$_3$CN is presented in Figure 9.

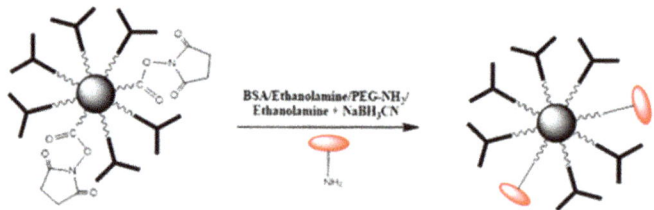

Figure 9. Deactivation of remaining R-CO-NHS groups using BSA/Ethanolamine/PEG-NH$_2$/Ethanolamine + NaBH$_3$CN.

The histogram presented in Figure 10 shows the absorbance reading for the supernatants containing the unreacted TNF-α after incubation with the nano-collector MNP@SiO$_2$-NH$_2$-CO-anti-TNF-α for the different deactivating mixtures.

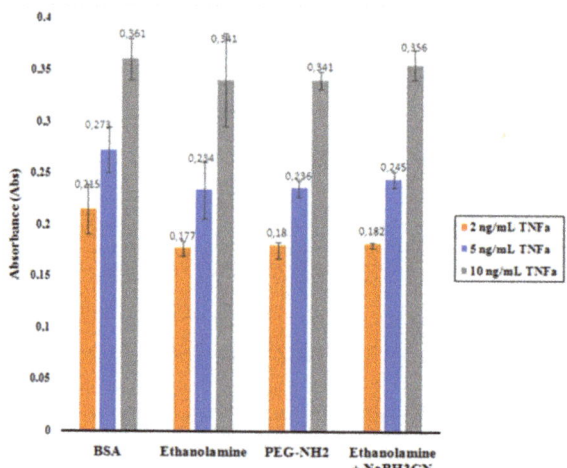

Figure 10. Impact of different deactivation mixtures (BSA, ethanolamine, PEG-NH$_2$, ethanolamine + NaBH$_3$CN) to the physisorption effect of TNF-α on 3D nano-collector MNP@SiO$_2$-NH$_2$-CO-anti-TNF-α. Error bars correspond to three replicates per sample.

It is observed that BSA appeared to be the best among the mixtures assessed. When 0.1% of BSA in PBS was used as deactivating mixture, TNF-α was less physisorbed to the nano-collector leading to a higher absorbance as the total amount of unreacted TNF-α remained relatively high. The ethanolamine

+ NaBH$_3$CN mixture also indicated similar behavior. However, in the case of ethanolamine and PEG-NH$_2$, the absorbance reading was lower as a result of a decreased concentration of unreacted TNF-α.

The bio-functionalization of MNPs with anti-TNF-α antibody at different concentrations was carried out. An illustration of the steps involved in the bio-functionalization of MNPs is presented in Figure 11. Firstly, the –COOH groups present on the anti-TNF-α antibody surface were activated using a mixture of EDC/NHS followed by the incubation of MNPs to the activated anti-TNF-α antibody to create the complex antibody-MNPs. Finally, the 3D nano-collector MNP@SiO$_2$-NH-anti-TNF-α was incubated with TNF-α.

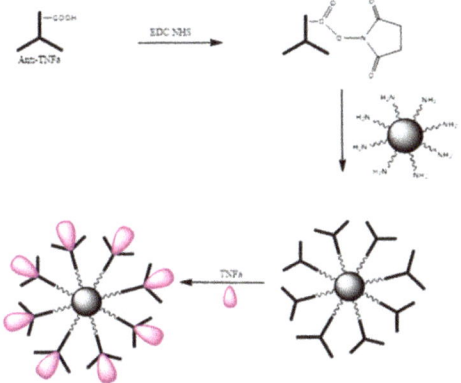

Figure 11. MNP@SiO$_2$-NH$_2$ bio-functionalization with anti-TNF-α antibody.

Based on the results shown in Figure 12, we can confirm again that the concentration obtained for the supernatant containing the unreacted amount of TNF-α after incubation with the nano-collector MNP@SiO$_2$-NH-anti-TNF-α is lower than the supernatant of TNF-α after incubation with non-activated anti-TNF-α antibody (without activation with EDC/NHS) used as a reference. Therefore, it can be concluded that the pre-concentration effect when incubating TNF-α with the complex anti-TNF-α-MNPs is much more important. The concentration of TNF-α decreased by 5%, 16%, and 17% when compared to the initial concentration of 2, 5, and 10 ng/mL, respectively.

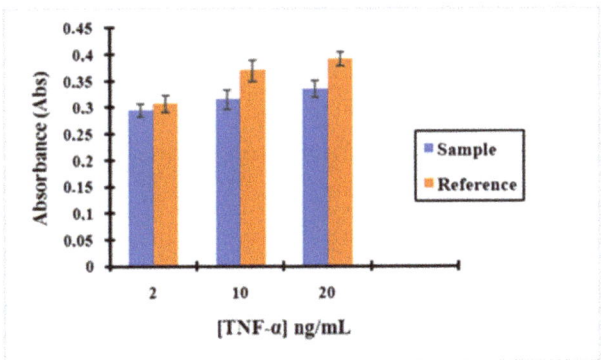

Figure 12. Bar graph showing (blue) the measurement of the absorbance after TNF-α incubation with the complex MNP@SiO$_2$-NH$_2$-CO-anti-TNF-αat three different concentrations of TNF-α: 2, 5, and 10 ng/L; (Orange) the measurement of the absorbance of TNF-α after incubation with no activated MNPs. Background fixed in artificial saliva at 1/500. Error bars correspond to three replicates per sample.

3. Materials and Methods

3.1. Materials

3.1.1. Instrumentation

A magnetic rack purchased from Sigma Aldrich (Lyon, Auvergne-Rhône-Alpes, France) and a tube revolver/rotator purchased from Thermo Fisher Scientific (Waltham, MA, U.S) were used during the bio-functionalization process of MNPs. All experiments for indirect measurements of the biomarkers were carried out at room temperature (22 ± 2 °C). The indirect measurements were carried out using UV-vis spectrophotometry (JENWAY 7205) purchased from Jeulin (Evreux, Normandy, France). Transmission electron microscopy (TEM) studies were carried out using JEOL JEM 1210 at 120 kV. Scanning electron microscopy (SEM) and energy dispersive X-ray spectroscopy (EDX) analysis was done using the QUANTA FEI 200 FEG-ESEM device. The solid sample was analyzed for this. Magnetization hysteresis cycle was measured using Quantum Design MPMS-XL system at 300K with a maximum of 60 kOe.

3.1.2. Chemicals and Reagents

Sodium chloride (NaCl, purity ≥ 99.5%), potassium chloride (KCl, purity 99.0–100.5%), sodium phosphate dibasic (Na_2HPO_4, PharmaGrade), magnesium nitrate ($Mg(NO_3)_2$, purity 98%), calcium chloride ($CaCl_2$, purity ≥ 93%), mucin (from pork stomach extract, type II), sodium hydroxide (NaOH, pellets, purity ≥ 98%), Urea, phosphate buffer solution (PBS) tablets, hydrochloric acid (HCl, ACS reagent, 37%), 1-ethyl-3-(dimethylaminopropyl) carbodiimide (EDC), N-hydroxysuccinimide (NHS), bovine serum albumin (BSA), ethanolamine, poly (ethylene glycol) methyl ether amine (PEG-NH_2), and sodium cyanoborohydride ($NaBH_3CN$) were purchased from Sigma-Aldrich (France). Anti-TNF-α antibody (Catalog number: MAB610-500), Anti-IL-10 antibody (Catalog number: MAB217-500), TNF-α (Catalog number: 210-TA), and IL-10 (Catalog number: 217-IL-050) were purchased from R&D Systems (France). Millipore Milli-Q nanopure water (resistivity > 18 MΩ cm) was produced by a Millipore Reagent Water System (France). The PBS buffer used in this study was prepared by dissolving PBS tablets in the nanopure water as indicated by the supplier by yielding 0.01 M phosphate buffer (pH 7.4) containing 0.0027 M potassium chloride and 0.137 M sodium chloride. All reagents used in the present work for the synthesis of magnetic nanoparticles and its functionalization were obtained from Aldrich Chemical Co and were used without further purification. Aqueous ammonia (30% as NH_3) was purchased from Panreac AppliChem and used as received.

3.2. Procedure of Synthesis of MNPs

The synthetic procedure has been modified and adapted to our needs. The procedure used for synthesis has been a modification of the procedure used by Hassani et al. for Stöber process and Krajl et al. for the amine and the consecutive COOH functionalization [50,51].

3.2.1. Synthesis of Fe_3O_4 Nanoparticles

First, 25 mL of a 15M solution of sodium hydroxide (NaOH) in distilled water was made. Then, a mixture of Iron (II) chloride tetrahydrate ($FeCl_2.4H_2O$) (2 g) and Iron (III) chloride hexahydrate ($FeCl_3.6H_2O$) (5.2 g) was added to 25 mL of distilled water. The NaOH solution previously made was added slowly to the mixture of the iron salts with vigorous stirring. Then, 1 mL of concentrated hydrochloric acid (HCl) was added dropwise along with vigorous stirring to make a black solid product. The resultant mixture was heated using an oil bath for 6 h at 80 °C. The black magnetite solid MNPs were isolated using an external magnet and washed three times with distilled water and then dried at 80 °C for overnight. The Iron oxide (Fe_3O_4) nanoparticles were formed. In the step for the addition of the NaOH, the base is added slowly to the iron salts to obtain smaller sized MNPs while if the iron salts are added directly into the base the sizes of the particle obtained are larger, so it was avoided.

3.2.2. Preparation of the $Fe_3O_4@SiO_2$ Core–Shells

First, 5 mmol of Fe_3O_4 as-synthesized was dispersed in a mixture of ethanol (100 mL) and distilled water (25 mL) for 30 min. Then, 3 mL of NH_3 was added followed by the dropwise addition of tetraethoxysilane (TEOS) (1.7 mL). This solution was stirred mechanically for 8 h at room temperature (r.t). Then, the product $Fe_3O_4@SiO_2$ was separated using an external magnet, was washed three times with distilled water and three times with ethanol, and then dried at 80 °C overnight.

3.2.3. Preparation of Carboxylic Acid and Amine Functionality on $Fe_3O_4@SiO_2$ Nanoparticles [$Fe_3O_4@SiO_2@NH_2$ and $Fe_3O_4@SiO_2@NH_2$-COOH]

First, 125 mg of $Fe_3O_4@SiO_2$ was suspended using an ultra-sound bath in 45 mL of ethanol. Then, 5 mL of 3-(2-aminoethylamino) propyl methyl dimethoxy silane (APMS) was dissolved in 20 mL ethanol (20 wt%), and was added to the suspension of $Fe_3O_4@SiO_2$ in ethanol and stirred vigorously. The pH of the solution was then maintained at 11 using tetramethyl ammonium hydroxide (TMAH). After that, the solution was stirred for 5 h at 50 °C. Then, the $Fe_3O_4@SiO_2$-NH_2 was obtained by magnetic separation, and was washed with distilled water thrice. Some of the particles were not isolated at this stage, and were kept in dispersed form, so sodium chloride (NaCl) (20 mg in 50 mL of deionized water) was added to the dispersion and with this, extra precipitation was achieved. After redispersion, these particles were magnetically decanted and washed thrice. Both crops were mixed together and dried at the oven at 80 °C. Then, 100 mg of $Fe_3O_4@SiO_2$-NH_2 was dispersed in 45 mL dry dimethylformamide (DMF) and 60 mg of succinic anhydride was added to this suspension. The mixture was then stirred with a magnetic stirrer for 22 h at room temperature r.t. The MNPs were then washed thrice with acetone, separated with a magnet, and dried under vacuum. The $Fe_3O_4@SiO_2$-NH_2-COOH MNPs were ready to be used.

3.2.4. Antibody (TNF-α) Grafted at the Surface of MNP@SiO_2-NH_2-COOH

(I)- 100 µL, 0.5% solid content of the mother solution containing MNPs@SiO_2-NH_2-COOH in water were washed following magnetic separation with 1 mL of 10 mM PBS. The washing was repeated three times to ensure all the residues were washed out. The activation of carboxylic acid groups present on the MNPs was achieved by incubation of MNPs in a 500 µL mixture of EDC/NHS (250 µL of 100 mM EDC + 250 µL of 100 mM NHS both prepared in 10 mM PBS) at r.t, and under soft stirring using tube revolver/rotator for 90 min. Then, the remaining EDC/NHS was removed and the MNPs were washed for three consecutive times using 1 mL of cold HCl 1 mM. Subsequently, 500 µL of purified anti-TNF-α antibody was added. Accordingly, three different concentrations of antibody (2, 10, and 20 ng/mL) were used to assess the impact of the concentration on the bio-functionalization efficacy. The activated mixture of MNPs + anti-TNF-α antibody was slowly stirred at r.t for 2 h 30 min. Finally, the complex MNPs@SiO_2-NH_2-CO-Antibody was obtained. It was further immobilized using the magnetic rack and the supernatants were collected and measured by UV-vis spectrophotometry.

(II)- For reference, the same procedure was repeated. However, the activation step with EDC/NHS was skipped. Consequently, the supernatants collected after incubation of anti-TNF-α antibody with MNPs correspond to the initial concentration of antibody minus the number of antibodies that were physisorbed to the MNPs. This phenomenon also occurs in (I)-. Therefore, the final concentration measured in the supernatants corresponding to (I) represents the initial concentration of anti-TNF-α antibody minus the amount of anti-TNF-α antibody covalently bonded to the MNPs minus the amount of anti-TNF-α antibody physisorbed. In summary, the absorbance read for (I) and (II) are as follows (Equations (4) and (5)):

$$Af(I) = A0 - A\ (MNP@SiO_2\text{-}NH_2\text{-}CO\text{-}anti\text{-}TNF\text{-}\alpha) - A\ (phy) \qquad (4)$$

$$Af(II) = A0 - A\,(phy) \tag{5}$$

where Af(I) and Af(II) correspond to the absorbance read of the supernatants obtained in (I) and (II), respectively. A0 represents the absorbance corresponding to the initial concentration of antibody, while A (phy) indicates the absorbance corresponding to the number of antibodies physisorbed onto the MNPs, and A (MNP@SiO$_2$-NH$_2$-CO-anti-TNF-α) to the absorbance corresponding to the number of antibodies covalently bonded to the MNPs. Figure S14 shows an illustration of the experimental procedure carried out for the bio-functionalization of MNPs with anti-TNF-α antibody.

3.2.5. Antibody Interleukin-10 (IL-10) Grafted at the Surface of MNP@SiO$_2$-NH$_2$-COOH

First, 100 µL, 0.5% solid content of the mother solution containing MNPs@SiO$_2$-NH$_2$-COOH in water were washed following magnetic separation with 1 mL of 10 mM PBS. The washing was done thrice to ensure all the residues were washed out. The activation of carboxylic acid groups present on the MNPs was achieved by incubation of MNPs in a 500 µL mixture of EDC/NHS (250 µL of 100 mM EDC + 250 µL of 100 mM NHS both prepared in 10 mM PBS) at r.t, and under soft stirring using tube revolver/rotator for 90 min. Then, the remaining EDC/NHS was removed and the MNPs were washed for three consecutive times using 1 mL of cold HCl 1 mM. Subsequently, 500 µL of purified Anti-IL-10antibody (10 ng/mL) was added. The activated mixture of MNPs + Anti-IL-10 was slowly stirred at r.t for 2 h 30 min. Finally, the complex MNPs@SiO$_2$-NH$_2$-CO-Anti-IL-10 was obtained.

3.2.6. Indirect Detection of TNF-α in PBS. Pre-Concentration of TNF-α onto the Complex MNP@SiO$_2$-NH$_2$-CO-anti-TNF-α

(I) The complex MNP@SiO$_2$-NH$_2$-CO-anti-TNF-α obtained was washed with 1 mL of 10 mM PBS. A magnetic rack was placed to immobilize the complex and allow the discharge of supernatants and the added PBS. The cleaning was repeated three times to ensure all the residues were washed out. Afterwards, the unreacted sites were deactivated using a solution of 0.1% BSA in PBS (500 µL in each Eppendorf). The mixture was stirred (33 rpm) using tube revolver/rotator at r.t for 45 min. Then, the unreacted BSA was removed and the complex was washed three times with 1 mL of 10 mM PBS. Again, magnetic separation was done. At this point, the complex was incubated with TNF-α at three different concentrations: 2, 5, and 10 ng/mL. Incubation was performed at r.t and stirred at 33 rpm for 45 min. Finally, after 45 min, the supernatants containing the unreacted TNF-α were collected and the absorbance was measured for the respective concentrations. Figure S15 shows an illustration of the experimental procedure carried out for the Pre-concentration of TNF-α using the complex MNP@SiO$_2$-NH$_2$-CO-anti-TNF-α.

(II) For the reference, two approaches were compared:

Reference (1) Supernatants containing unreacted TNF-α after incubation with MNPs that were NOT ACTIVATED with EDC/NHS during the functionalization with anti-TNF-α. Since -COOH groups present on the MNPs were not activated, we do not expect any covalent bonding anti-TNF-α-MNPs. Therefore, we only expect interaction of TNF-α with those antibodies physisorbed on the MNPs.

Reference (2) Supernatants containing unreacted TNF-α after incubation with the initial MNPs that have only been washed four times with 1mL of 10 mM PBS. In this case, the only physisorption of TNF-α on the MNPs is expected.

3.2.7. Indirect Detection of IL-10 in Artificial Saliva. Pre-Concentration of IL-10 onto the Complex MNP@SiO$_2$-NH$_2$-CO-Anti-IL-10

The complex MNP@SiO$_2$-NH$_2$-CO-Anti-IL-10 obtained was washed with 1 mL of 10 mM PBS. A magnetic rack was placed to immobilize the complex and allow the discharge of supernatants and the added PBS. The cleaning was done thrice to ensure all the residues were washed out. Afterwards, the unreacted sites were deactivated using a solution of 0.1% BSA in PBS (500 µL in each Eppendorf). The mixture was stirred (33 rpm) using tube revolver/rotator at room temperature for 45 min. Then, the unreacted BSA was removed and the complex was washed three times with 1 mL of

10 mM PBS. Again, magnetic separation was done. At this point, the complex was incubated with IL-10 at three different concentrations: 10, 40, and 100 ng/mL. Incubation was performed at r.t and stirred at 33 rpm for 45 min. Finally, after 45 min, the supernatants containing the unreacted IL-10 were collected and the absorbance was measured for the respective concentrations.

3.2.8. Bio-Functionalization of MNPs Decorated with Amine Groups "MNP@SiO$_2$-NH$_2$"

First, 100 µL of the anti-TNF-α antibody at 10 ng/mL were activated by the incubation in a 500 µL mixture of EDC/NHS (250 µL of 100 mM EDC + 250 µL of 100 mM NHS both prepared in 10 mM PBS) at r.t, and under soft stirring for 90 min. Then, the activated anti-TNF-α antibody was added to the MNP@SiO$_2$-NH$_2$ (100 µL, solid content 0.5%) at r.t, and under soft stirring, for 2 h 30 min to create the complex MNPs@SiO$_2$-NH-Antibody. Afterwards, MNPs were washed for three consecutive times using 1 mL of cold HCl 1 mM, followed by washing for three times using 1 mL of PBS 10 Mm. Finally, the complex was used for the pre-concentration with different concentrations of TNF-α (2, 10, 20 ng/mL) and the supernatants were collected and measured by UV-vis spectrophotometry.

3.3. Calibration Curve of Anti-TNF-α Antibody and TNF-α in PBS

In the first stage, a calibration curve was obtained in PBS in order to determine the optimal range of concentration at which both TNF-α and Anti-TNF-α can be analyzed by UV-vis spectrophotometry.

3.3.1. Calibration Study of Anti-TNF-α Antibody

Firstly, Anti-TNF-α antibody was prepared in PBS 10 mM at three different concentrations: 20, 100, and 200 ng/mL; 10 mM solution of PBS was used as blank. All the measurements were carried out in duplicates.

As shown in Figure S16, a peak for Anti-TNF-α was found at $\lambda = 206$ nm. However, we also observed that the absorbance is saturated at concentrations of antibody higher than 100 ng/mL. Therefore, the experiment was repeated using a dilution factor Anti-TNF-α antibody/PBS 1/10. Figure S17 shows the UV-vis spectrum of Anti-TNF-α at concentrations of 2, 10, and 20 ng/mL.

Here, the absorbance increased linearly with the concentration of antibody in conformity with the Beer–Lambert's law ($A = \varepsilon c l$). The saturation effect disappeared and therefore the concentration range was fixed between 2 and 20 ng/mL. Measurements were repeated three times and the calibration curve obtained is shown in Figure S18.

3.3.2. Calibration Study of TNF-α

Likewise, a calibration study was carried out to assess the concentration range at which the antigen TNF-α is measurable by UV-vis spectrophotometry. For this purpose, TNF-α was prepared in PBS 10 mM at different concentrations: 2, 5, and 10 ng/mL. A 10mM solution of PBS was used as blank. All the measurements were carried out in duplicates. A peak for TNF-α was found at $\lambda = 205$ nm. In this case, no saturation was observed and the concentration range of antigen was fixed from 2 to 10 ng/mL. The calibration curve obtained is shown in Figure S19.

4. Conclusions

We would like to conclude that MNPs with Fe$_3$O$_4$ core has been synthesized and coated with a layer of SiO$_2$ using the Stöber process. A good SiO$_2$ layer homogeneity has been achieved using an overhead mechanical stirrer with a glass bar containing two blades in a beaker with a Teflon lid to prevent alcohol evaporation while other processes have been performed using typical magnetic stirrers with a magnetic bead. The SiO$_2$ coating has been decorated with –NH$_2$ and –COOH free end groups. The MNPs thus produced were Fe$_3$O$_4$@SiO$_2$, Fe$_3$O$_4$@SiO$_2$-NH$_2$, and Fe$_3$O$_4$@SiO$_2$-NH$_2$-COOH that show a gradual increase in size with the additional functionalized layers. Colloidal stability of the MNPs have also been studied using different surfactants and observing the change in the zeta-potential values of the colloids. The –COOH and –NH$_2$ MNPs have been further functionalized with anti-TNF-α and anti-IL-10 antibodies using (EDC) and (NHS) reagents. The magnetic conjugates produced can be easily extracted using an external magnet. These MNPs have been prepared for testing the tumor necrosis factor alpha (TNF-α), while the others have been prepared for Interleukin-10 (IL-10). To have the highest selectivity towards TNF-α, and IL-10, the non-functionalized MNPs surface sites have been neutralized with the BSA protein. The anti-TNF-α and anti-IL-10 antibodies MNPs have been incubated with TNF-α and IL-10, respectively, in PBS, and after magnetic separation the supernatant solution has been tested for TNF-α and IL-10, respectively, in PBS. They have been measured by UV-vis spectrophotometry where a lower absorbance reading indicates better MNPs capture. The supernatant containing unreacted TNF-α that had been incubated with MNPs whose –COOH reagents has been used as the main reference. It had not been activated with EDC/NHS, although anti-TNF-α or anti-IL-10 antibodies had been added (named Reference 1). Although these did not presumably have any bonded anti-TNF-α antibody or anti-IL-10 antibody, they may indeed have some anti-TNF-α antibody or anti-IL-10 antibody physisorbed, thus interacting with TNF-α or IL-10. Further, there could have been the physisorption on the MNPs by the very same TNF-α and IL-10. To test this, TNF-α and IL-10 were incubated with MNPs before EDC/NHS activation (named Reference 2). Without any exception, the MNPs incorporating the antibodies gave the lowest value of TNF-α and IL-10 in solution, which implies their highest uptake by the MNPs, whereas the highest value of TNF-α and IL-10 was for Reference 2. All this implies that antibodies can be adsorbed, in part, on the surface of silica, a factor that has to be taken in consideration. All these studies were done in PBS, but it was sought to use saliva as the body fluid. To apply the UV-vis spectrophotometry technique it was observed that the saliva had to be diluted with PBS to record the amount of TNF-α and IL-10. Dilution was necessary to 1:500 with PBS. By using Fe$_3$O$_4$@SiO$_2$-NH$_2$-CO-anti-TNF-α antibody or anti-IL-10 antibody the pre-concentration process for TNF-α and IL-10 was clear.

When the coupling with the anti-TNF-α antibody or anti-IL-10 antibody occurs, not all activated COOH generate the desired CO-anti-TNF-α or anti-IL-10; some remain "unreacted" and to avoid later reaction with TNF-α or IL-10 that would obscure the reaction, it is necessary to use one deactivating agent. Additionally, these deactivating agents also fill the cavities in the MNPs preventing their occupation by physisorbed antibodies or antigens. Among the several compounds/mixtures utilized like BSA, ethanolamine, ethanolamine+ sodium cyanoborohydride (NaBH3CN), and poly(ethylene glycol) methyl ether amine (PEG-NH2), the BSA and ethanolamine + NaBH3CN were found to be the most efficient. To conclude, in this work, we demonstrate the successful synthesis of two different kind of MNPs (Fe3O4@SiO2-NH2-COOH; Fe3O4@SiO2-NH2) as well as the characterization of these MNPs using different techniques. Then, the high selectivity of these MNPs to pre-concentrate two kind of Heart failure biomarkers: TNF-α (2–20 ng/mL) and IL-10 (10–100 ng/mL) in a complex medium (artificial saliva) after five hours of biofunctionalization process was studied.

In this paper, other than the use of the new set-up using mechanical stirrer with glass blades having a Teflon cover for homogeneous coating of the Silica layer, it reports one of the first studies done using saliva as a diagnostic fluid for heart failure research. Furthermore, colloidal stability of MNPs has been studied here with different surfactants and by increasing the viscosity of the medium by using a surfactant. Preliminary results of how the stability varies with the change in viscosity of the medium and how difficult it is to extract with an external magnet is reported here. This work tries to point out the little concern that exists on the numerous studies of MNPs about taking into consideration the time required for magnetic extraction, or the medium and how it affects the magnetic extraction and the stability of the MNPs for long time shell storage in case they are intended to be used in automatic machines without trained personnel. Further, it deals with the kind of stirrers needed for synthesizing MNPs and that simple UV-vis spectroscopy could be used to determine the loading of MNPs. This paper aims towards showing certain basic fundamental things needed for using MNPs, rather than just reporting a synthesis of MNPs and the detection of antigens. Some basic fundamentals of magnetic nanoparticles, which are mostly excluded from the papers dealing with such topics have been addressed here.

Supplementary Materials: The following are available online at http://www.mdpi.com/1420-3049/25/17/3968/s1, It includes the Results and Discussion Images (Figure S1 to S19, SEM Image, ED pattern, EDX spectrum, IR studies, UV-vis spectra, calibration curves, reaction mechanism, time studies for magnetic extraction and image of MNPs).

Author Contributions: All authors have contributed to the development of this research, but particularly on the subject of conceptualization, F.T. and A.E. (Abdelhamid Errachid); methodology and validation, A.S. (Arpita Saha) and H.B.H.; formal analysis, A.E. (Abdelhamid Elaissari); resources, A.E. (Abdelhamid Errachid), C.V., and F.T.; writing—original draft preparation, A.S. (Arpita Saha), H.B.H., and F.T.; writing—review and editing, A.S. (Arpita Saha), H.B.H., F.T., A.S. (Abhishek Saini), J.G.-G., N.Z., C.V., A.E. (Abdelhamid Elaissari), and A.E. (Abdelhamid Errachid); visualization, A.S. (Arpita Saha); supervision, F.T. and A.E. (Abdelhamid Errachid); project administration, A.E. (Abdelhamid Errachid), C.V., and F.T.; funding acquisition, A.E. (Abdelhamid Errachid), C.V., and F.T. All authors have read and agreed to the published version of the manuscript.

Funding: We acknowledge the funding through the European Union's Horizon 2020 research and innovation programme entitled (An integrated POC solution for non-invasive diagnosis and therapy monitoring of Heart Failure patients, KardiaTool) under grant agreement No 768686 and from CAMPUS FRANCE program under grant agreement PHC PROCOPE 40544QH. This work has been supported by the Spanish Ministerio de Economia y Competitividad (CTQ2016-75150-R), CSIC through 2019AEP132 and Generalitat de Catalunya (2014/SGR/149).

Acknowledgments: A.S. (Arpita Saha) thanks Generalitat de Catalunya for FI predoctoral grant (2016FI_B 00214). A.S. (Abhishek Saini) and A.S. (Arpita Saha) were enrolled in the PhD program of UAB.

Conflicts of Interest: The authors declare no conflict of interest. The funders had no role in the design of the study; in the collection, analyses, or interpretation of data; in the writing of the manuscript, or in the decision to publish the results.

References

1. Pankhurst, A.; Connolly, J.; Jones, S.K.; Dobson, J. Applications of magnetic nanoparticles in biomedicine. *J. Phys. D Appl. Phys.* **2003**, *36*, R167. [CrossRef]
2. Berry, C.C.; Curtis, A.S.G. Functionalisation of magnetic nanoparticles for applications in biomedicine. *J. Phys. D Appl. Phys.* **2003**, *36*, R198. [CrossRef]
3. Berry, C.C. Possible exploitation of magnetic nanoparticle-cell interaction for biomedical applications. *J. Mater. Chem.* **2005**, *15*, 543–547. [CrossRef]
4. Mornet, S.; Portier, J.; Duguet, E. A method for synthesis and functionalization of ultrasmall superparamagnetic covalent carriers based on maghemite and dextran. *J. Magn. Magn. Mater.* **2005**, *293*, 127–134. [CrossRef]
5. Parton, E.; De Palma, R.; Borghs, G. Biomedical applications using magnetic nanoparticles. *Solid State Technol.* **2007**, *50*, 47.
6. Herrera, A.P.; Barrera, C.; Rinaldi, C. Synthesis and functionalization of magnetite nanoparticles with aminopropylsilane and carboxymethyldextran. *J. Mater. Chem.* **2008**, *18*, 3650–3654. [CrossRef]
7. Varadan, V.K.; Chen, L.; Xie, J. *Nanomedicine: Design and Applications of Magnetic Nanomaterials, Nanosensors and Nanosystems*; John Wiley and Sons Pub.: New York, NY, USA, 2008.
8. Lascialfari, A.; Sangregorio, C. Magnetic Nanoparticles in Biomedicine: Recent Advances. *Chimica Oggi* **2011**, *29*, 20–23.
9. Kim, D.H.; Nikles, D.E.; Johnson, D.T.; Brazel, C.S. Heat Generation of Aqueously Dispersed CoFe2O4 Nanoparticles as Heating Agents for Magnetically Activated Drug Delivery and Hyperthermia. *J. Magn. Magn. Mater.* **2008**, *320*, 2390–2396. [CrossRef]
10. Chertok, B.; Moffat, B.A.; David, A.E.; Yu, F.; Bergemann, C.; Ross, B.D.; Yang, V.C. Iron Oxide Nanoparticles as a Drug Delivery Vehicle for MRI Monitored Magnetic Targeting of Brain Tumors. *Biomaterials* **2008**, *29*, 487–496. [CrossRef]
11. Jain, T.K.; Richey, J.; Strand, M.; Leslie-Pelecky, D.L.; Flask, C.A.; Labhasetwar, V. Magnetic Nanoparticles with Dual Functional Properties: Drug Delivery and Magnetic Resonance Imaging. *Biomaterials* **2008**, *29*, 4012–4021. [CrossRef]

12. Dulinska-Litewka, J.; Łazarczyk, A.; Hałubiec, P.; Szafranski, O.; Karnas, K.; Karewicz, A. Superparamagnetic Iron Oxide Nanoparticles—Current and Prospective Medical Applications. *Materials* **2019**, *12*, 617. [CrossRef] [PubMed]
13. Ulman, A. Formation and Structure of Self-Assembled Monolayers. *Chem. Rev.* **1996**, *96*, 1533–1554.
14. Stöber, W.; Fink, A.; Bohn, E. Controlled Growth of Monodispersed Silica Spheres in the Micron Size Range. *J. Colloid Interf. Sci.* **1968**, *26*, 62–69.
15. Jamshaid, T.; Zine, N.; Errachid El-Salhi, A.; Ahmad, N.M.; Elaissari, A. *Soft Nanoparticles for Biomedical Applications*; Royal Society of Chemistry: London, UK, 2014; pp. 312–341. ISBN 978-1-84973-811-8.
16. Tangchaikeeree, T.; Polpanich, D.; Bentaher, A.; Baraket, A.; Errachid, A.; Agusti, G.; Elaissari, A.; Jangpatarapongsa, K. Combination of PCR and dual nanoparticles for detection of Plasmodium falciparum. *Colloids Surf. B.* **2017**, *159*, 888–897. [CrossRef] [PubMed]
17. Jamshaid, T.; Eissa, M.M.; Lelong, Q.; Bonhommé, A.; Augsti, G.; Zine, N.; Errachid, A.; Elaissari, A. Tailoring of carboxyl-decorated magnetic latex particles using seeded emulsion polymerization. *Polym. Adv. Technol.* **2017**, *28*, 1088–1096. [CrossRef]
18. Bray, M.; Yang, Y.; Fare, C.; Chai, C.; Jessieville, F.; Barakat, A.; Arachnid, A.; Zhang, A.D.; Jaffrezic-Renault, A. Boron-doped Diamond Electrodes Modified with Fe3O4@Au Magnetic Nanocomposites as Sensitive Platform for Detection of a Cancer Biomarker, Interleukin-8. *Electroanalysis* **2017**, *28*, 1810–1816.
19. Jamshaid, T.; Neto, E.T.T.; Elissa, M.M.; Zine, N.; Hiroiuqui Kunita, M.; Errachid, A.; Elaissari, A. Magnetic particles: From preparation to lab-on-a-chip, biosensors, microsystems and microfluidics applications. *TrAC-Trend Anal. Chem.* **2016**, *79*, 344–362. [CrossRef]
20. Amstad, E.; Textor, M.; Reimhult, E. Stabilization and functionalization of iron oxide nanoparticles for biomedical applications. *Nanoscale* **2011**, *3*, 2819–2843. [CrossRef]
21. Berry, C.C. Functionalization of magnetic nanoparticles for applications in biomedicine. *J. Phys. D Appl. Phys.* **2009**, *42*, 224003. [CrossRef]
22. Faraji, M.; Yamini, Y.; Rezaee, M. Magnetic nanoparticles: Synthesis, stabilization, functionalization, characterization and applications. *JICS* **2010**, *7*, 1–37. [CrossRef]
23. Hao, R.; Xing, R.; Xu, Z.; Hou, Y.; Gao, S.; Sun, S. Synthesis, functionalization, and biomedical applications of multifunctional magnetic nanoparticles. *Adv. Mater.* **2010**, *22*, 2729–2742. [CrossRef] [PubMed]
24. Eissa, M.M.; Rahman, M.M.; Zine, N.; Jaffrezic-Renault, N.; Errachid, A.; Fessi, H.; Elaissari, A. Reactive magnetic poly(divinylbenzene-co-glycidyl methacrylate) colloidal particles for specific antigen detection using microcontact printing technique. *Acta Biomater.* **2013**, *9*, 5573–5582. [CrossRef] [PubMed]
25. Lu, A.H.; Salabas, E.L.; Schüth, F. Magnetic nanoparticles: Synthesis, protection, functionalization and applications. *Angew. Chem. Int. Ed.* **2007**, *46*, 1222–1244. [CrossRef] [PubMed]
26. Xu, C.; Sun, S. Monodisperse magnetic nanoparticles for biomedical applications. *Polym. Int.* **2007**, *56*, 821–826. [CrossRef]
27. Gan, N.; Yang, X.; Xie, D.; Wu, Y.; Wen, W. A disposable organophosphorus pesticides enzyme biosensor based on magnetic composite nano-particles modified screen printed carbon electrode. *Sensors* **2010**, *10*, 625–638. [CrossRef]
28. Justino, C.I.L.; Rocha-Santos, T.A.P.; Cardoso, S.; Duarte, A.C. Strategies for enhancing the analytical performance of nanomaterial-based sensors. *Trends Anal. Chem.* **2013**, *47*, 27–36. [CrossRef]
29. Justino, C.I.L.; Rocha-Santos, T.A.; Duarte, A.C. Review of analytical figures of merit of sensors and biosensors in clinical applications. *Trends Anal. Chem.* **2010**, *29*, 1172–1183. [CrossRef]
30. Li, J.; Gao, H.; Chen, Z.; Wei, X.; Yang, C.F. An electrochemical immunosensor for carcinoembryonic antigen enhanced by self-assembled nanogold coatings on magnetic nanoparticles. *Anal. Chim. Acta* **2010**, *665*, 98–104. [CrossRef]
31. Xin, Y.; Fu-Bing, X.; Hong-Wei, L.; Feng, W.; Di-Zhao, C.; Zhao-Yang, W. A novel H_2O_2 biosensor based on Fe3O4-Au magnetic nanoparticles coated horseradish peroxidase and graphene sheets-Nafion film modified screen-printed carbon electrode. *Electrochim. Acta* **2013**, *109*, 750–755. [CrossRef]
32. Chen, D.; Deng, J.; Liang, J.; Xie, J.; Hue, C.; Huang, K. A core-shell molecularly imprinted polymer grafted onto a magnetic glassy carbon electrode as a selective sensor for the determination of metronidazole. *Sens. Actuators B Chem.* **2013**, *183*, 594–600. [CrossRef]

33. Prakash, A.; Chandra, S.; Bahadur, D. Structural, magnetic, and textural properties of iron oxide-reduced graphene oxide hybrids and their use for the electrochemical detection of chromium. *Carbon* **2012**, *50*, 4209–4219. [CrossRef]
34. Hu, Y.; Zang, Z.; Zhang, H.; Luo, L.; Yao, S. Selective and sensitive molecularly imprinted sol-gel film-based electrochemical sensor combining mercaptoacetic acid-modified PbS nanoparticles with Fe_3O_4@Au-multi-walled carbon nanotubes-chitosan. *J. Solid State Electrochem.* **2012**, *16*, 857–867. [CrossRef]
35. Arvand, M.; Hassannezhad, M. Magnetic core–shell Fe_3O_4@SiO_2/MWCNT nanocomposite modified carbon paste electrode for amplified electrochemical sensing of uric acid. *Mater. Sci. Eng. C* **2014**, *36*, 160. [CrossRef] [PubMed]
36. Chen, X.; Zhu, J.; Chen, Z.; Xu, C.; Wang, Y.; Yao, C. A novel bienzyme glucose biosensor based on three-layer Au-Fe_3O_4@SiO_2 magnetic nanocomposite. *Sens. Actuators B Chem.* **2011**, *159*, 220–228. [CrossRef]
37. Halima, H.B.; Zine, N.; Gallardo-González, J.; Aissari, A.E.; Sigaud, M.; Alcacer, A.; Bausells, J.; Errachid, A. A Novel Cortisol Biosensor Based on the Capacitive Structure of Hafnium Oxide: Application for Heart Failure Monitoring. In Proceedings of the 2019 20th International Conference on Solid-State Sensors, Actuators and Microsystems & Eurosensors XXXIII (TRANSDUCERS & EUROSENSORS XXXIII), Berlin, Germany, 23–27 June 2019; p. 1067.
38. Drobitch, R.K.; Svensson, C.K. Therapeutic drug monitoring in saliva. *Clin. Pharm.* **1992**, *23*, 365–379. [CrossRef]
39. Jasim, H.; Olausson, P.; Hedenberg-Magnusson, B.; Ernberg, M.; Ghafouri, B. The proteomic profile of whole and glandular saliva in healthy pain-free subjects. *Sci. Rep.* **2016**, *6*, 39073. [CrossRef]
40. Xu, R.; Cui, B.; Duan, X.; Zhang, P.; Zhou, X.; Yuan, Q. Saliva: Potential diagnostic value and transmission of 2019-nCoV. *Int. J. Oral. Sci.* **2020**, *12*, 1–6. [CrossRef]
41. Fakheran, O.; Dehghannejad, M.; Khademi, A. Saliva as a diagnostic specimen for detection of SARS-CoV-2 in suspected patients: A scoping review. *Infect. Dis. Poverty* **2020**, *9*, 927. [CrossRef]
42. Azzi, L.; Carcano, G.; Gianfagna, F.; Grossi, P.; Gasperina, D.D.; Genoni, A.; Fasano, M.; Sessa, F.; Tettamanti, L.; Carinci, F.; et al. Saliva is a reliable tool to detect SARS-CoV-2. *J. Infect.* **2020**, *81*, e45–e50. [CrossRef]
43. Segal, A.; Wong, D.T. Salivary diagnostics: Enhancing disease detection and making medicine better. *Eur. J. Dent. Educ.* **2008**, *12*, 22–29.
44. Lee, Y.H.; Wong, D.T. Saliva: An emerging biofluid for early detection of diseases. *Am. J. Dent.* **2009**, *22*, 241–248. [PubMed]
45. Javaid, M.A.; Ahmed, A.S.; Durand, R.; Tran, S.D. Saliva as a diagnostic tool for oral and systemic diseases. *J. Oral Biol. Craniofac. Res.* **2016**, *6*, 67–76. [CrossRef] [PubMed]
46. Golas, P.L.; Louie, S.; Lowry, G.V.; Matyjaszewski, K.; Tilton, R.D. Comparative study of polymeric stabilizers for magnetite nanoparticles using ATRP. *Langmuir* **2010**, *26*, 16890–16900. [CrossRef] [PubMed]
47. Yu, W.; Xie, H. A review on nanofluids: Preparation, stability mechanisms, and applications. *J. Nanomater.* **2012**, *2012*, 1–17. [CrossRef]
48. Matlochová, A.; Plachá, D.; Rapantová, N. The application of nanoscale materials in groundwater remediation. *PJoES* **2013**, *22*, 1401–1410.
49. Svoboda, J. *Encyclopedia of Materials: Science and Technology*; Elsevier: New York, NY, USA, 2005.
50. Hassani, H.; Zakerinasab, B.; Nasseri, M.A.; Shavakandi, M. The preparation, characterization and application of COOH grafting on ferrite-silica nanoparticles. *Rsc Adv.* **2016**, *6*, 17560–17566. [CrossRef]
51. Kralj, S.; Drofenik, M.; Makovec, D. Controlled surface functionalization of silica-coated magnetic nanoparticles with terminal amino and carboxyl groups. *J. Nanopart. Res.* **2010**, *13*, 2829–2841. [CrossRef]
52. O'Handley, R.C. *Modern Magnetic Materials: Principles and Applications*; Wiley: New York, NY, USA, 1999.

Sample Availability: Not available.

© 2020 by the authors. Licensee MDPI, Basel, Switzerland. This article is an open access article distributed under the terms and conditions of the Creative Commons Attribution (CC BY) license (http://creativecommons.org/licenses/by/4.0/).

MDPI
St. Alban-Anlage 66
4052 Basel
Switzerland
Tel. +41 61 683 77 34
Fax +41 61 302 89 18
www.mdpi.com

Molecules Editorial Office
E-mail: molecules@mdpi.com
www.mdpi.com/journal/molecules

www.ingramcontent.com/pod-product-compliance
Lightning Source LLC
LaVergne TN
LVHW070654100526
838202LV00013B/960